ANSWER BOOK
FOR
CALCULUS

Third Edition

Michael Spivak

Publish or Perish, Inc.
HOUSTON, TEXAS
Ⓑ

ANSWER BOOK FOR CALCULUS
Third Edition

Manufactured in the United States of America
ISBN 0-914098-90-X

1. (ii)

$$(x - y)(x + y) = [x + (-y)](x + y) = x(x + y) + (-y)(x + y)$$
$$= x(x + y) - [y(x + y)] = x^2 + xy - [yx + y^2]$$
$$= x^2 + xy - y^2 = x^2 - y^2.$$

(iv)

$$(x - y)(x^2 + xy + y^2) = x(x^2 + xy + y^2) - [y(x^2 + xy + y^2)]$$
$$= x^3 + x^2 y + xy^2 - [yx^2 + xy^2 + y^3] = x^3 - y^3.$$

(v)

$$(x - y)(x^{n-1} + x^{n-2}y + \cdots + xy^{n-2} + y^{n-1})$$
$$= x(x^{n-1} + x^{n-2}y + \cdots + xy^{n-2} + y^{n-1})$$
$$- [y(x^{n-1} + x^{n-2}y + \cdots + xy^{n-2} + y^{n-1})]$$
$$= x^n + x^{n-1}y + \cdots + x^2 y^{n-2} + xy^{n-1}$$
$$- [x^{n-1}y + x^{n-2}y^2 + \cdots + xy^{n-1} + y^n]$$
$$= x^n - y^n.$$

Using the notation of Chapter 2, this proof can be written as follows:

$$(x - y) \cdot \sum_{j=0}^{n-1} x^j y^{n-1-j} = x\left(\sum_{j=0}^{n-1} x^j y^{n-1-j}\right) - \left[y\left(\sum_{j=0}^{n-1} x^j y^{n-1-j}\right)\right]$$
$$= x^n + \sum_{j=0}^{n-2} x^{j+1} y^{n-1-j} - \left[\sum_{j=1}^{n-1} x^j y^{n-j} + y^n\right]$$
$$= x^n + \sum_{j=0}^{n-2} x^{j+1} y^{n-1-j} - \left[\sum_{k=0}^{n-2} x^{k+1} y^{n-(k+1)} + y^n\right]$$

(letting $k = j - 1$)

$$= x^n - y^n.$$

A formal proof requires such a scheme, in which the expression $(x^{n-1} + x^{n-2}y + \cdots + xy^{n-2} + y^{n-1})$ is replaced by the inductively defined symbol $\sum_{j=0}^{n-1} x^j y^{n-1-j}$.
Along the way we have used several other manipulations which can, if necessary, be justified by inductive arguments.

3. (iv) $(a/b)(c/d) = (ab^{-1})(cd^{-1}) = (ac)(b^{-1}d^{-1}) = (ac)(bd)^{-1}$ (by (iii)) $= (ac)/(bd).$

1

(vi) If $ab^{-1} = cd^{-1}$, then $(ab^{-1})bd = (cd^{-1})bd$, or $ad = bc$. Conversely, if $ad = bc$, then $(ad)d^{-1}b^{-1} = (bc)d^{-1}b^{-1}$, or $ab^{-1} = cd^{-1}$. If $ab^{-1} = ba^{-1}$, then $a^2 = b^2$, so by Problem 1(iii), $a = b$ or $a = -b$. Conversely, if $a = b$, then $a/b = b/a = 1$ and if $a = -b$, then $a/b = b/a = -1$.

4. (ii) All x.

(iv) $x > 3$ or $x < 1$.

(vi) $x > [-1 + \sqrt{5}]/2$ or $x < [-1 - \sqrt{5}]/2$.

(viii) All x, since $x^2 + x + 1 = [x + (1/2)]^2 + 3/4$.

(x) $x > \sqrt{2}$ or $x < \sqrt[3]{2}$.

(xii) $x < 1$.

(xiv) $x > 1$ or $x < -1$.

5. (ii) $b - a$ is in P, so $-a - (-b)$ is in P.

(iv) $b - a$ is in P and c is in P, so $c(b - a) = bc - ac$ is in P.

(vi) If $a > 1$, then $a > 0$, so $a^2 > a \cdot 1$, by part (iv).

(viii) If $a = 0$ or $c = 0$, then $ac = 0$, but $bd > 0$, so $ac < bd$. Otherwise we have $ac < bc < bd$ by applying part (iv) twice.

(x) If $a < b$ were false, then either $a = b$ or $a > b$. But if $a = b$, then $a^2 = b^2$, and if $a > b \geq 0$, then $a^2 > b^2$, by part (ix).

6. (a) From $0 \leq x < y$ and Problem 5(viii) we have $x^2 < y^2$ [as in Problem 5(ix)]. Then from $0 \leq x < y$ and $x^2 < y^2$ we have $x^3 < y^3$. We can continue in this way to prove that $x^n < y^n$ for $n = 2, 3, \ldots$ (a rigorous proof uses induction, covered in the next chapter).

(b) If $0 \leq x < y$, then $x^n < y^n$ by part (a). If $x < y \leq 0$, then $0 \leq -y < -x$, so $(-y)^n < (-x)^n$ by part (a); this means that $-y^n < -x^n$ (since n is odd) and hence $x^n < y^n$. Finally, if $x < 0 \leq y$, then $x^n < 0 \leq y^n$ (since n is odd). Thus, in all cases, if $x < y$, then $x^n < y^n$.

(c) This follows immediately from part (b), since $x < y$ would imply that $x^n < y^n$, while $y < x$ would imply that $y^n < x^n$.

(d) Similarly, if n is even, then using part (a) instead of part (b) we see that if $x, y \geq 0$ and $x^n = y^n$, then $x = y$. Moreover, if $x, y \leq 0$ and $x^n = y^n$, then $-x, -y \geq 0$ and $(-x)^n = (-y)^n$, so again $x = y$. The only other possibility is that one of x and y is positive, the other negative. In this case x and $-y$ are both positive or both negative. Moreover $x^n = (-y)^n$, since n is even, so it follows from the previous cases that $x = -y$.

7. If $a < b$, then

$$a = \frac{a+a}{2} < \frac{a+b}{2} < \frac{b+b}{2} = b.$$

If $0 < a < b$, then $a^2 < ab$ by Problem 5(iv), so $a < \sqrt{ab}$ by Problem 5(x). Moreover, $(a - b)^2 > 0$, so

$$a^2 + b^2 > 2ab,$$
$$a^2 + 2ab + b^2 > 4ab,$$
$$(a + b)^2 > 4ab,$$

so $a + b > 2\sqrt{ab}$. Moreover, for all a, b we have $(a - b)^2 \geq 0$, and thus $(a+b)^2 \geq 4ab$, which implies that $a + b \geq 2\sqrt{ab}$ for $a, b \geq 0$.

8. Two applications of P′12 show that if $a < b$ and $c < d$, then $a+c < b+c < b+d$, so $a + c < b + d$ by P′11. In particular, if $0 < b$ and $0 < d$, then $0 < b + d$, which proves P11. It follows, in addition, that if $a < 0$, then $-a > 0$; for if $-a < 0$ were true, then $0 = a + (-a) < 0$, contradicting P′10. Consequently, any number a satisfies precisely one of the conditions $a = 0$, $a > 0$, $a < 0$, the last being equivalent to $-a > 0$. This proves P10. Finally, P′13 shows that if $0 < a$ and $0 < c$, then $0 < ac$, which proves P12.

9. (ii) $|a| + |b| - |a + b|$.

(iv) $x^2 - 2xy + y^2$.

10. (ii)

$$x - 1 \quad \text{if } x \geq 1;$$
$$1 - x \quad \text{if } 0 \leq x \leq 1;$$
$$1 + x \quad \text{if } -1 \leq x \leq 0;$$
$$-1 - x \quad \text{if } x \leq -1.$$

(iv)

$$a \text{ if } a \geq 0;$$
$$3a \text{ if } a \leq 0.$$

11. (ii) $-5 < x < 11$.

(iv) $x < 1$ or $x > 2$ (the distance from x to 1 plus the distance from x to 2 equals 1 precisely when $1 \leq x \leq 2$).

(vi) No x.

(viii) If $x > 1$ or $x < -2$, then the condition becomes $(x - 1)(x + 2) = 3$, or $x^2 + x - 5 = 0$, for which the solutions are $(-1 + \sqrt{21})/2$ and $(-1 - \sqrt{21})/2$.

Since the first is > 1 and the second is < -2, both are solutions to the equation $|x - 1| \cdot |x + 2| = 3$. For $-2 < x < 1$ the condition becomes $(1 - x)(x + 2) = 3$ or $x^2 + x + 1 = 0$, which has no solutions.

12. (ii) $|1/x| \cdot |x| = |(1/x) \cdot x|$ (by (i)) $= |1| = 1$, so $|1/x| = 1/|x|$.

(iv) $|x - y| = |x + (-y)| \leq |x| + |-y| = |x| + |y|$.

(vi) Interchanging x and y in part (v) gives $|y| - |x| \leq |x - y|$. Combining this with part (v) yields $|(|x| - |y|)| \leq |x - y|$.

13. If $x \leq y$, then $|y - x| = y - x$, so $x + y + |y - x| = x + y + y - x = 2y$, which is $2 \max(x, y)$. Interchanging x and y proves the formula when $x \geq y$, and the same type of argument works for $\min(x, y)$. Also

$$\max(x, y, z) = \max(x, \max(y, z))$$

$$= \frac{x + \dfrac{y + z + |y - z|}{2} + \left| \dfrac{y + z + |y - z|}{2} - x \right|}{2}$$

$$= \frac{|y - z| + y + z + 2x + |y + z + |y - z| - 2x|}{4}.$$

14. (a) If $a \geq 0$, then $|a| = a = -(-a) = |-a|$, since $-a \leq 0$. The equality is proved for $a \leq 0$ by replacing a by $-a$.

(b) If $|a| \leq b$, then clearly $b \geq 0$. Now $|a| \leq b$ means that $a \leq b$ if $a \geq 0$, and surely $a \leq b$ if $a \leq 0$. Similarly, $|a| \leq b$ means $-a \leq b$, and hence $-b \leq a$, if $a \leq 0$, and surely $-b \leq a$ if $a \geq 0$. So $-b \leq a \leq b$.

Conversely, if $-b \leq a \leq b$, then $|a| = a \leq b$ if $a \geq 0$, while $|a| = -a \leq b$ if $a \leq 0$.

(c) From $-|a| \leq a \leq |a|$ and $-|b| \leq b \leq |b|$ it follows that

$$-(|a| + |b|) \leq a + b \leq |a| + |b|,$$

so $|a + b| \leq |a| + |b|$.

15. If $x \neq y$, then

$$x^2 + xy + y^2 = \frac{x^3 - y^3}{x - y}.$$

Problem 6(b) shows that the quotient on the right is always positive (since $x^3 - y^3 > 0$ if $x - y > 0$ and $x^3 - y^3 < 0$ if $x - y < 0$). Moreover, if $x = y \neq 0$, then $x^2 + xy + y^2 = 3x^2 > 0$. The other inequality is proved similarly, using the factorization for $x^5 - y^5$.

16. (a) If
$$x^2 + y^2 = (x + y)^2 = x^2 + 2xy + y^2,$$

then $xy = 0$, so $x = 0$ or $y = 0$. If
$$x^3 + y^3 = (x + y)^3 = x^3 + 3x^2y + 3xy^2 + y^3,$$

then $3xy(x + y) = 0$, so $x = 0$ or $y = 0$ or $x = -y$.

(b) The first equation implies that
$$4x^2 + 8xy + 4y^2 \geq 0.$$

Suppose that we also had
$$4x^2 + 6xy + 4y^2 \leq 0.$$

Subtracting the second from the first would give $2xy \geq 0$. If neither x nor y is 0, this means that we must have $2xy > 0$; but this implies that $4x^2 + 6xy + y^2 > 0$, a contradiction.

Moreover, it is clear that if one of x and y is 0, but not the other, then we also have $4x^2 + 6xy + 4y^2 > 0$.

(c) If
$$x^4 + y^4 = (x + y)^4 = x^4 + 4x^3y + 6x^2y^2 + 4xy^3 + y^4,$$

then
$$0 = 4x^3y + 6x^2y^2 + 4xy^3 = xy(4x^2 + 6xy + 4y^2),$$

so $x = 0$ or $y = 0$, or $4x^2 + 6xy + 4y^2 = 0$. But by part (b), the last equation implies that x and y are both 0. Thus we must always have $x = 0$ or $y = 0$.

(d) If
$$x^5 + y^5 = (x + y)^5 = x^5 + 5x^4y + 10x^3y^2 + 10x^2y^3 + 5xy^4 + y^5,$$

then
$$0 = 5x^4y + 10x^3y^2 + 10x^2y^3 + 5xy^4$$
$$= 5xy(x^3 + 2x^2y + 2xy^2 + y^3),$$

so $xy = 0$ or
$$x^3 + 2x^2y + 2xy^2 + y^3 = 0.$$

Subtracting this equation from
$$(x + y)^3 = x^3 + 3x^2y + 3xy^2 + y^3$$

we obtain
$$(x + y)^3 = x^2y + xy^2 = xy(x + y).$$

So either $x + y = 0$ or $(x+y)^2 = xy$; the latter condition implies that $x^2 + xy + y^2 = 0$, so $x = 0$ or $y = 0$ by Problem 15. Thus $x = 0$ or $y = 0$ or $x = -y$.

17. (a) Since

$$2x^2 - 3x + 4 = 2\left(x - \frac{3}{4}\right)^2 + 4 - \frac{9}{8}$$

$$= 2\left(x - \frac{3}{4}\right)^2 + \frac{23}{8},$$

the smallest possible value is $23/8$, when $(x - 3/4)^2 = 0$, or $x = 3/4$.

(b) We have

$$x^2 - 3x + 2y^2 + 4y + 2 = \left(x - \frac{3}{2}\right)^2 + 2(y + 1)^2 - \frac{9}{4},$$

so the smallest possible value is $-9/4$, when $x = 3/2$ and $y = -1$.

(c) For each y we have

$$x^2 + 4xy + 5y^2 - 4x - 6y + 7 = x^2 + 4(y - 1)x + 5y^2 - 6y + 7$$

$$= [x + 2(y - 1)]^2 + 5y^2 - 6y + 7 - 4(y - 1)^2$$

$$= [x + 2(y - 1)]^2 + (y + 1)^2 + 2,$$

so the smallest possible value is 2, when $y = -1$ and $x = -2(y - 1) = 4$.

18. (a) is a straightforward check.

(b) We have

$$x^2 + bx + c = \left(x + \frac{b}{2}\right)^2 + \left(c - \frac{b^2}{4}\right) \geq c - \frac{b^2}{4};$$

but $c - b^2/4 > 0$, so $x^2 + bx + c > 0$ for all x.

(c) Apply part (b) with y for b and y^2 for c: we have $b^2 - 4c = y^2 - 4y^2 < 0$ for $y \neq 0$, so $x^2 + xy + y^2 > 0$ for all x, if $y \neq 0$ (and surely $x^2 + xy + y^2 > 0$ for all $x \neq 0$ if $y = 0$).

(d) α must satisfy $(\alpha y)^2 - 4y^2 < 0$, or $\alpha^2 < 4$, or $|\alpha| < 2$.

(e) Since

$$x^2 + bc + c = \left(x + \frac{b}{2}\right)^2 + \left(c - \frac{b^2}{4}\right) \geq c - \frac{b^2}{4},$$

and since $x^2 + bx + c$ has the value $c - b^2/4$ when $x = -b/2$, the minimum value is $c - b^2/4$. Since

$$ax^2 + bx + c = a\left(x^2 + \frac{b}{a}x + \frac{c}{a}\right),$$

the minimum value is

$$a \left(\frac{c}{a} - \frac{b^2}{4a^2} \right) = c - \frac{b^2}{4a}.$$

19. (a) The proofs when $x_1 = \lambda y_1$ and $x_2 = \lambda y_2$, or $y_1 = y_2 = 0$, are straightforward. If there is no such λ, then the equation

$$\lambda^2(y_1{}^2 + y_2{}^2) - 2\lambda(x_1 y_1 + x_2 y_2) + (x_1{}^2 + y_1{}^2) = 0$$

has *no* solution λ, so by Problem 18(a) we must have

$$\left[\frac{2(x_1 y_1 + x_2 y_2)}{(y_1{}^2 + y_2{}^2)} \right]^2 - \frac{4(x_1{}^2 + y_1{}^2)}{(y_1{}^2 + y_2{}^2)} < 0,$$

which yields the Schwarz inequality.

(b) We have $2xy \le x^2 + y^2$, since $0 \le (x - y)^2 = x^2 - 2xy + y^2$. Thus

(1)
$$\frac{2x_1 y_1}{\sqrt{x_1{}^2 + x_2{}^2} \sqrt{y_1{}^2 + y_2{}^2}} \le \frac{x_1{}^2}{(x_1{}^2 + x_2{}^2)} + \frac{y_1{}^2}{(y_1{}^2 + y_2{}^2)},$$

(2)
$$\frac{2x_1 y_1}{\sqrt{x_1{}^2 + x_2{}^2} \sqrt{y_1{}^2 + y_2{}^2}} \le \frac{x_2{}^2}{(x_1{}^2 + x_2{}^2)} + \frac{y_2{}^2}{(y_1{}^2 + y_2{}^2)};$$

addition yields

$$\frac{2(x_1 y_1 + x_2 y_2)}{\sqrt{x_1{}^2 + x_2{}^2} \sqrt{y_1{}^2 + y_2{}^2}} \le 2.$$

(c) The equality is a straightforward computation. Since $(x_1 y_2 - x_2 y_1)^2 \ge 0$, the Schwarz inequality follows immediately.

(d) The proof in part (a) already yields the desired result.

In part (b), equality holds only if it holds in (1) and (2). Since $2xy = x^2 + y^2$ only when $0 = (x - y)^2$, i.e., $x = y$, this means that

$$\frac{x_i}{\sqrt{x_1{}^2 + x_2{}^2}} = \frac{y_i}{\sqrt{y_1{}^2 + y_2{}^2}} \qquad \text{for } x = 1, 2,$$

so we can choose $\lambda = \sqrt{x_1{}^2 + x_2{}^2} / \sqrt{y_1{}^2 + y_2{}^2}$.

In part (c), equality holds only when $x_1 y_2 - x_2 y_1 = 0$. One possibility is $y_1 = y_2 = 0$. If $y_1 \ne 0$, then $x_1 = (x_1/y_1)y_1$ and also $x_1 = (x_1/y_1)y_2$; similarly, if $y_2 = 0$, then $\lambda = x_2/y_2$.

20, 21, 22. See Chapter 5.

23. According to Problem 21, we have $|x/y - x_0/y_0| < \varepsilon$ if

$$|x - x_0| < \min \left(\frac{\varepsilon}{2(1/|y_0| + 1)}, 1 \right)$$

and

$$\left| \frac{1}{y} - \frac{1}{y_0} \right| < \frac{\varepsilon}{2(|x_0| + 1)},$$

and the latter is true, according to Problem 22, if

$$|y - y_0| < \min \left(\frac{|y_0|}{2}, \frac{\varepsilon |y_0|^2}{4(|x_0| + 1)} \right).$$

24. (a) For $k = 1$ the equation reads $a_1 + a_2 = a_1 + a_2$. If the equation holds for k, then

$$(a_1 + \cdots + a_{k+1}) + a_{k+2} = [(a_1 + \cdots + a_k) + a_{k+1}] + a_{k+2}$$
$$= (a_1 + \cdots + a_k) + (a_{k+1} + a_{k+2}) \qquad \text{by P1}$$
$$= a_1 + \cdots + a_k + (a_{k+1} + a_{k+2})$$
$$\text{since the equation holds for } k$$
$$= a_1 + \cdots + a_{k+2}$$
$$\text{by the definition of } a_1 + \cdots + a_{k+2}.$$

(b) For $k = 1$ the equation reduces to the definition of $a_1 + \cdots + a_k$. If the equation is true for some $k < n$, then

$$(a_1 + \cdots + a_{k+1}) + (a_{k+2} + \cdots + a_n)$$
$$= ([a_1 + \cdots + a_k] + a_{k+1}) + (a_{k+2} + \cdots + a_n)$$
$$\text{by part (a)}$$
$$= (a_1 + \cdots + a_k) + (a_{k+1} + (a_{k+2} + \cdots + a_n))$$
$$\text{by P1}$$
$$= (a_1 + \cdots + a_k) + (a_{k+1} + \cdots + a_n)$$
$$\text{by the definition of } a_{k+1} + \cdots + a_n$$
$$= a_1 + \cdots + a_n \qquad \text{by assumption.}$$

(c) The proof is by "complete induction" on k (see Chapter 2). The assertion is clear for $k = 1$. Assume that it is true for all $l < k$. Then

$$s(a_1, \ldots, a_k) = s'(a_1, \ldots, a_l) + s''(a_{l+1}, \ldots, a_k)$$
$$= (a_1 + \cdots + a_l) + (a_{l+1} + \cdots + a_k) \qquad \text{by assumption}$$
$$= a_1 + \cdots + a_k \qquad \text{by part (b).}$$

25. P2, P3, P4, P6, P7, P8 are obvious from a glance at the tables. There are eight cases for P1, and even this number can be reduced: because P2 is true, it is clear that $a + (b + c) = (a + b) + c$ if a, b, or c is 0, so only the case $a = b = c = 1$ must be checked. Similarly for P5. Finally, P9 is true for $a = 0$, since $0 \cdot b = 0$ for all b, and for $a = 1$, since $1 \cdot b = b$ for all b.

CHAPTER 2

1. (ii) Since $1^3 = 1^2$, the formula is true for $n = 1$. Suppose that the formula is true for k. Then

$$(1 + \cdots + k + [k+1])^2 = (1 + \cdots + k)^2 + 2(1 + \cdots + k)(k+1) + (k+1)^2$$

$$= 1^3 + \cdots + k^3 + 2\frac{k(k+1)}{2}(k+1) + (k+1)^2$$

$$= 1^3 + \cdots + k^3 + (k^3 + 2k^2 + k) + (k^2 + 2k + 1)$$

$$= 1^3 + \cdots + k^3 + (k+1)^3,$$

so the formula is true for $k + 1$.

2. (ii)

$$\sum_{i=1}^{n}(2i-1)^2 = 1^2 + 3^2 + \cdots + (2n-1)^2$$

$$= [1^2 + 2^2 + \cdots + (2n)^2] - [2^2 + 4^2 + 6^2 + \cdots + (2n)^2]$$

$$= [1^2 + 2^2 + \cdots + (2n)^2] - 4[1^2 + 2^2 + 3^2 + \cdots + (n)^2]$$

$$= \frac{2n(2n+1)(4n+1)}{6} - \frac{4n(n+1)(2n+1)}{6}$$

$$= \frac{2n(2n+1)[4n+1-2(n+1)]}{6}$$

$$= \frac{n(2n+1)(2n-1)}{3}.$$

3. (a)

$$\binom{n}{k-1} + \binom{n}{k} = \frac{n!}{(k-1)!(n-k+1)!} + \frac{n!}{k!(n-k)!}$$

$$= \frac{kn!}{k!(n+1-k)!} + \frac{(n+1-k)n!}{k!(n+1-k)!}$$

$$= \frac{(n+1)n!}{k!(n+1-k)!} = \binom{n+1}{k}.$$

(b) Clearly $\binom{1}{1}$ is a natural number. Suppose that $\binom{n}{p}$ is a natural number for all $p \leq n$. Since

$$\binom{n+1}{p} = \binom{n}{p-1} + \binom{n}{p} \qquad \text{for } p \leq n,$$

it follows that $\binom{n+1}{p}$ is a natural number for all $p \leq n$, while $\binom{n+1}{n+1}$ is also a natural number. So $\binom{n+1}{p}$ is a natural number for all $p \leq n+1$.

(c) There are $n(n-1)\cdots(n-k+1)$ k-tuples of distinct integers each chosen from $1,\ldots,n$, since the first can be picked in n ways, the next in $n-1$ ways, etc. Now each set of exactly k integers can be arranged in $k!$ k-tuples, so there are $n(n-1)\cdots(n-k+1)/k! = \binom{n}{k}$ such sets.

(d) The binomial theorem is clear for $n = 1$. Suppose that

$$(a+b)^n = \sum_{j=0}^{n} \binom{n}{j} a^{n-j} b^j.$$

Then

$$(a+b)^{n+1} = (a+b)(a+b)^n = (a+b)\sum_{j=0}^{n} \binom{n}{j} a^{n-j} b^j$$

$$= \sum_{j=0}^{n} \binom{n}{j} a^{n+1-j} b^j + \sum_{j=0}^{n} \binom{n}{j} a^{n-j} b^{j+1}$$

$$= \sum_{j=0}^{n} \binom{n}{j} a^{n+1-j} b^j + \sum_{j=1}^{n+1} \binom{n}{j-1} a^{n+1-j} b^j$$

(we have replaced j by $j-1$ in the second sum)

$$= \sum_{j=0}^{n+1} \binom{n+1}{j} a^{n+1-j} b^j \qquad \text{by part (a),}$$

so the binomial theorem is true for $n+1$.

(e) (i)

$$2^n = (1+1)^n = \sum_{j=0}^{n} \binom{n}{j}.$$

(ii)

$$0 = (1+-1)^n = \sum_{j=0}^{n} (-1)^j \binom{n}{j}.$$

(iii) Subtracting (ii) from (i) we obtain

$$2 \sum_{l \text{ odd}} \binom{n}{l} = 2^n.$$

(iv) Add (i) and (ii).

4. (a) Since

$$(1+x)^n (1+x)^m = (1+x)^{n+m}$$

we have

$$\sum_{k=0}^{n} \binom{n}{k} x^k \cdot \sum_{j=0}^{m} \binom{m}{j} x^j = \sum_{l=0}^{n+m} \binom{n+m}{l} x^l.$$

But the coefficient of x^l on the left is clearly

$$\sum_{k=0}^{l} \binom{n}{k}\binom{m}{l-k},$$

one term of the sum occurring for each pair k, $j = l - k$.

(b) Let $m, l = n$ in part (a) [note that $\binom{n}{k} = \binom{n}{n-k}$].

6. (ii) From

$$(k + 1)^5 - k^5 = 5k^4 + 10k^3 + 10k^2 + 5k + 1 \qquad k = 1, \ldots, n$$

we obtain

$$(n + 1)^5 - 1 = 5\left(\sum_{k-1}^{n} k^4\right) + 10\left(\sum_{k=1}^{n} k^3\right) + 10\left(\sum_{k=1}^{n} k^2\right) + 5\left(\sum_{k=1}^{n} k\right) + n,$$

so

$$\sum_{k=1}^{n} k^4 = \frac{(n + 1)^5 - 1 - 10\left(\frac{n^4}{4} + \frac{n^3}{2} + \frac{n^2}{4}\right) - 10\frac{n(n+1)(n+2)}{6} - 5\frac{n(n+1)}{2} - n}{5}$$

$$= \frac{n^5}{5} + \frac{n^4}{2} + \frac{n^3}{3} - \frac{n}{30}.$$

(iv) From

$$\frac{1}{k^2} - \frac{1}{(k + 1)^2} = \frac{2k + 1}{k^2(k + 1)^2} \qquad k = 1, \ldots, n$$

we obtain

$$1 - \frac{1}{(n + 1)^2} = \sum_{k=1}^{n} \frac{2k + 1}{k^2(k + 1)^2}.$$

7. The proof is by complete induction on p. The statement is true for $p = 1$, since

$$\sum_{k=1}^{n} k = \frac{n(n + 1)}{2} = \frac{n^2}{2} + n.$$

Suppose that the statement is true for all natural numbers $\leq p$. The binomial theorem yields the equations

$$(k + 1)^{p+1} - k^{p+1} = (p + 1)k^p + \text{terms involving lower powers of } k.$$

Adding for $k = 1, \ldots, n$, we obtain

$$\frac{(n + 1)^{p+1}}{p + 1} = \sum_{k=1}^{n} k^p + \text{terms involving } \sum_{k=1}^{n} k^r \text{ for } r < p.$$

By assumption, we can write each $\sum_{k=1}^{n} k^r$ as an expression involving powers n^s with $s \leq p$. It follows that

$$\sum_{k=1}^{n} k^p = \frac{(n+1)^{p+1}}{p+1} + \text{terms involving powers of } n \text{ less than } p+1.$$

10. Suppose A contains 1, and that A contains $n+1$ if it contains n. If A does not contain all natural numbers, then the set B of natural numbers *not* in A is not $\emptyset$. So B has a smallest member n_0. Now $n_0 \neq 1$, since A contains 1, so we can write $n_0 = (n_0 - 1) + 1$, where $n_0 - 1$ is a natural number. Now $n_0 - 1$ is *not* in B, so $n_0 - 1$ *is* in A. By hypothesis, n_0 must be in A, so n_0 is not in B, a contradiction. (By the way, the assertion that a natural number $n \neq 1$ can be written $n = m + 1$ for some other natural number m, can itself be proved by induction.)

11. Clearly 1 is in B. If k is in B, then $1, \ldots, k$ are all in A, so $k+1$ is in A, so $1, \ldots, k+1$ are in A, so $k+1$ is in B. By (ordinary) induction, $B = \mathbf{N}$, so also $A = \mathbf{N}$.

14. (a) If $\sqrt{2} + \sqrt{6}$ were rational, then $(\sqrt{2} + \sqrt{6})^2$ would certainly be rational. So $8 + 2\sqrt{12} = 8 + 4\sqrt{3}$ would be rational, so $\sqrt{3}$ would be rational, which is false.

(b) Similarly, if $\sqrt{2} + \sqrt{3}$ were rational, then its square $5 + 2\sqrt{6}$ would be rational, so $\sqrt{6}$ would be rational, which is false.

15. (a) The assertion is true for $m = 1$. If it is true for m, then

$$(p + \sqrt{q})^{m+1} = (p + \sqrt{q})(a + b\sqrt{q}) = (ap + bq) + (a + pb)\sqrt{q},$$

and $ap + bq$ and $a + bp$ are rational.

(b) The assertion is true for $m = 1$. If it is true for m, then

$$(p - \sqrt{q})^{m+1} = (p - \sqrt{q})(a - b\sqrt{q}) = (ap + bq) - (a + pb)\sqrt{q},$$

whereas $(p + \sqrt{q})^{m+1} = (ap + bq) + (a + pb)\sqrt{q}$ by part (a).

16. (a) The inequality $(m + 2n)^2/(m + n)^2 > 2$ is equivalent to

$$m^2 + 4mn + 4n^2 > 2m^2 + 4mn + 2n^2,$$

or simply $2n^2 > m^2$.

The second inequality is equivalent to

$$n^2[(m + 2n)^2 - 2(m + n)^2] < (2n^2 - m^2)(m + n)^2,$$

or

$$n^2(2n^2 - m^2) < (2n^2 - m^2)(n^2 + [2mn + m^2]),$$

or

$$0 < (2n^2 - m^2)(2mn + m^2).$$

(b) Reverse all inequality signs in the solution for part (a).

(c) Let $m_1 = m + 2n$ and $n_1 = m + n$, and then choose

$$m' = m_1 + 2n_1 = 3m + 4n,$$
$$n' = m_1 + \ n_1 = 2m + 3n.$$

17. (a) Suppose that every number $< n$ can be written as a product of primes. If $n > 1$ is not a prime, then $n = ab$ for $a, b < n$. By assumption, a and b are each products of primes, so $n = ab$ is also.

(b) If $\sqrt{n} = a/b$, then $nb^2 = a^2$, so the factorization into primes of nb^2 and of a^2 must be the same. Now every prime appears an even number of times in the factorization of a^2, and of b^2, so the same must be true of the factorization of n. This implies that n is a square.

(c) Repeat the same argument, using the fact that every prime occurs a multiple of k times in a^k and b^k.

(d) If $p_1, \ldots, p_n$ were the only primes, then $(p_1 \cdot p_2 \cdots p_n) + 1$ could not be a prime, since it is larger than all of them (and is not 1), so it must be divisible by a prime. But $p_1, \ldots, p_n$ clearly do not divide it, a contradiction. (Although this is a proof by contradiction, it can be used to obtain some positive information: If $p_1, \ldots, p_n$ are the first n primes, then the $(n + 1)^{\text{st}}$ prime is $\leq (p_1 \cdot p_2 \cdots p_n) + 1$. It is not necessarily true, however, that the number $(p_1 \cdot p_2 \cdots p_n) + 1$ is a prime; for example, $(2 \cdot 3 \cdot 5 \cdot 7 \cdot 11 \cdot 13) + 1 = 30{,}031 = 59 \cdot 509$.)

18. (a) Suppose $x = p/q$ where p and q are natural numbers with no common factor. Then

$$\frac{p^n}{q^n} + a_{n-1}\frac{p^{n-1}}{q^{n-1}} + \cdots + a_0 = 0,$$

so

$(*)$ $$p^n + a_{n-1}p^{n-1}q + \cdots + a_0 q^n = 0.$$

Now if $q \neq \pm 1$, then q has some prime number as a factor. This prime factor divides every term of $(*)$ other than p^n, so it must divide p^n also. Therefore it divides p, a contradiction. So $q = \pm 1$, which means that x is an integer.

(b) If

$$x = \sqrt{6} - \sqrt{2} - \sqrt{3},$$

then

$$x^2 = 6 + (\sqrt{2} + \sqrt{3})^2 - 2\sqrt{6}(\sqrt{2} + \sqrt{3})$$
$$= 11 + 2\sqrt{6}[1 - (\sqrt{2} + \sqrt{3})],$$

so

$$(x^2 - 11)^2 = 24[1 - (\sqrt{2} + \sqrt{3})]^2$$
$$= 24[1 + (\sqrt{2} + \sqrt{3})^2 - 2(\sqrt{2} + \sqrt{3})]$$
$$= 24[6 + 2(\sqrt{6} - \sqrt{2} - \sqrt{3})]$$
$$+ 24[6 + 2x].$$

It follows from part (a) that either x is irrational or else x is an integer. But it is easy to check that

$$0 < \sqrt{2} + \sqrt{3} - \sqrt{6} < 1$$

(the inequalities $\sqrt{6} < \sqrt{2} + \sqrt{3}$ and $\sqrt{2} + \sqrt{3} < 1 + \sqrt{6}$ are easily checked by squaring them), so x is not an integer.

(c) Writing the various powers of $x = 2^{2/6} + 2^{3/6}$ in terms of the powers of $\eta = 2^{1/6}$, we obtain the following table for the coefficients.

	η^0	η^1	η^2	η^3	η^4	η^5
x^0	1					
x^1			1	1		
x^2	2				1	2
x^3	2	6	6	2		
x^4			2	8	12	8
x^5	40	40	20	4	2	10
x^6	12	24	60	80	60	24

We can then find numbers $a_0, \ldots, a_5$ such that

$$x^6 + a_5 x^5 + a_4 x^4 + a_3 x^3 + a_2 x^2 + a_1 x + a_0 = 0$$

by solving the equations $a_0 + 2a_2 + 2a_3 + 40a_5 + 12 = 0$, etc. It turns out that

$$x^6 - 6x^4 - 4x^3 + 12x^2 - 24x - 4 = 0.$$

Part (a) implies that either x is irrational or else x is an integer, and it is easy to see that x is not an integer, because $1.4 < \sqrt{2} < 1.5$ and $1.2 < \sqrt[3]{2} < 1.3$, so $2.6 < \sqrt{2} + \sqrt[3]{2} < 2.8$.

This is one of those problems where a little learning, though perhaps a dangerous thing, could save a lot of work: The proper equation for x can also be found by noting that $\sqrt{2} + \sqrt[3]{2}$ clearly satisfies the equation

$$[(x - \sqrt{2})^3 - 2] \cdot [(x + \sqrt{2})^3 - 2] = 0;$$

when the left side is multiplied out we obtain

$$(x - 2)^3 + 4 - 2 \cdot [(x - \sqrt{2})^3 + (x + \sqrt{2})^3]$$
$$= (x - 2)^3 + 4 - 2 \cdot [2x^3 + 12x] \qquad \text{(the odd powers of } x \text{ cancel out)}$$
$$= x^6 - 6x^4 - 4x^3 + 12x^2 - 24x - 4.$$

Of course, this method depends on the observation that the equation for $x = \sqrt{2} + \sqrt[3]{2}$ should also have $-\sqrt{2} + \sqrt[3]{2}$ as a root (a hint as to why this should be true will be found in Problem 25-8).

20. Since

$$\frac{\left(\dfrac{1 + \sqrt{5}}{2}\right)^1 - \left(\dfrac{1 - \sqrt{5}}{2}\right)^1}{\sqrt{5}} = \frac{\sqrt{5}}{\sqrt{5}} = 1,$$

$$\frac{\left(\dfrac{1 + \sqrt{5}}{2}\right)^2 - \left(\dfrac{1 - \sqrt{5}}{2}\right)^2}{\sqrt{5}} = \frac{\sqrt{5}}{\sqrt{5}} = 1,$$

the assertion is true for $n = 1$ and $n = 2$. Now suppose that the assertion is true for all $k < n$, where $n \geq 3$. Then it is true, in particular, for $n - 1$ and $n - 2$, so

$$a_n = a_{n-1} + a_{n-2}$$

$$= \frac{\left(\dfrac{1 + \sqrt{5}}{2}\right)^{n-2} - \left(\dfrac{1 - \sqrt{5}}{2}\right)^{n-2} + \left(\dfrac{1 + \sqrt{5}}{2}\right)^{n-1} - \left(\dfrac{1 - \sqrt{5}}{2}\right)^{n-1}}{\sqrt{5}}$$

$$= \frac{\left(\dfrac{1 + \sqrt{5}}{2}\right)^{n-2}\left(1 + \dfrac{1 + \sqrt{5}}{2}\right) - \left(\dfrac{1 - \sqrt{5}}{2}\right)^{n-2}\left(1 + \dfrac{1 - \sqrt{5}}{n}\right)}{\sqrt{5}}$$

$$= \frac{\left(\dfrac{1 + \sqrt{5}}{2}\right)^{n-2}\left(\dfrac{1 + \sqrt{5}}{2}\right)^2 - \left(\dfrac{1 - \sqrt{5}}{2}\right)^{n-2}\left(\dfrac{1 - \sqrt{5}}{2}\right)^2}{\sqrt{5}}$$

$$= \frac{\left(\dfrac{1 + \sqrt{5}}{2}\right)^{n} - \left(\dfrac{1 - \sqrt{5}}{2}\right)^{n}}{\sqrt{5}}.$$

21. (a) As before, the proof is trivial if all $y_i = 0$ or if there is some number λ with $x_i = \lambda y_i$ for all i. Otherwise,

$$0 < \sum_{i=1}^{n} (\lambda y_i - x_i)^2$$

$$= \lambda^2 \left(\sum_{i=1}^{n} y_i^{\,2} \right) - 2\lambda \left(\sum_{i=1}^{n} x_i y_i \right) + \sum_{i=1}^{n} x_i^{\,2},$$

so Problem 1-18 again gives the result.

(b) Using $2xy \le x^2 + y^2$ with

$$x = \frac{x_i}{\sqrt{\displaystyle\sum_{i=1}^{n} x_i^{\,2}}}, \qquad y = \frac{y_i}{\sqrt{\displaystyle\sum_{i=1}^{n} y_i^{\,2}}}$$

we obtain

(1)
$$\frac{2x_i y_i}{\sqrt{\displaystyle\sum_{i=1}^{n} x_i^{\,2}} \sqrt{\displaystyle\sum_{i=1}^{n} y_i^{\,2}}} \le \frac{x_i^{\,2}}{\displaystyle\sum_{i=1}^{n} x_i^{\,2}} + \frac{y_i^{\,2}}{\displaystyle\sum_{i=1}^{n} y_i^{\,2}}.$$

Adding we obtain

$$\frac{\displaystyle\sum_{i=1}^{n} 2x_i y_i}{\sqrt{\displaystyle\sum_{i=1}^{n} x_i^{\,2}} \sqrt{\displaystyle\sum_{i=1}^{n} y_i^{\,2}}} \le \frac{\displaystyle\sum_{i=1}^{n} x_i^{\,2}}{\displaystyle\sum_{i=1}^{n} x_i^{\,2}} + \frac{\displaystyle\sum_{i=1}^{n} y_i^{\,2}}{\displaystyle\sum_{i=1}^{n} y_i^{\,2}} = 2.$$

Again, equality holds only if it holds in (1) for all i, which means that

$$\frac{x_i}{\sqrt{\displaystyle\sum_{i=1}^{n} x_i^{\,2}}} = \frac{y_i}{\sqrt{\displaystyle\sum_{i=1}^{n} y_i^{\,2}}}$$

for all i. If all y_i are not 0, this means that $x_i = \lambda y_i$ for

$$\lambda = \frac{\sqrt{\displaystyle\sum_{i=1}^{n} x_i^{\,2}}}{\sqrt{\displaystyle\sum_{i=1}^{n} y_i^{\,2}}}.$$

(c) This is the most interesting proof—it depends on the equality

$$\sum_{i=1}^{n} x_i{}^2 \cdot \sum_{i=1}^{n} y_i{}^2 = \left(\sum_{i=1}^{n} x_i y_i\right)^2 + \sum_{i<j}(x_i y_j - x_j y_i)^2.$$

To check this equality, note that

$$\sum_{i=1}^{n} x_i{}^2 \cdot \sum_{i=1}^{n} y_i{}^2 = \sum_{i=1}^{n} x_i{}^2 y_i{}^2 + \sum_{i \neq j} x_i{}^2 y_j{}^2$$

$$\left(\sum_{i=1}^{n} x_i y_i\right)^2 = \sum_{i=1}^{n}(x_i y_i)^2 + \sum_{i \neq j} x_i y_i x_j y_j.$$

The difference is

$$\sum_{i \neq j}(x_i{}^2 y_j{}^2 - x_i y_i x_j y_j) = 2\sum_{i<j}(x_i{}^2 y_j{}^2 + x_j{}^2 y_i{}^2 - x_i y_i x_j y_j)$$

$$= 2\sum_{i<j}(x_i y_j - x_j y_i)^2.$$

If equality holds in the Schwarz inequality, then all $x_i y_j = x_j y_i$. If some $y_i \neq 0$, say $y_1 \neq 0$, then $x_i = \dfrac{x_1}{y_1} y_i$ for all i, so we can let $\lambda = x_1/y_1$.

22. (a) We have to prove that

$$A_n(a_1 + a_2 - A_n) \geq a_1 a_2$$

or

$$0 \geq A_n{}^2 - (a_1 + a_2)A_n + a_1 a_2$$
$$= (A_n - a_1)(A_n - a_2),$$

which is indeed true, since $a_1 < A_n < a_2$. If fact, we actually have $\bar{a}_1 \bar{a}_2 > a_1 a_2$.

This shows that $G_n \leq \bar{G}_n$, the geometric mean of $\bar{a}_1, \bar{a}_2, \ldots, \bar{a}_n$, while the arithmetic mean $\bar{A}_n$ is the same as A_n. So it suffices to prove that $\bar{G}_n \leq \bar{A}_n = A_n$. In other words, we can assume that one of the numbers (namely $\bar{a}_1$) actually equals the arithmetic mean. But now we can repeat this process and see that it suffices to prove the inequality when *two* of the numbers equal the arithmetic mean. Continuing enough times, it suffices to prove the inequality when all numbers are equal, in which case it is clearly true, and in fact, is an equality. This is clearly the only case where we have equality, since at the very first stage we get $\bar{G}_n < G_n$ if some $a_i \neq A_n$.

(b) We know that $G_n \leq A_n$ when $n = 2^1$. Suppose that $G \leq A_n$ for $n = 2^k$ and let $m = 2^{k+1} = 2n$. Then

$$G_m = \sqrt[m]{a_1 \cdots a_m} = \sqrt{\sqrt[n]{a_1 \cdots a_n} \sqrt[n]{a_{n+1} \cdots a_m}}$$

$$\leq \frac{\sqrt[n]{a_1 \cdots a_n} + \sqrt[n]{a_{n+1} \cdots a_m}}{2} \qquad \text{using } G_2 \leq A_2$$

$$\leq \frac{\dfrac{a_1 + \cdots + a_n}{n} + \dfrac{a_{n+1} + \cdots + a_m}{n}}{2} \qquad \text{by assumption}$$

$$= \frac{a_1 + \cdots + a_m}{2n}$$

$$= A_m.$$

(c) Applying (b) to these 2^m numbers yields, for $k = 2^m - n$,

$$(a_1 \cdots a_n)(A_n)^k \leq \left[\frac{a_1 + \cdots + a_n + kA_n}{2^m} \right]^{2^m}$$

$$= \left[\frac{nA_n + kA_n}{2^m} \right]^{2^m} = (A_n)^{2^m},$$

so

$$a_1 \cdots a_n \leq (A_n)^{2^m - k} = (A_n)^n.$$

23. Since $a^{n+1} = a^n \cdot a = a^n \cdot a^1$, the first equation is true for $m = 1$. Suppose that $a^{n+m} = a^n \cdot a^m$. Then

$$a^{n+(m+1)} = a^{(n+m)+1} = a^{n+m} \cdot a \qquad \text{by definition}$$

$$= (a^n \cdot a^m) \cdot a$$

$$= a^n \cdot (a^m \cdot a)$$

$$= a^n \cdot a^{m+1} \qquad \text{by definition,}$$

so the first equation is true for $m + 1$.

Since $(a^n)^1 = a^n = a^{n \cdot 1}$, the second equation is true for $m = 1$. Suppose that $(a^n)^m = a^{nm}$. Then

$$(a^n)^{m+1} = (a^n)^m \cdot a^n \qquad \text{by definition}$$

$$= a^{nm} \cdot a^n$$

$$= a^{nm+n} \qquad \text{by (i)}$$

$$= a^{n(m+1)}.$$

24. Since

$$1 \cdot (b + c) = b + c \qquad \text{by definition}$$
$$= 1 \cdot b + 1 \cdot c \qquad \text{by definition,}$$

the first result is true for $a = 1$. Suppose that $a \cdot (b + c) = a \cdot b + a \cdot c$ for all b and c. Then

$$(a + 1) \cdot (b + c) = a \cdot (b + c) + (b + c) \qquad \text{by definition}$$
$$= (a \cdot b + a \cdot c) + (b + c)$$
$$= (a \cdot b + b) + (a \cdot c + c) \qquad \text{by P1 and P4}$$
$$= (a + 1) \cdot b + (a + 1) \cdot c \qquad \text{by definition.}$$

The equation $a \cdot 1 = a$ is true for $a = 1$ by definition. Suppose that $a \cdot 1 = a$. Then

$$(a + 1) \cdot 1 = a \cdot 1 + 1 \cdot 1 \qquad \text{by definition}$$
$$= a + 1.$$

For $b = 1$, the equation $a \cdot b = b \cdot a$ follows from $a \cdot 1 = a$, which has just been proved, and $1 \cdot a = a$, which is true by definition. Suppose that $a \cdot b = b \cdot a$. Then

$$a \cdot (b + 1) = a \cdot b + a \cdot 1$$
$$= a \cdot b + a$$
$$= b \cdot a + a$$
$$= (b + 1) \cdot a \qquad \text{by definition.}$$

25. (a) (i) is clear.

(ii) This is clear, because 1 is positive, and if k is positive, then $k + 1$ is positive.

(iii) Clearly 1 is in this set. If condition (2) failed for this set, then there would be some k in the set with $k + 1 = 1/2$. But his is false, since $k = -1/2$ is not positive.

(iv) This set contains 4 but not $4 + 1$.

(v) Since 1 is in A and B, also 1 is in C. If k is in C, then k is in both A and B, so $k + 1$ is in A and B, so $k + 1$ is in C.

(b) (i) 1 is a natural number because 1 is in every inductive set, by definition of inductive sets.

(ii) If k is a natural number, then k is in every inductive set. So $k + 1$ is in every inductive set. So $k + 1$ is a natural number.

26. If there is only $n = 1$ ring, it can clearly be moved onto spindle 3 in $1 = 2^1 - 1$ moves. Assume the result for k rings. Then given $k + 1$ rings,

> (a) move the top k rings onto spindle 2 in $2^k - 1$ moves,
>
> (b) move the bottom ring onto spindle 3,
>
> (c) move the top k rings back onto spindle 3 in $2^k - 1$ moves.

This takes $2(2^k - 1) + 1 = 2^{k+1} - 1$ moves. If $2^k - 1$ moves is the minimum possible for k rings, then $2^{k+1} - 1$ is the minimum for $k + 1$ rings, since the bottom ring can't be moved at all until the top k rings are moved somewhere, taking at least $2^k - 1$ moves, the bottom ring has to be moved to spindle 3, taking at least 1 move, and then the other rings have to be placed on top of it, taking at least another $2^k - 1$ moves.

27. Everyone resigned on the seventeenth luncheon meeting.

The reasoning is as follows (for the sake of sanity, "he or she" shall be rendered as "he" throughout). First suppose there were only 2 professors, Prof. A and Prof. B, each knowing of the error in the other's work, but unaware of any error in his own. Then neither is surprised by Prof. X's statement, but each *expects the other to be surprised*, and to resign at the first luncheon meeting next year. When this doesn't happen, each (being a mathematics professor capable of logical deduction) realizes that this can only be because he has also made an error. So at the next meeting, both resign.

Next consider the case of 3 professors, Profs. A, B and C. Prof. C knows that Prof. A is aware of an error in Prof. B's work (either because Prof. A found the error and informed him, or because he found the error and informed Prof. A). Similarly, he knows that Prof. B knows that there is an error in Prof. A's work. But Prof. C thinks he has made no errors, so as far as he is concerned, the situation vis-a-vis Profs. A and B is precisely that analyzed in the previous paragraph (Prof. C is assuming, of course, that no one believes an error to exist when one doesn't). So Prof. C expects both Prof. A and Prof. B to resign at the second meeting. Of course, Profs. A and B similarly expect the other two to resign at the second meeting. When no one resigns, everyone realizes that he has made an error, so all resign at the third meeting.

Now you can turn this into a proof by induction (can't you?).

28. Again it is a good idea to start with the case when the department consists only of Profs. A and B. Now, of course, both professors know that some one has published an incorrect result, but Prof. A thinks that Prof. B *doesn't* know, and vice-versa. Once Prof. X makes his announcement, Prof. A knows that Prof. B knows. And that's why he expects Prof. B to resign at the next meeting.

In the case of three professors, the situation is more complicated. Each knows that some one has made an error, and moreover each knows that the others know—for example, Prof. C knows that Prof. A knows, since he and Prof. A have discussed the error in Prof. B's work, and he knows similarly that Prof. B knows. But Prof. C *doesn't* think that Prof. A knows that Prof. B knows. So Prof. X's announcement changes things: now Prof. C knows that Prof. A knows that Prof. B knows.

Well, you can see what happens in general. This seems to prove that statements like "A knew that B knew that C knew that ... " actually make sense.

CHAPTER 3

1. (ii) $x/(x+1)$ (for $x \neq 0, -1$).

(iv) $1/(1+x+y)$ (for $x+y \neq -1$).

(vi) For all c, since $f(c \cdot 0) = f(0)$.

2. (ii) Rational y between -1 and 1, and all y with $|y| > 1$.

(iv) All w with $0 \leq w \leq 1$.

3. (ii) $\{x : -1 \leq x \leq 1\}$.

(iv) $\{-1, 1\}$.

4. (ii) $\sin^2 y$.

(iv) $\sin t^3$.

5. (ii) $s \circ P$.

(iv) $S \circ s$.

(vi) $s \circ (P + P \circ S)$.

(viii) $P \circ S \circ s + s \circ S + P \circ s \circ (S + s)$.

6. (a) Let

$$f_i(x) = \frac{\prod\limits_{\substack{j=1 \\ j \neq i}}^{n} (x - x_j)}{\prod\limits_{\substack{j=1 \\ j \neq i}}^{n} (x_i - x_j)}.$$

(b) Let

$$f(x) = \sum_{i=1}^{n} a_i f_i(x)$$

$$= \sum_{i=1}^{n} a_i \cdot \prod_{\substack{j=1 \\ j \neq i}}^{n} \frac{(x - x_j)}{x_i - x_j}.$$

7. (a) If the degree of f is 1, then f is of the form

$$f(x) = cx + d = c(x - a) + (d + ac),$$

so we can let $g(x) = c$ and $b = d+ac$. Suppose that the result is true for polynomials of degree $\leq k$. If f has degree $k+1$, then f has the form

$$f(x) = a_{k+1}x^{k+1} + \cdots + a_1 x + a_0.$$

Now the polynomial function $h(x) = f(x) - a_{k+1}(x - a)$ has degree $\leq k$, so we can write

$$f(x) - a_{k+1}(x - 1) = (x - a)g(x) + b,$$

or

$$f(x) = (x - a)[g(x) + a_{k+1}] + b,$$

which is the required form.

(b) By part (a), we can write $f(x) = (x - a)g(x) + b$. Then

$$0 = f(a) = (a - a)g(a) + b = b,$$

so $f(x) = (x - a)g(x)$.

(c) Suppose f has n roots $a_1, \ldots, a_n$. Then by part (b) we can write $f(x) = (x - a)g_1(x)$ where the degree of $g_1(x)$ is $n - 1$. Now

$$0 = f(a_2) = (a_2 - a_1)g_1(a_2),$$

so $g_1(a_2) = 0$, since $a_2 \neq a_1$. Thus we can write

$$f(x) = (x - a_1)(x - a_2)g_2(x),$$

where the degree of g_2 is $n - 2$. Continuing in this way, we find that

$$f(x) = (x - a_1)(x - a_2) \cdots (x - a_n)c$$

for some number $c \neq 0$. It is clear that $f(a) \neq 0$ if $a \neq a_1, \ldots, a_n$. So f can have at most n roots.

(d) If $f(x) = (x - 1)(x - 2) \cdots (x - n)$, then f has n roots. If n is even, then $f(x) = x^n + 1$ has no roots. If n is odd, then $f(x) = x^n$ has only one root, namely 0.

8. If

$$x = f(f(x)) = \frac{a\left(\dfrac{ax + b}{cx + d}\right) + b}{c\left(\dfrac{ax + b}{cx + d}\right) + d}$$

for all x, then

$$(ac + cd)x^2 + (d^2 - a^2)x - ab - bd = 0 \qquad \text{for all } x,$$

so

$$ac + cd = 0,$$
$$ab + bd = 0,$$
$$d^2 - a^2 = 0.$$

It follows that $a = d$ or $a = -d$. One possibility is $a = d = 0$, in which case $f(x) = b/(cx)$, which satisfies $f(f(x)) = x$ for all $x \neq 0$. If $a = d \neq 0$, then $b = c = 0$, so $f(x) = x$. The third possibility is $a + d = 0$, so that $f(x) = (ax + b)/(cx - a)$, which satisfies $f(f(x)) = x$ for all $x \neq a/c$ (strictly speaking we should add the proviso that $f(x) \neq a/c$ for $x \neq a/c$, which means that

$$\frac{ax + b}{cx - a} \neq \frac{a}{c},$$

or $a^2 + bc \neq 0$).

9. (a)

$$C_{A \cap B} = C_A \cdot C_B,$$
$$C_{R-A} = 1 - C_A,$$
$$C_{A \cup B} = C_A + C_B - C_A \cdot C_B.$$

(b) Let $A = \{x : f(x) = 1\}$.

(c) $f = f^2$ if and only if $f(x) = 0$ or 1 for all x; so part (b) may be applied.

10. (a) Those functions f satisfying $f(x) \geq 0$ for all x.

(b) Those functions f with $f(x) \neq 0$ for all x.

(c) Those functions b and c satisfying $(b(t))^2 - 4c(t) \geq 0$ for all t.

(d) $b(t)$ must $= 0$ whenever $a(t) = 0$. If $a(t) \neq 0$ for all t, then there is a unique such function, namely $x(t) = a(t)/b(t)$. If $a(t) = 0$ for some t, then $x(t)$ can be chosen arbitrarily, so there are infinitely many such x.

11. (d) Let $H(1)$, $H(2)$, $H(13)$, $H(36)$, $H(\pi/3)$ and $H(47)$ have the values already prescribed, and let $H(x) = 0$ for $x \neq 1, 2, 13, 36, \pi/3, 47$. Since, in particular, $H(0) = 0$, the equation $H(H(x)) = H(x)$ holds for all x.

(e) Let $H(1) = 7$, $H(7) = 7$, $H(17) = 18$, $H(18) = 18$, and $H(x) = 0$ for $x \neq 1, 7, 17, 18$.

13. (a) Let

$$E(x) = \frac{f(x) + f(-x)}{2}, \qquad O(x) = \frac{f(x) - f(x)}{2}.$$

(b) If $f = E + O$, where E is even and O is odd, then

$$f(x) = E(x) + O(x),$$
$$f(-x) = E(x) - O(x).$$

Solving, we obtain the above expressions for $E(x)$ and $O(x)$.

14. $\max(f, g) = (f + g + |f - g|)/2$; $\min(f, g) = (f + g - |f - g|)/2$. (See Problem 1-13.)

15. (a) $f = \max(f, 0) + \min(f, 0)$ because

$$f(x) = \max(f(x), 0) + \min(f(x), 0) \qquad \text{for all } x,$$

the equation $a = \max(a, 0) + \min(a, 0)$ holding for all numbers a.

(b) For each x, choose numbers $g(x), h(x) \geq 0$ with $f(x) = g(x) - h(x)$. Since we can choose each pair $g(x)$ and $h(x)$ in infinitely many ways, there are infinitely many such functions g and h.

16. (a) The result is true for $n = 1$. If $f(x_1 + \cdots + x_n) = f(x_1) + \cdots + f(x_n)$ for all $x_1, \ldots, x_n$, then

$$\begin{aligned}
f(x_1 + \cdots + x_{n+1}) &= f([x_1 + \cdots + x_n] + x_{n+1}) \\
&= f(x_1 + \cdots + x_n) + f(x_{n+1}) \\
&= f(x_1) + \cdots + f(x_n) + f(x_{n+1}).
\end{aligned}$$

(b) Let $c = f(1)$. Now for any natural number n,

$$f(n) = f(\underbrace{1 + \cdots + 1}_{n \text{ times}}) = \underbrace{f(1) + \cdots + f(1)}_{n \text{ times}} = cn.$$

Since

$$f(x) + f(0) = f(x + 0) = f(x),$$

it follows that $f(0) = 0$. Then since

$$f(x) + f(-x) = f(x + (-x)) = f(0) = 0,$$

it follows that $f(-x) = f(x)$. In particular, for any natural number n,

$$f(-n) = -f(n) = -cn = c(-n).$$

Moreover,

$$\underbrace{f\left(\frac{1}{n}\right) + \cdots + f\left(\frac{1}{n}\right)}_{n \text{ times}} = f\left(\underbrace{\frac{1}{n} + \cdots + \frac{1}{n}}_{n \text{ times}}\right) = f(1) = c,$$

so

$$f\left(\frac{1}{n}\right) = c \cdot \frac{1}{n},$$

and consequently

$$f\left(\frac{1}{-n}\right) = f\left(-\frac{1}{n}\right) = -f\left(\frac{1}{n}\right) = -c \cdot \frac{1}{n} = c\left(\frac{1}{-n}\right).$$

Finally, any rational number can be written m/n for m a natural number, and n an integer; and

$$f\left(\frac{m}{n}\right) = f\underbrace{\left(\frac{1}{n} + \cdots + \frac{1}{n}\right)}_{m \text{ times}} = \underbrace{f\left(\frac{1}{n}\right) + \cdots + f\left(\frac{1}{n}\right)}_{m \text{ times}}$$

$$= mc \cdot \frac{1}{n} = c \cdot \frac{m}{n}.$$

17. **(a)** Since $f(a) = f(a \cdot 1) = f(a) \cdot f(1)$ and $f(a) \neq 0$ for some a, we have $f(1) = 1$.

(b) According to Problem 16, $f(x) = f(1)x = x$ for all rational x.

(c) If $c > 0$, then $c = d^2$ for some d, so $f(c) = f(d^2) = (f(d))^2 \geq 0$. Moreover, we cannot have $f(c) = 0$, since this would imply that

$$f(x) = f\left(c \cdot \frac{a}{c}\right) = f(c) \cdot f\left(\frac{a}{c}\right) = 0 \qquad \text{for all } a.$$

(d) If $x > y$, then $x - y > 0$, so $f(x) - f(y) > 0$, by part (c).

(e) Suppose that $f(x) > x$ for some x. Choose a rational number r with $x < r < f(x)$. Then, by parts (b) and (d),

$$f(x) < f(r) = r < f(x),$$

a contradiction. Similarly, it is impossible that $f(x) < x$. (There is a minor detail here which requires justification. See Problem 8-5.)

18. If either $f = 0$ or $g = 0$ holds, and also either $h = 0$ or $k = 0$, then the equation certainly holds. If not, then there is some x with $f(x) \neq 0$, and some y with $g(y) \neq 0$. Then $0 \neq f(x)g(y) = h(x)k(y)$, so we also have $h(x) \neq 0$ and $k(y) \neq 0$. Letting $\alpha = h(x)/f(x)$, we have $g(y') = \alpha k(y')$ for all y'. Moreover $\alpha = g(y)/k(y)$, so we also have $h(x') = \alpha f(x')$ for all y'. Moreover $\alpha = g(y)/k(y)$, so we also have $h(x') = \alpha f(x')$ for all x'. Thus we have $g = \alpha k$ and $h = \alpha f$ for some number $\alpha \neq 0$.

19. **(a)** **(i)** If $f(x) + g(y) = xy$ for all x and y, then, in particular,

$$f(x) + g(0) = 0 \qquad \text{for all } x.$$

So $f(x) = -g(0)$ for all x, and

$$-g(0) + g(y) = xy \qquad \text{for all } y;$$

setting $x = 0$ we obtain $g(y) = g(0)$. So we must have

$$0 = -g(0) + g(0) = xy \qquad \text{for all } x \text{ and } y,$$

which is absurd.

(ii) Setting $y = 0$, we obtain $f(x) = x/g(0)$. Similarly, setting $x = 0$, we obtain $g(y) = y/f(0)$. So

$$\frac{x}{g(0)} \cdot \frac{y}{f(0)} = x + y \qquad \text{for all } x \text{ and } y.$$

Choosing $y = 0$ we obtain $x = 0$ for all x, which is absurd.

(b) Let f and g be the same constant function. (Arguments similar to those used in part (a) show that these are the only possible choices.)

20. (a) Let $f(x) = x$.

(b) For every natural number n we have

$$|f(y) - f(x)| = \left| \sum_{k=1}^{n} f\left(x + \frac{k}{n}[y - x]\right) - f\left(x + \frac{k-1}{n}[y - x]\right) \right|$$

$$\leq \sum_{k=1}^{n} \left| f\left(x + \frac{k}{n}[y - x]\right) - f\left(x + \frac{k-1}{n}[y - x]\right) \right|$$

$$\leq \sum_{k=1}^{n} \frac{1}{n^2}(y - x)^2$$

$$= \frac{(y - x)^2}{n}.$$

Therefore $f(y) = f(x)$ for all x and y.

22. (a) If $f(x) = f(y)$, then $g(x) = h(f(x)) = h(f(y)) = g(y)$.

(b) If $z = f(x)$, define $h(z) = g(x)$. This definition makes sense, because if $z = f(x')$, then $g(x) = g(x')$ by part (a). For z not of the form $f(x)$, define h any old way (or leave it undefined). Then for all x in the domain of f we have $g(x) = h(f(x))$.

23. (a) Suppose $x \neq y$. Then $g(x) = g(y)$ would imply that $x = f(g(x)) = f(g(y)) = y$, a contradiction.

(b) $b = f(g(b))$, so let $a = g(b)$.

24. (a) The hypothesis can be stated as follows: If $x = y$, then $g(x) = g(y)$. The conclusion now follows from Problem 22(b), applied to g and I.

(b) For each x, choose some number a such that $x = f(a)$. Call this number $g(x)$. Then $f(g(x)) = x = I(x)$ for all x.

25. It suffices to find a function f such that $f(x) \neq f(y)$ if $x \neq y$, but such that not every number is of the form $f(x)$, because by Problem 24(a) there will be a function g with $g \circ f = I$, and by Problem 23(b) there will not be a function g with $f \circ g = I$. One such function is

$$f(x) = \begin{cases} x, & x \le 0 \\ x+1, & x > 0; \end{cases}$$

no number between 0 and 1 is of the form $f(x)$.

26. $h \circ f \circ g = h \circ (f \circ g) = h \circ I = h$, and also $h \circ f \circ g = (h \circ f) \circ g = I \circ g = g$.

27. (a) The condition $f \circ g = g \circ f$ means that $g(x) + 1 = g(x+1)$ for all x. There are many such g. In fact, g can be defined arbitrarily for $0 \le x < 1$, and its values for other x determined from this equation.

(b) If $f(x) = c$ for all x, then $f \circ g = g \circ f$ if and only if $c = f(g(x)) = g(f(x)) = g(x)$, i.e., $c = g(c)$.

(c) If $f \circ g = g \circ f$ for all g, then in particular this is true for all constant functions $g(x) = c$. It follows from part (b) that $f(c) = c$ for all c.

28. (a) is a straightforward check.

(b) Let f be a function with $f(x) = 0$ for some x, but not all x. Then $f \ne 0$, but there is clearly no function g with $f(x) \cdot g(x) = 1$ for all x.

(c) Let f and g be the two functions which are 0 except at x_0 and x_1, with $f(x_0) = 1$, $f(x_1) = 0$ and $g(x_0) = 0$, $g(x_1) = 1$. Neither is 0, so f or $-f$ would have to be in P, and likewise g or $-g$. But $(\pm f)(\pm g) = 0$, which contradicts P12.

(d) P'11, P'12 and P'13 are true. P'10 is false; although at most one of the conditions holds, it is not necessarily true that at least one holds. For example, if $f(x) > 0$ for some x and < 0 for other x, then neither $f = 0$, $f < 0$, nor $0 < f$ is true.

(e) The first inequality is not necessarily true. In fact, if $h(x) = -x$, then $f < g$ actually implies that $h \circ f > h \circ g$. The second inequality is true, since $f(h(x)) < g(h(x))$ for all x.

CHAPTER 4

1. (i) $(2, 4)$

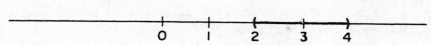

(ii) $[2, 4]$

(iii) $(a - \varepsilon, a + \varepsilon)$

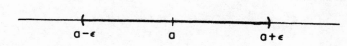

(iv) $\left(-\sqrt{3/2}, -\sqrt{1/2}\right) \cup \left(\sqrt{1/2}, \sqrt{3/2}\right)$.

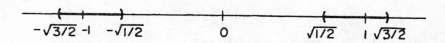

(v) $(-2, 2)$.

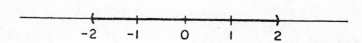

(vi) $\emptyset$ if $a \leq 0$;
$\mathbf{R}$ is $a \geq 1$;
$\left(-\infty, -\sqrt{(1/a) - 1}\,\right] \cup \left[\sqrt{(1/a) - 1}, \infty\right)$ if $0 < a < 1$.

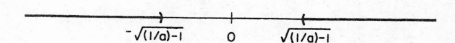

(vii) $(-\infty, 1] \cup [1, \infty)$.

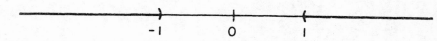

(viii) $(-1, 1) \cup (2, \infty)$.

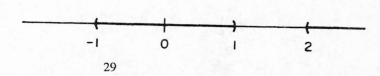

2. (a) Since $0 \leq x \leq b$, we have $0 \leq x/b \leq 1$, and $x = (x/b) \cdot b$; so choose $t = x/b$. Clearly t represents the ratio in which x divides the interval $[0, b]$. The midpoint of $[0, b]$ is $b/2$.

(b) If x is in $[a, b]$, so that

$$a \leq x \leq b,$$

then

$$0 \leq x - a \leq b - a,$$

so that $x - a$ is in $[0, b - a]$. It follows from part (a) that for some t with $0 \leq t \leq 1$ we have

$$x - a = t(b - a)$$

or

$$x = a + t(b - a) = (1 - t)a + tb.$$

The midpoint of $[a, b]$ is

$$a + \frac{b - a}{2} = \frac{a + b}{2}.$$

The point $1/3$ of the way from a to b is

$$a + \frac{b - a}{3} = \frac{2}{3}a + \frac{1}{3}b.$$

(c) and **(d)** are clear.

3. (i)   **(ii)**

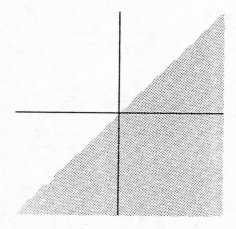

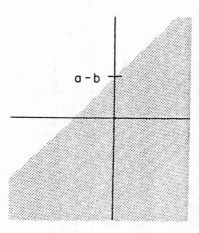

(iii)

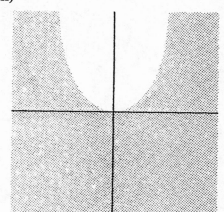

(iv)

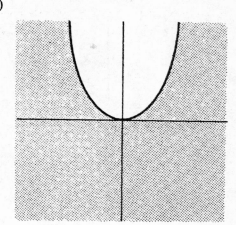

(v)

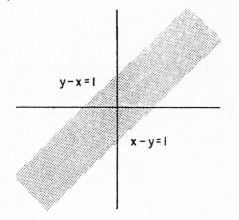

(vi)

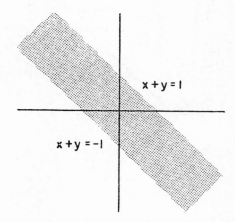

(vii)

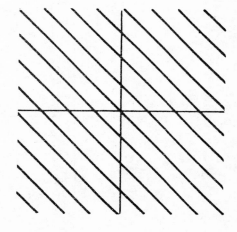

(viii)

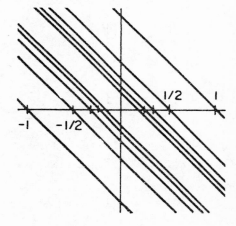

(ix)

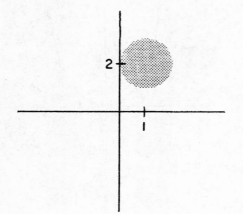

(x)

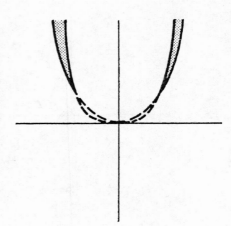

4. (i)

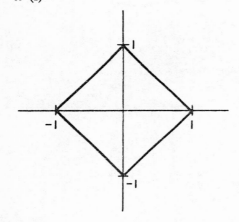

(ii)

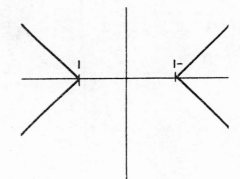

(iii), (iv)

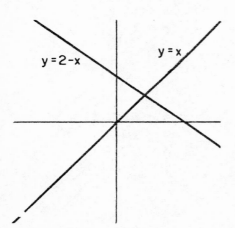

(v)

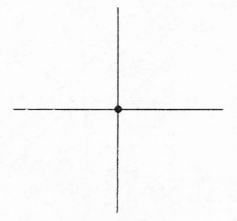

(vi)

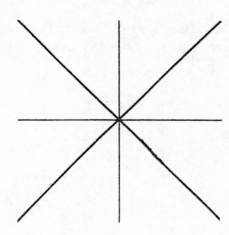

(vii) $x^2 - 2x + y^2 = (x - 1)^2 + y^2 - 1$

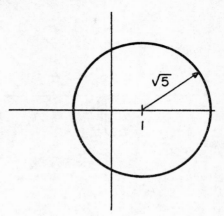

(viii)

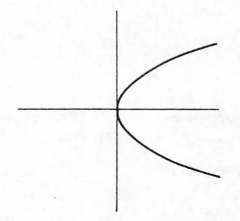

5. (i)

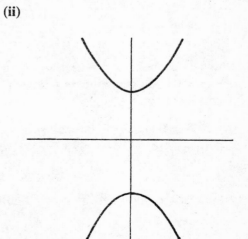

(ii)

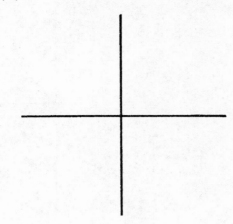

(iii)   **(iv)**

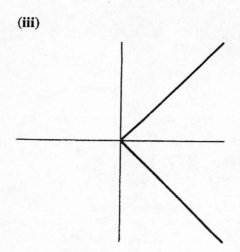 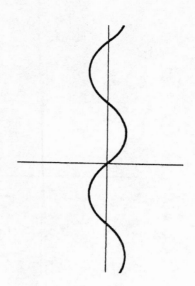

8. (a) The angle POQ is a right angle if and only if $(PQ)^2 = (PO)^2 + (OQ)^2$.

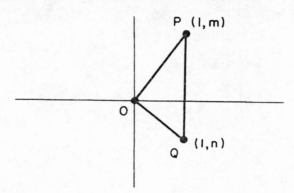

This means that

$$(m-n)^2 = m^2 + 1 + n^2 + 1,$$

which is equivalent to $-2mn = 2$, or $mn = -1$. This proves the result when $b = c = 0$. The general case follows from this special case, since perpendicularity depends only on the slope.

(b) If $B \neq 0$ and $B' \neq 0$, these straight lines are the graphs of

$$f(x) = (-A/B)x - C/A,$$
$$g(x) = (-A'/B')x - C/A;$$

so, by part (a), the lines are perpendicular if and only if

$$\left(-\frac{A}{B}\right) \cdot \left(-\frac{A'}{B'}\right) = -1,$$

which is equivalent to $AA' + BB' = 0$. If $B = 0$ (and consequently $A \neq 0$), then the first line is vertical, so the second is perpendicular to it if and only if $A' = 0$, which happens precisely when $AA' + BB' = 0$. Similarly if $B' = 0$.

9. (a) This inequality is equivalent to the squared inequality,

$$(x_1 + y_1)^2 + (x_2 + y_2)^2 \leq (x_1^2 + x_2^2) + (y_1^2 + y_2^2) + 2\sqrt{x_1^2 + x_2^2}\sqrt{y_1^2 + y_2^2},$$

which is easily seen to be equivalent to the Schwarz inequality.

(b) In part (a), replace

$$
\begin{array}{lll}
x_1 & \text{by} & x_2 - x_1, \\
x_2 & \text{by} & y_2 - y_1, \\
y_1 & \text{by} & x_3 - x_2, \\
y_2 & \text{by} & y_3 - y_2
\end{array}
$$

Geometrically, this inequality says that the length of one side of a triangle is less than the sum of the lengths of the other two. (Notice that the additional information about the Schwarz inequality which was presented in Problem 1-19(d) shows that $\leq$ can be replaced by $<$ in the triangle inequality except when (x_1, y_1), (x_2, y_2) and (x_3, y_3) lie on a straight line.)

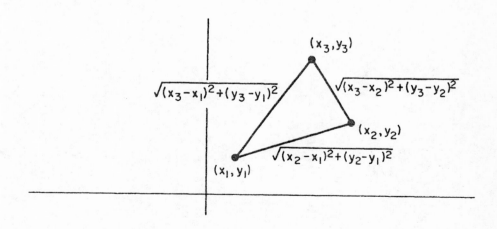

10. (The following figures do not indicate any particular points, since they were drawn using the method of Chapter 11, rather than by plotting points.)

(i) This function is odd.

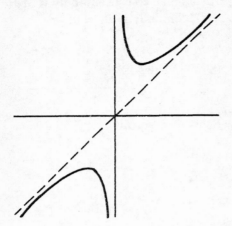

(ii) This function is odd.

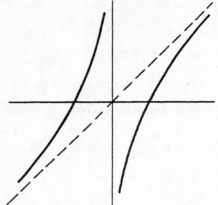

(iii) This function is even.

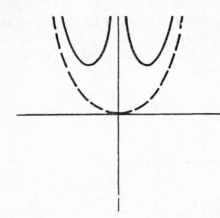

(iv) This function is even.

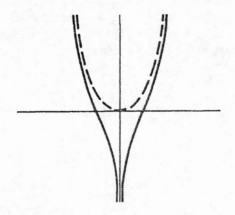

11. (i) The graph of f is symmetric with respect to the vertical axis.

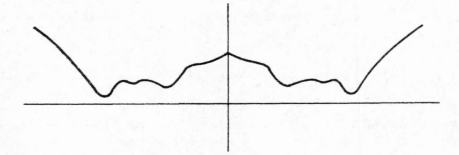

(ii) The graph of f is symmetric with respect to the origin. Equivalently, the part of the graph to the left of the vertical axis is obtained by reflecting first through the vertical axis, and then through the horizontal axis.

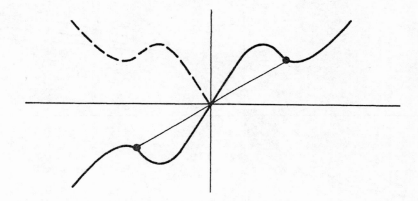

(iii) The graph of f lies above or on the horizontal axis.

(iv) The graph of f repeats the part between 0 and a over and over.

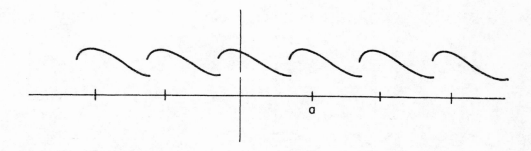

12. When n is odd, the domain of f is **R**, but when n is even, the domain of f is $[0, \infty)$.

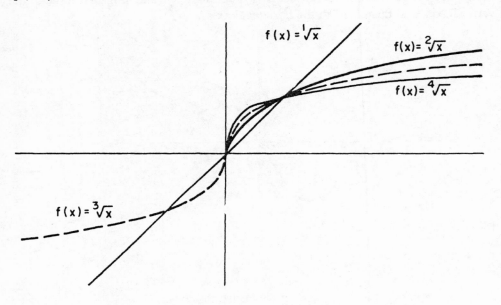

13. The graphs of $f(x) = |x|$ and $f(x) = |\sin x|$ contain "corners".

(a)

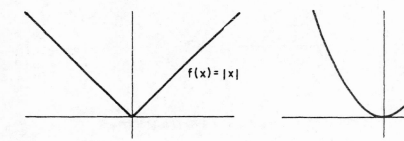

(b)

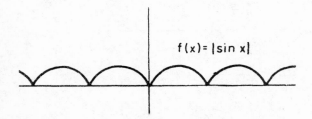
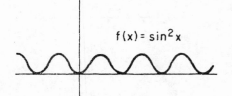

14. (i) The graph of g is the graph of f moved up c units.

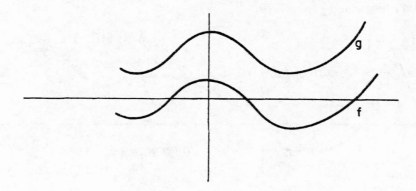

(ii) The graph of g is the graph of f moved over c units to the *left* (if $c > 0$).

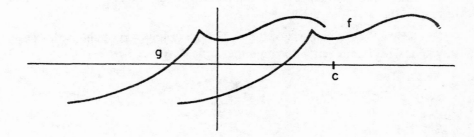

(iii) The height of the graph of f is multiplied by a factor of c everywhere. If $c = 0$, this means that $g = 0$; if $c > 0$, distances from the horizontal are increased in the same direction; if $c < 0$, distances are increased, but directions are changed.

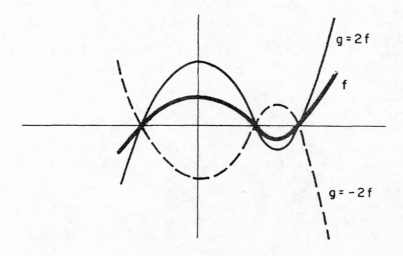

(iv) The graph of f is compressed by a factor of c if $c > 0$; if $c < 0$, the compression is combined with reflection through the vertical axis. If $c = 0$, then g is a constant function, $g(x) = f(0)$.

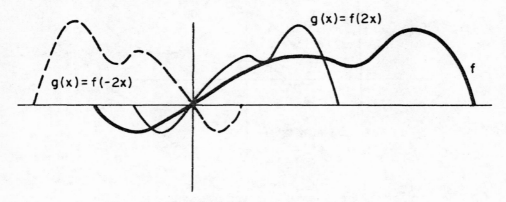

(v) "Everything that happens far out happens near 0, and vice versa", amply illustrated by the graph of $g(x) = \sin(1/x)$.

(vi) The graph of g consists of the part of the graph to the right of the vertical axis, together with its reflection through the vertical axis.

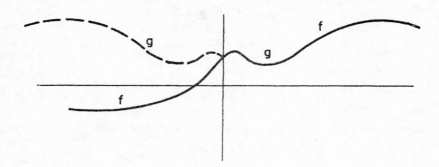

(vii) The graph of g is obtained by flipping up any parts of the graph of f which lie below the horizontal axis.

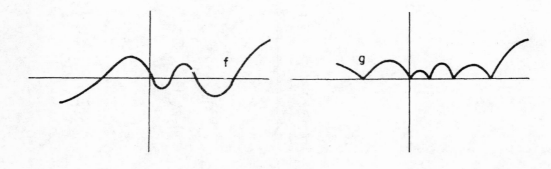

(viii) The graph of g is obtained by "cutting off" the part of the graph of f which lies below the horizontal axis.

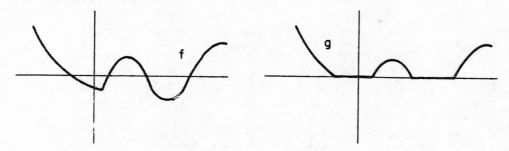

(ix) The graph of g is obtained by "cutting off" the part of the graph of f which lies above the horizontal axis.

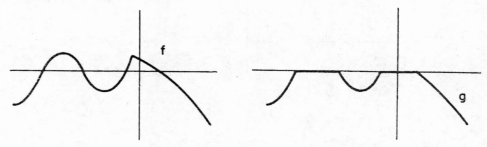

(x) The graph of g is obtained by "cutting off" the part of the graph of f which lies below the horizontal line at height 1.

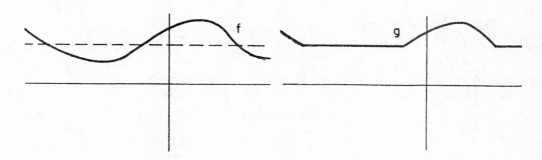

15. Since

$$f(x) = ax^2 + bx + c = a\left(x^2 + \frac{b}{a}x + \frac{c}{a}\right)$$

$$= a\left[\left(x + \frac{b}{2a}\right)^2 + \left(\frac{c}{a} - \frac{b^2}{4a}\right)\right],$$

the graph looks like the figure below.

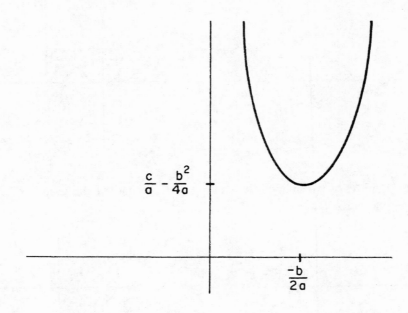

$$\frac{c}{a} - \frac{b^2}{4a}$$

$$\frac{-b}{2a}$$

16. Suppose $C = 0$, so that we have the equation

$$Ax^2 + Bx + Dy + E = 0.$$

If $D \neq 0$, this is equivalent to

$$y = -\frac{A}{D}x^2 - \frac{B}{D}x - \frac{E}{D},$$

so the set of all (x, y) satisfying this equation is the same as the graph of $f(x) = (-A/D)x^2 - (B/D)x - (E/D)$, which is a parabola, by Problem 15. [If $D = 0$, we have the equation $Ax^2 + Bx + E = 0$, $(A \neq 0)$, which may have zero, one or two solutions for x; in this case the set of all (x, y) satisfying the equation is either $\emptyset$, one straight line, or two parallel straight lines.] Similarly, if $A = 0$, then we again have a parabola [compare Problem 5(i)]. When $A, C \neq 0$ we can write the equation as

$$A\left(x + \frac{B}{2A}\right)^2 + C\left(y + \frac{D}{2C}\right)^2 = F$$

for some F.

When $A = C > 0$ we have a circle [compare page 65 of the text], unless $F = 0$, in which case we have a point (a "circle of radius 0"), or $F < 0$, in which case we have $\emptyset$. In general, when $A, C > 0$ we have an ellipse not necessarily centered at the origin (or a point, or $\emptyset$). There is no need to consider separately the case $A, C < 0$, since we have the same situation, replacing F by $-F$.

When A and C have different signs we have a hyperbola for $F \neq 0$ (which way it points depends on the signs of A, C and F). For $F = 0$ we have the equation

$$x + \frac{B}{2A} = \pm \sqrt{\frac{-C}{A}} \left(y + \frac{D}{2C} \right)$$

which gives two intersecting lines (a "degenerate hyperbola").

17. (i)

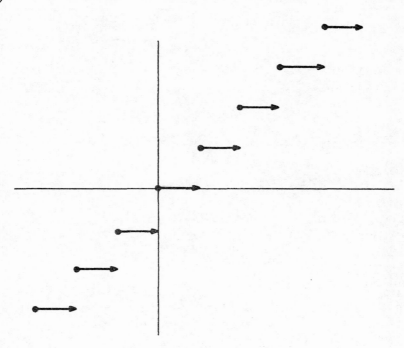

(ii)

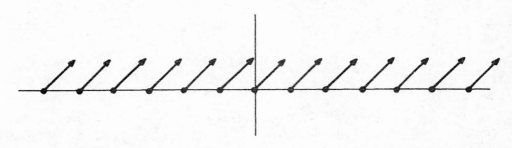

(iii)

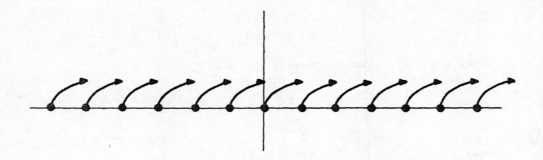

(iv)

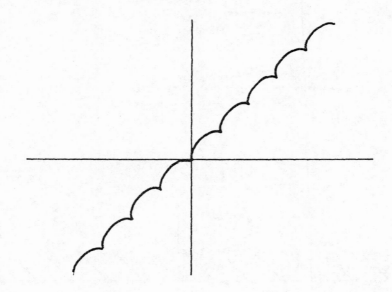

(v)

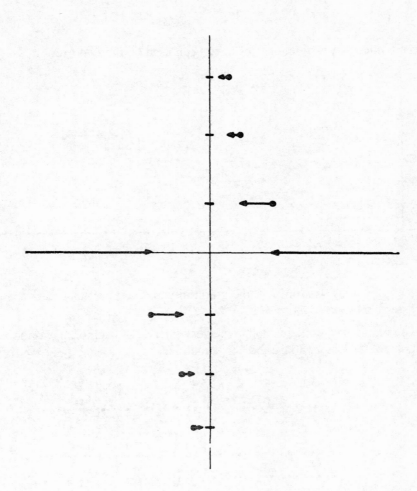

(vi) Notice that the domain of f is $\{x : -1 \leq x \leq 1 \text{ and } x \neq 0\}$.

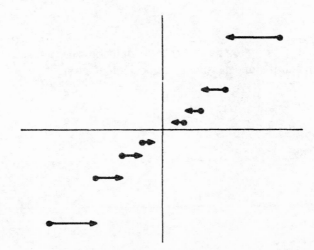

18. See pages 500 and 502 of the text.

19. (i) Notice that different scales have been used on the two axes.

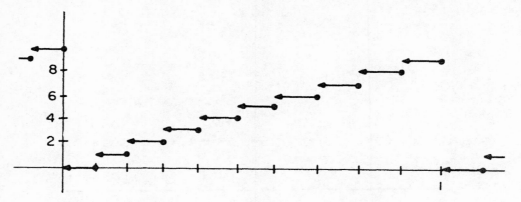

(ii) The graph of f is similar to the graph in part (i), except that there are ten sets of ten steps between n and $n + 1$.

(iii) The graph of f contains points in every interval of each of the horizontal lines at distance 0, 1, 2, ... , above the horizontal axis.

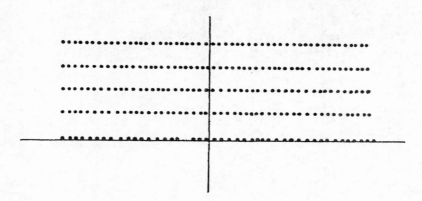

(iv) The graph of f contains points in every interval of the horizontal axis and of the horizontal line at distance 1 above the horizontal axis.

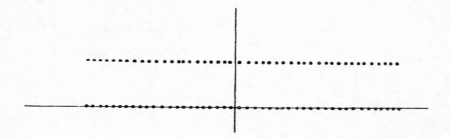

(v) The figure below shows a (rough) picture of the part of the graph of f which lies over $[6/10, 1]$.

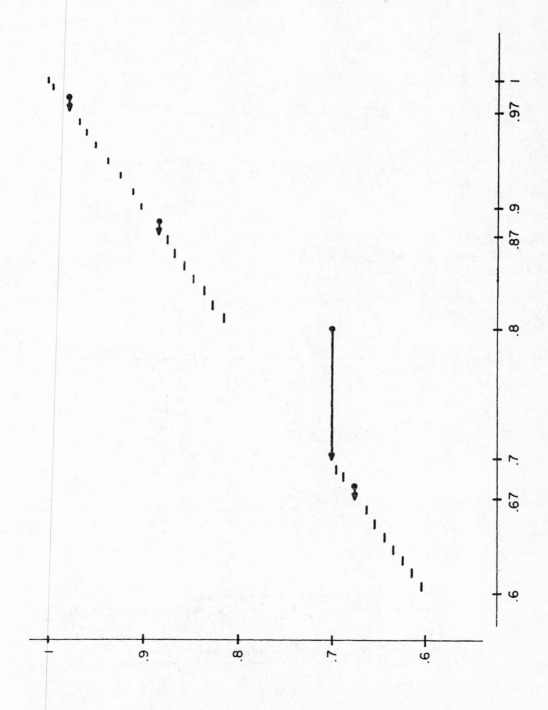

(vi) The figure below shows a (rough) picture of the graph of f. Notice that different scales have been used on the two axes.

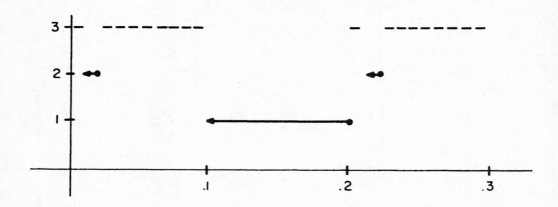

20. See page 97 of the text.

22. (a) The first part is a straightforward computation. By Problem 1-18, the minimum of these numbers is

$$d^2 + c^2 - \frac{(-2md - 2c)^2}{4(m^2 + 1)} = \frac{4m^2d^2 + 4d^2 + 4m^2c^2 + 4c^2 - (4m^2d^2 + 8mcd + 4c^2)}{4(m^2 + 1)}$$

$$= \frac{d^2 + m^2c^2 - 2mcd}{m^2 + 1} = \frac{(cm - d)^2}{m^2 + 1}.$$

(b) The distance from (c, d) to the graph of f is the same as the distance from $(c, d - b)$ to the graph of $g(x) = mx$. By part (a), this is

$$\frac{|cm - d + b|}{\sqrt{m^2 + 1}}.$$

23. (a)

$x' =$ distance from (x, y) to the graph of $f(x) = -x$ if (x, y) lies above this grpah (i.e., if $x + y \geq 0$), and the negative of this distance if $x + y \leq 0$.

$y' =$ distance from (x, y) to the graph of $f(x) = x$ if (x, y) lies above this grpah (i.e., if $x - y < 0$), and the negative of this when $x - y \geq 0$.

By Problem 22, these distances are given by

$$\frac{|-x-y|}{\sqrt{2}} = \left| \frac{x}{\sqrt{2}} + \frac{y}{\sqrt{2}} \right|,$$

$$\frac{|x-y|}{\sqrt{2}} = \left| \frac{x}{\sqrt{2}} - \frac{y}{\sqrt{2}} \right|,$$

from which the desired formulas follow.

(b) Since

$$\frac{x'}{\sqrt{2}} = \frac{x}{2} + \frac{y}{2},$$

$$\frac{y'}{\sqrt{2}} = -\frac{x}{2} + \frac{y}{2},$$

we have $(x'/\sqrt{2})^2 - (y'/\sqrt{2})^2 = 1$ if and only if

$$
\begin{aligned}
1 &= \left(\frac{x}{2} + \frac{y}{2} \right)^2 - \left(-\frac{x}{2} + \frac{y}{2} \right)^2 \\
&= \frac{x^2}{4} + \frac{y^2}{4} + \frac{xy}{2} - \left(\frac{x^2}{4} + \frac{y^2}{4} - \frac{xy}{2} \right) \\
&= xy.
\end{aligned}
$$

1. (a) The first formula is basically just the definition of $\sin\theta$ and $\cos\theta$. For the second formula, note that $R_\theta(0, 1)$ makes an angle of $\theta + 90°$ with the first axis, so

$$R_\theta(0, 1) = R_{\theta+90}(1, 0) = (\cos(\theta + 90°), \sin(\theta + 90°))$$
$$= (-\sin\theta, \cos\theta).$$

(b) Let the rotation R_θ be applied to Figure 3 on page 76. Then v moves to $R_\theta(v)$, and w moves to $R_\theta(w)$. Moreover, the (dashed) lines parallel to v and w become lines parallel to $R_\theta(v)$ and $R_\theta(w)$, respectively. This means that the intersection of those two lines, i.e., $v + w$, must move to the intersection of the two lines parallel to $R_\theta(v)$ and $R_\theta(w)$, i.e., $R_\theta(v) + R_\theta(w)$. This shows that

$$R_\theta(v + w) = R_\theta(v) + R_t(w).$$

To prove the second equation, simply note that since $a \cdot w$ lies along the line through the origin and w, it follows that $R_\theta(a \cdot w)$ must lie along the line through the origin and $R_\theta(w)$. Moreover, since the length of $a \cdot w$ is a times the length of w, the length of $R_\theta(a \cdot w)$ must also be a times the length of $R_\theta(w)$.

(c) We have

$$R_\theta(x, y) = R_\theta(x \cdot (1, 0) + y \cdot (0, 1))$$
$$= R_\theta(x \cdot (1, 0)) + R_\theta(y \cdot (0, 1))$$
$$= x \cdot R_\theta(1, 0) + y \cdot R_\theta(0, 1)$$
$$= x \cdot (\cos\theta, \sin\theta) + y \cdot (-\sin\theta, \cos\theta)$$
$$= (x\cos\theta, x\sin\theta) + (-y\sin\theta, y\cos\theta)$$
$$= (x\cos\theta - y\sin\theta, x\sin\theta + y\cos\theta).$$

(d) For $\theta = -45°$ we have

$$(x', y') = R_\theta(x, y) = (x\cos(-45°) - y\sin(-45°), x\sin(-45°) + y(-\cos 45°)).$$

Substituting

$$\sin(-45°) = -\frac{1}{\sqrt{2}} \qquad \cos(-45°) = \frac{1}{\sqrt{2}}$$

we get

$$(x', y') = \left(\frac{1}{\sqrt{2}}x + \frac{1}{\sqrt{2}}y, -\frac{1}{\sqrt{2}}x + \frac{1}{\sqrt{2}}y\right),$$

and thus the desired formulas for x' and y'.

2. (a) If w satisfies this equation, then so do all multiples $a \cdot w$. To solve

$$v_1 w_1 + v_2 w_2 = 0,$$

where v_1 and v_2 are fixed, we can assign w_1 arbitrarily, and then obtain

$$w_2 = -\frac{v_1 w_1}{v_2}.$$

Multiplying w_1 by a factor simply multiplies w_2 by the same factor, so the solutions are precisely the multiples of the one we obtain for any particular w_1.

This works provided $v_2 \neq 0$. If $v_2 = 0$, so that v is a multiple of $(1, 0)$, then it is easy to see that the possible w's consist of all multiples of $(0, 1)$, and vice-versa.

(b) These are all straightforward computations from the definition.

(c) Since $v \cdot v = v_1{}^2 + v_2{}^2$, this is obvious. The norm

$$\|v\| = \sqrt{v \cdot v} = \sqrt{v_1{}^2 + v_2{}^2}$$

is just the distance from v to the origin.

(d) This is simply Problem 4-9: The squared inequality is equivalent to the Schwarz inequality (Problem 1-19); equality holds in this squared inequality only when $v = 0$ or $w = 0$ or $w = a \cdot v$ for some a. For the original inequality it is then easy to see that equality holds only when $a > 0$.

(e) We have

$$\|v + w\|^2 = (v + w) \cdot (v + w) = v \cdot v + 2v \cdot w + w \cdot w$$
$$\|v - w\|^2 = (v - w) \cdot (v - w) = v \cdot v - 2v \cdot w + w \cdot w.$$

Subtracting the second from the first we get

$$\|v + w\|^2 - \|v - w\|^2 = 4(v \cdot w).$$

3. (a) We have

$$R_\theta(v) \cdot R_\theta(w) = (v_1 \cos\theta - v_2 \sin\theta, \, v_1 \sin\theta + v_2 \cos\theta) \cdot$$
$$(w_1 \cos\theta - w_2 \sin\theta, \, w_1 \sin\theta + w_2 \cos\theta)$$
$$= v_1 w_1 \cos^2\theta + v_1 w_1 \sin^2\theta + v_2 w_2 \sin^2\theta + v_2 w_2 \cos^2\theta$$
$$+ \sin\theta \cos\theta [-v_1 w_2 - w_1 v_2 + v_1 w_2 + w_1 v_2]$$
$$= v_1 w_1 + v_2 w_2 = v \cdot w.$$

(b) The formula for $e \cdot w$ is a straightforward calculation. For the vectors $v = a \cdot e$ and $u = b \cdot w$ we then have, using Problem 2(b),

$$v \cdot u = (a \cdot e) \cdot u = a \cdot (e \cdot u)$$
$$= a \cdot (e \cdot (b \cdot w)) = a \cdot (b \cdot (e \cdot w))$$
$$= ab \cdot (e \cdot w),$$

which gives the formula

$$v \cdot w = \|v\| \cdot \|w\| \cdot \cos\theta$$

when v a multiple of (1,0).

For the general case, choose ϕ to be the angle from the first axis to v, so that $v = R_\phi(v')$ for some v' pointing along the first axis, and let $w = R_\phi(w')$. Since rotation doesn't change lengths, we have

$$\|v\| = \|v'\|, \qquad \|w\| = \|w'\|;$$

moreover, the angle θ' between v' and w' is the same as the angle θ between v and w. Then by part (a) we have

$$\begin{aligned} v \cdot w &= R_\phi(v') \cdot R_\phi(w') \\ &= v' \cdot w' \\ &= \|v'\| \cdot \|w'\| \cdot \cos\theta' \\ &= \|v\| \cdot \|w\| \cdot \cos\theta. \end{aligned}$$

4. Using the "point-slope" form (Problem 4-6) the line L is the graph of

$$f(x) = \frac{w_2}{w_1}(x - v_1) + v_2.$$

Solving $f(x) = 0$, we find the desired first coordinate of B, and thus the formula for the area of the parallelogram, which has that base and height w_2.

5. (a) For $v_2 = 0$ the formula for det reduces to $v_1 w_2$, and v_1 (> 0) is the base of the parallelogram; the height is w_2 (and hence the area is $v_1 w_2 = \det$) for $w_2 > 0$, while the height is $-w_2$ (and hence the area is $-v_1 w_2 = -\det$) for $w_2 < 0$.

(b)

$$\begin{aligned} \det(R_\theta v, R_\theta w) &= \det\big((v_1\cos\theta - v_2\sin\theta, \, v_1\sin\theta + v_2\cos\theta), \\ &\qquad\qquad (w_1\cos\theta - w_2\sin\theta, \, w_1\sin\theta + w_2\cos\theta)\big) \\ &= \big[v_1\cos\theta - v_2\sin\theta\big] \cdot \big[w_1\sin\theta + w_2\cos\theta\big] \\ &\quad - \big[v_1\sin\theta + v_2\cos\theta\big] \cdot \big[w_1\cos\theta - w_2\sin\theta\big] \\ &= v_1 w_2 - v_2 w_1 = \det(v, w). \end{aligned}$$

For any v and w, we can write $v = R_\theta(v')$ for some v' that points along the positive horizontal axis; then $w = R_\theta(w')$ for some w', and w lies above the horizontal axis when the rotation from v to w is counterclockwise, and below the axis when the rotation is clockwise. The area of the parallelogram spanned by v and w is the same as that spanned by $v' = R_\theta(v)$ and $w' = R_\theta(w)$, which by part (a) is therefore $\pm\det(v', w')$, depending on whether the rotation is clockwise or counterclockwise. But we have just seen that this is $\pm\det(v, w)$.

6. These are all straightforward computations from the definition.

7. As in Problem 3, we first check the formula when v is a multiple of $(0, 1)$. Then choose ϕ so that $v = R_\phi(v')$ for some v' pointing along the first axis, and let $w = R_\phi(w')$; we again have

$$\|v\| = \|v'\|, \qquad \|w\| = \|w'\|;$$

moreover, the angle θ' between v' and w' is the same as the angle θ between v and w. Then by the formula in Problem 5(b) we have

$$
\begin{aligned}
\det(v, w) &= \det(R_\phi(v'), R_\phi(w')) \\
&= \det(v', w') \\
&= \|v'\| \cdot \|w'\| \cdot \sin\theta' \\
&= \|v\| \cdot \|w\| \cdot \sin\theta.
\end{aligned}
$$

1. The point (x, y, z) is in the cylinder if and only if

$$x^2 + y^2 = C^2.$$

Choosing coordinates in the plane P as on page 81, we see that the points in the intersection of P and the cylinder are those satisfying

$$(\alpha x + \beta)^2 + y^2 = C^2.$$

The possibilities are $\emptyset$, a straight line, two parallel straight lines, or an ellipse (or circle).

2. (a) Consider the plane containing the line L_1 from z to F_1 and the line L: it intersects the sphere S_1 in a circle C. Since S_1 is tangent to the plane P at F_1 it follows that L_1 is tangent to C at F_1, and L is also tangent to C. The desired result now follows from the fact that the two line segments tangent to a circle from an outside point have the same length.

(b) Similarly, the length of the line from z to F_2 is the length of the vertical line L' from z to C_2. But L and L' together form a vertical straight line from the plane of C_1 to the plane of C_2. Hence the distance from z to F_1 plus the distance from z to F_2 is always exactly the distance between these two planes.

3. The proof is similar, except that now the sum will always be the length of a straight line generator of the cone between the planes of the two circles.

CHAPTER 4, Appendix 3

1. The points with polar coordinates (r_1, θ_1) and (r_2, θ_2) are

$$(r_1 \cos \theta_1, r_1 \sin \theta_1) \qquad \text{and} \qquad (r_2 \cos \theta_2, r_2 \sin \theta_2)$$

and the distance d between them is given by

$$\begin{aligned}
d^2 &= (r_2 \cos \theta_1 - r_1 \cos \theta_1)^2 + (r_2 \sin \theta_2 - r_1 \sin \theta_1)^2 \\
&= r_2{}^2 (\cos^2 \theta_2 + \sin^2 \theta_2) + r_1{}^2 (\cos^2 \theta_1 + \sin^2 \theta_1) \\
&\quad - 2 r_1 r_2 [\cos \theta_1 \cos \theta_2 + \sin \theta_1 \sin \theta_2] \\
&= r_1{}^2 + r_2{}^2 - 2 r_1 r_2 \cos(\theta_1 - \theta_2).
\end{aligned}$$

This is just the "law of cosines".

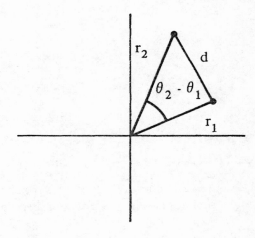

2. (i) For each point (x, y) on the graph of f, with

$$x = f(\theta) \cos \theta, \qquad y = f(\theta) \sin \theta$$

we also have the point (x', y') with

$$\begin{aligned}
x' &= f(-\theta) \cos(-\theta) = f(\theta) \cos \theta = x, \\
y' &= f(-\theta) \sin(-\theta) = -f(\theta) \sin \theta = -y.
\end{aligned}$$

The point $(x', y') = (x, -y)$ is the reflection of (x, y) through the horizontal axis, so the graph of f in polar coordinates is symmetric with respect to this axis.

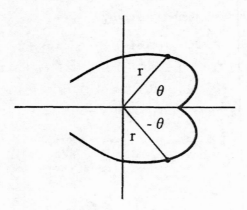

(ii) Similarly, if f is odd, then

$$x' = f(-\theta)\cos(-\theta) = -f(\theta)\cos\theta = -x,$$
$$y' = f(-\theta)\sin(-\theta) = -f(\theta)(-\sin\theta) = y.$$

The point $(x', y') = (-x, y)$ is the reflection of (x, y) through the vertical axis, so the graph of f in polar coordinates is symmetric with respect to the vertical axis.

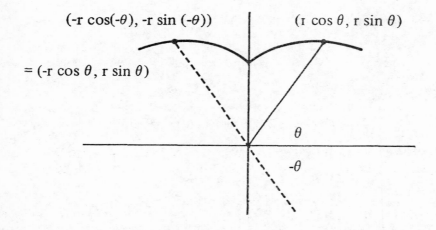

(iii) The graph of f in polar coordinates is symmetric with respect to the origin.

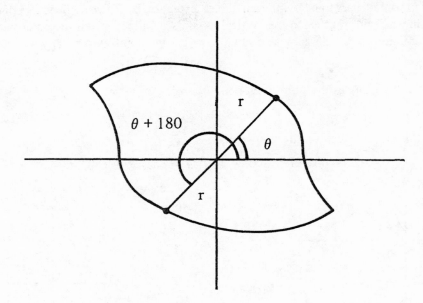

3. (i) $r = a \sin \theta$ implies $r^2 = ar \sin \theta$, so if $(x, y) = (r \cos \theta, r \sin \theta)$, then $x^2 + y^2 = ay$, or

$$x^2 + \left(y - \frac{a}{2}\right)^2 = \left(\frac{a}{2}\right)^2,$$

so (x, y) lies on the circle of radius $a/2$ with center $(0, a/2)$. [Conversely, if (x, y) satisfies $x^2 + y^2 = ay$ and (r, θ) are polar coordinates for (x, y), so that $x = r \cos \theta$ and $y = r \sin \theta$, then $r^2 = ar \sin \theta$. This implies that $r = a \sin \theta$, except when $r = 0$. In this case we have the point $(x, y) = (0, 0)$, which also lies on the graph of $r = a \sin \theta$, since it has polar coordinates $r = \theta = 0$.]

(ii) If $a = 0$, we have the equation $r = 0$, which is the single point $(0, 0)$. Suppose $a \neq 0$. Now $r = a \sec \theta = a/\cos \theta$ implies that

$$r \cos \theta = a,$$

so if $(x, y) = (r \cos \theta, r \sin \theta)$, then

$$x = a,$$

and (x, y) lies on the vertical line through $(a, 0)$. Notice that we must exclude points with $\cos \theta = 0$, but they can't be on this vertical line anyway, since $a \neq 0$. [Conversely, if (r, θ) are polar coordinates for a point (x, y) on this line, then

$$a = r \cos \theta, \qquad y = r \sin \theta,$$

and in particular $r = a/\cos \theta$ ($\cos \theta \neq 0$, since $a \neq 0$).]

(iii) Figure (a) shows the part of the graph from $\theta = 0$ to $\theta = 90°$. Figure (b) shows the part from $\theta = -90°$ to $\theta = 90°$. It is symmetric with respect to the horizontal axis, since cos is even. Finally, Figure (c) shows the whole graph, a four-leaf clover. The graph appears to be symmetric under a rotation by 90°, and hence, in particular, symmetric with respect to the vertical axis also. In fact, when the point with polar coordinate $(\cos 2\theta, \theta)$ is rotated by 90° we get the point with polar coordinates

$$(\cos 2\theta, \theta + 90°).$$

This is the same as the point with polar coordinates

$$(-\cos 2\theta, \theta + 90° + 180°)$$

and this point is also on the graph, since

$$\cos\big(2(\theta + 90° + 180°)\big) = \cos(2\theta + 180°)$$
$$= -\cos 2\theta.$$

(a)

(b)

(c)

Although Figure (c) shows θ going from $-90°$ to $270°$, it could just as well show θ going from $0°$ to $360°$. Note that if we do not allow negative values for r, the graph will contain only the left and right leaves.

(iv) Figure (a) shows the part of the graph from $\theta = 0°$ to $\theta = 60°$. Figure (b) shows the part from $\theta = -60°$ to $\theta = 60°$. It is symmetric with respect to the horizontal axis, as before. Finally, Figure (c) shows the whole graph. It is symmetric under a rotation by 120°, for if the point with polar coordinates $(\cos 3\theta, \theta)$ is rotated by 120° we get the point with polar coordinates

$$(\cos 3\theta, \theta + 120°)$$

and this is on the graph, since

$$\cos 3\theta = \cos(3(\theta + 120°)).$$

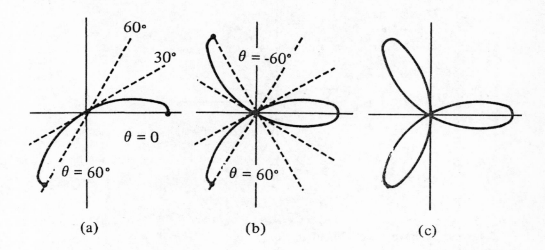

(a) (b) (c)

Notice that in this case we will get the whole graph even if we allow only $r > 0$:

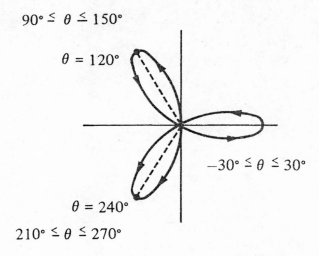

The proof of symmetry with respect to rotation through 120° didn't involve replacing θ by $\theta + 180°$, as in the previous example.

(v) The graph is the same as in (iii). (However, now we obtain 4 leaves no matter what conventions we adopt about the sign of r, since $r \geq 0$ in any case.)

(vi) The graph has 6 leaves (each leaf in (iv) arises from an interval on which $r \leq 0$ as well as from one on which $r \geq 0$).

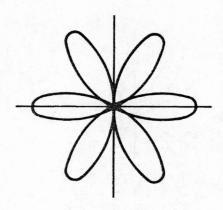

4. (i) and **(ii)** have already been given.

(iii)
$$r^3 = r^2 \cos 2\theta = r^2 \cos^2 \theta - r^2 \sin^2 \theta = x^2 - y^2$$

so

$$(x^2 + y^2)^{3/2} = x^2 - y^2.$$

5. As before, the distance r from (x, y) to O is given by

$$(1) \qquad\qquad r^2 = x^2 + y^2,$$

while the distance s to $\mathbf{f}$ is given by

$$s^2 = (x + 2\varepsilon a)^2 + y^2.$$

Now writing the condition

$$r - s = 2a$$

as

$$r - 2a = s$$

and squaring, we get the same equation as before,

$$(2) \qquad\qquad 4a^2 - 4ar + r^2 = x^2 + 4\varepsilon a x + 4\varepsilon^2 a^2 + y^2,$$

so subtracting (1) from (2) again gives

$$a - r = \varepsilon x + \varepsilon^2 a,$$

and thus

$$(3) \qquad\qquad r = \Lambda - \varepsilon x, \qquad \text{for } \Lambda = (1 - \varepsilon^2)a,$$

and once again

(4)
$$r = \frac{\Lambda}{1 + \varepsilon \cos \theta}.$$

It remains to consider the points satisfying

$$s - r = 2a,$$

or

$$r + 2a = s$$

Squaring we now obtain

(2′)
$$r^2 + 4ar + 4a^2 = x^2 + 4\varepsilon ax + 4\varepsilon^2 a^2 + y^2.$$

Subtracting (1) from (2′) gives

$$a + r = \varepsilon x + \varepsilon^2 a,$$

or

$$r = (\varepsilon^2 - 1)a + \varepsilon x$$
$$= -(\Lambda - \varepsilon x),$$

which is simply the negative of the r found previously; thus, the other branch of the hyperbola is obtained by choosing $-\Lambda$ for Λ.

6. The distance from the line to (x, y) is just

$$a - x = a - r \cos \theta;$$

thus our condition is

$$r = a - r \cos \theta,$$

or equivalently

$$a = r(1 + \cos \theta).$$

7. Squaring (3) and substituting $x^2 + y^2$ for r^2, we get

$$x^2 + y^2 = \Lambda^2 - 2\varepsilon \Lambda x + \varepsilon^2 x^2,$$

which gives the desired equation,

$$(1 - \varepsilon^2)x^2 + y^2 = \Lambda^2 - 2\Lambda \varepsilon x.$$

Problem 4-16 shows that this is a circle or ellipse when $1 - \varepsilon^2 > 0$ is positive, i.e., when $\varepsilon < 1$ (remember that we have already specified $\varepsilon > 0$), a hyperbola when $1 - \varepsilon^2 < 0$, i.e., when $\varepsilon > 1$, and a parabola when $1 - \varepsilon^2 = 0$, i.e., when $\varepsilon = 1$.

8. (a) The graph is the heart-shaped curve shown below.   (Hence the name "cardioid" = heart-shaped).

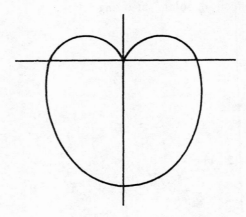

(b) The point with polar coordinates (r, θ) is also the point with polar coordinates $(-r, \theta + 180°)$. So the graph of $r = 1 - \sin = \theta$ is also the graph of

$$-r = 1 - \sin(\theta + 180°) = 1 + \sin \theta.$$

(c) Since $r = 1 - \sin \theta$ we have

$$r^2 = r - r \sin \theta$$

or

$$x^2 + y^2 = \sqrt{x^2 + y^2} - y.$$

(Notice that if we start with $r = -1 - \sin \theta$, then we obtain the same result since now $r < 0$, so $r = -\sqrt{x^2 + y^2}$.)

The squared equation

$$(x^2 + y^2 + y)^2 = x^2 + y^2$$

might seem to have the extraneous solutions

$$x^2 + y^2 = -\sqrt{x^2 + y^2} - y,$$

but for $x \neq 0$ this has no solutions, for we have

$$-y \leq |y| < \sqrt{x^2 + y^2},$$

and hence

$$-y - \sqrt{x^2 + y^2} < 0.$$

9. (i) The graph is shown below (the dashed line is the cardioid $r = 1 - \sin\theta$).

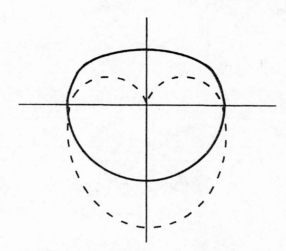

(ii) The graph is shown below, together with the cardioid (dashed line).

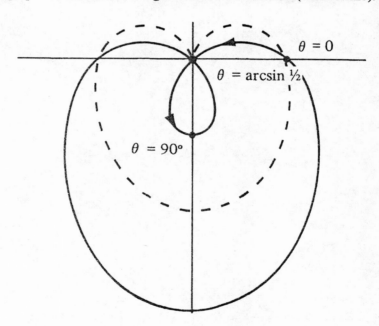

(iii) This graph has the same shape as (i): Since

$$\cos\theta = -\sin(\theta - 90°)$$

we can write

$$r = 2 - \sin(\theta - 90°)$$
$$= 2\left(1 - \tfrac{1}{2}\sin(\theta - 90°)\right),$$

which shows that the graph is twice as large as the curve in (i) and rotated 90° counter-clockwise.

10. (a)

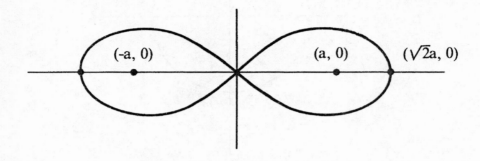

(b) Since

$$r^4 = 2a^2 r^2 \cos 2\theta = 2a^2 r^2 (\cos^2 \theta - \sin^2 \theta)$$

we obtain

$$(x^2 + y^2)^2 = 2a^2 (x^2 - y^2).$$

(c) If $P = (x, y)$, then

$$
\begin{aligned}
(d_1 d_2)^2 &= [(x - a)^2 + y^2] \cdot [(x + a)^2 + y^2] \\
&= [(x - a)(x + a)]^2 + y^2 [(x + a)^2 + (x - a)^2] + y^4 \\
&= (x^2 - a^2)^2 + y^2 [2x^2 + 2a^2] + y^4 \\
&= x^4 - 2a^2 x^2 + a^4 + 2x^2 y^2 + 2a^2 y^2 + y^4 \\
&= (x^2 + y^2)^2 - 2a^2 (x^2 - y^2) + a^4,
\end{aligned}
$$

so $d_1 d_2 = a^2$ if and only if

$$(x^2 + y^2)^2 = 2a^2 (x^2 - y^2).$$

(d) For $b < a$ we obtain two curves, inside the two portions of the lemniscate, as in the figure below.

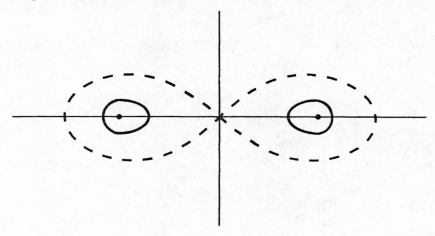

For $b < a$ we obtain a single curve surrounding the lemniscate. It happens to be indented for $b < a\sqrt{2}$.

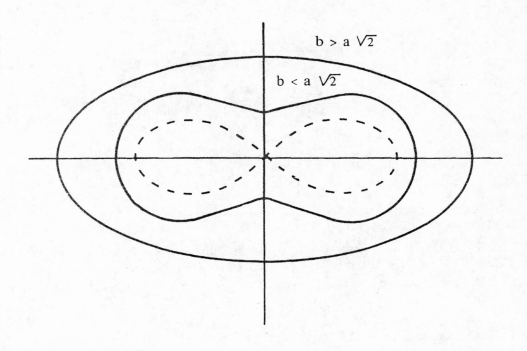

CHAPTER 5

1. (i)

$$\lim_{x \to 1} \frac{x^2 - 1}{x + 1} = \frac{0}{2} = 0.$$

(iii)

$$\lim_{x \to 3} \frac{x^3 - 8}{x - 2} = \frac{27 - 8}{3 - 2} = 19.$$

(v)

$$\lim_{y \to x} \frac{x^n - y^n}{x - y} = \lim_{y \to x} \frac{y^n - x^n}{y - x} = nx^{n-1}, \qquad \text{by (iv).}$$

2. (i)

$$\lim_{x \to 1} \frac{1 - \sqrt{x}}{1 - x} = \lim_{x \to 1} \frac{\left(1 - \sqrt{x}\right)\left(1 + \sqrt{x}\right)}{\left(1 - x\right)\left(1 + \sqrt{x}\right)}$$

$$= \lim_{x \to 1} \frac{1 - x}{\left(1 - x\right)\left(1 + \sqrt{x}\right)}$$

$$= \lim_{x \to 1} \frac{1}{1 + \sqrt{x}} = \frac{1}{2}.$$

(ii)

$$\lim_{x \to 0} \frac{1 - \sqrt{1 - x^2}}{x} = \lim_{x \to 0} \frac{\left(1 - \sqrt{1 - x^2}\right)\left(1 + \sqrt{1 - x^2}\right)}{x\left(1 + \sqrt{1 + x^2}\right)}$$

$$= \lim_{x \to 0} \frac{1 - (1 - x^2)}{x\left(1 + \sqrt{1 + x^2}\right)} = \lim_{x \to 0} \frac{x}{1 + \sqrt{1 - x^2}} = 0.$$

(iii)

$$\lim_{x \to 0} \frac{1 - \sqrt{1 - x^2}}{x^2} = \lim_{x \to 0} \frac{1}{1 + \sqrt{1 + x^2}} = \frac{1}{2}.$$

3. (iv) Let $\delta = \varepsilon$, since $|x/(1 + \sin^2 x)| \leq |x|$.

(vi) If $\varepsilon > 1$, let $\delta = 1$. Then $|x - 1| < \delta$ implies that $0 < x < 2$, so $0 < \sqrt{x} < 2$, so $|\sqrt{x} - 1| < 1$. If $\varepsilon < 1$, then $(1 - \varepsilon)^2 < x < (1 + \varepsilon)^2$ implies that $|\sqrt{x} - 1| < \varepsilon$, so it suffices to choose δ so that $(1 - \varepsilon)^2 \leq 1 - \delta$ and $1 + \delta \leq (1 + \varepsilon)^2$. Thus we can choose $\delta = 2\varepsilon - \varepsilon^2$.

4. (i) (ii) (iii) All numbers a which are not integers.

(iv) All a.

(v) All a with $a \neq 0$ and $a \neq 1/n$ for any integer n.

(vi) All a with $|a| < 1$ and $a \neq 1/n$ for any integer n.

5. (a) (i) All a not of the form $n + k/10$ for integers n and k.

(ii) All a not of the form $n + k/100$ for integers n and k.

(iii), (iv) No a.

(v) All numbers a whose decimal expansion does not end $7999\ldots$.

(b) The answers are the same as in part (a) (although the description of the numbers in terms of their new "decimal expansions" may be different).

6. (ii) We need

$$|f(x) - 2| < \min\left(1, \frac{\varepsilon}{2(|4| + 1)}\right) \quad \text{and} \quad |g(x) - 4| < \frac{\varepsilon}{2(|2|) + 1},$$

so we need

$$0 < |x - 2| < \min\left(\sin^2\left(\frac{[\min(1, \varepsilon/10)]^2}{9}\right) + \min(1, \varepsilon/10), [\min(1, \varepsilon/6)]^2\right)$$
$$= \delta.$$

(iv) We need

$$\left|\frac{1}{g(x)} - \frac{1}{4}\right| < \frac{\varepsilon}{2(|2| + 1)} \quad \text{and} \quad |f(x) - 2| < \min\left(1, \frac{\varepsilon}{2(|1/4| + 1)}\right),$$

so we need

$$0 < |x - 2|$$
$$< \min\left(\left[\min\left(2, \frac{8\varepsilon}{2(|2|) + 1)}\right)\right]^2, \sin^2\left(\frac{[\min(1, 2\varepsilon/5)]^2}{9}\right) + \min(1, 2\varepsilon/5)\right)$$
$$= \delta.$$

7. Let $f(x) = \sqrt{|x|}$ with $a = 0$ and $l = 0$. Then for $\varepsilon < 1$ we have $\left|\sqrt{|x|} - 0\right| < \varepsilon$ when $0 < |x - 0| < \varepsilon^2$; but if $0 < |x - 0| < \varepsilon^2/2$, it does not follow that $\left|\sqrt{|x|} - 0\right| < \varepsilon/2$ (instead we must let $0 < |x - 0| < (\varepsilon/2)^2$).

8. (a) Yes. For example, if $g = 1 - f$, then $\lim_{x \to a} [f(x) + g(x)]$ exists even if $\lim_{x \to a} f(x)$ [and consequently $\lim_{x \to a} g(x)$] does not exist; and if $g = 1/f$ where $f(x) \neq 0$ for all $x \neq a$, then $\lim_{x \to a} f(x)g(x)$ does exist even if $\lim_{x \to a} f(x)$ and $\lim_{x \to a} g(x)$ do not exist [for example, if $f(x) = 1/(x - a)$ for $x \neq 0$, and $g(x) = x - a$].

(b) Yes, since $g = (f + g) - f$.

(c) No. (This is just another way of stating part (b).)

(d) No. The argument analogous to part (b), that $g = (f \cdot g)/f$, will not work if $\lim\limits_{x \to a} f(x) = 0$, and this is precisely the case in which one can find a counterexample. For example, let $f(x) = x - a$, and let $g(x) = 0$ for x rational and 1 for x irrational. Then $\lim\limits_{x \to a} g(x)$ does not exist, but $\lim\limits_{x \to a} f(x)g(x) = 0$, since $|f(x)g(x) - 0| \leq |f(x)|$.

11. Intuitively, this is true because we only have to consider x's satisfying $0 < |x - a| < \delta'$, where we can pick $\delta' < \delta$. In fact, if $\lim\limits_{x \to a} f(x) = l$, and $\varepsilon > 0$, there is a δ' such that if $0 < |x - a| < \delta'$, then $|f(x) - l| < \varepsilon$. Now there is also a $\delta' < \delta$ with this property (namely, $\min(\delta, \delta')$). Since $f(x) = g(x)$ for all x with $0 < |x - a| < \delta$, we also have $f(x) = g(x)$ for all x with $0 < |x - a| < \delta'$, so the conclusion $|f(x) - l| < \varepsilon$ can just as well be written $|g(x) - l| < \varepsilon$. This shows that $\lim\limits_{x \to a} g(x) = l$.

12. (a) Intuitively, $f(x)$ cannot be made close to a number $> \lim\limits_{x \to a} g(x)$ because $f(x) \leq g(x)$ and $g(x)$ is close to $\lim\limits_{x \to a} g(x)$. A rigorous proof is by contradiction. Suppose that $l = \lim\limits_{x \to a} f(x) > \lim\limits_{x \to a} g(x) = m$. Let $\varepsilon = l - m > 0$. Then there is a $\delta > 0$ such that if $0 < |x - a| < \delta$, then $|l - f(x)| < \varepsilon/2$ and $|m - g(x)| < \varepsilon/2$.

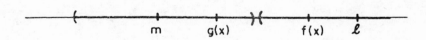

Thus for $0 < |x - a| < \delta$ we have

$$g(x) < m + \frac{\varepsilon}{2} = l - \frac{\varepsilon}{2} < f(x),$$

contradicting the hypothesis.

(b) It suffices to assume that $f(x) \leq g(x)$ for all x satisfying $0 < |x - a| < \delta$, for some $\delta > 0$.

(c) No. For example, let $f(x) = 0$ and let $g(x) = |x|$ for $x \neq 0$, and $g(0) = 1$. Then $\lim\limits_{x \to 0} f(x) = 0 = \lim\limits_{x \to 0} g(x)$.

13. Intuitively, g is squeezed between f and h, which approach the same number:

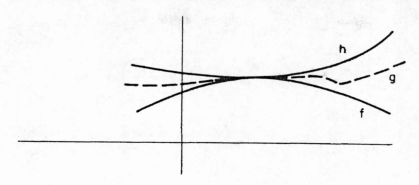

Let $l = \lim\limits_{x \to a} f(x)$. Given $\varepsilon > 0$, there is a $\delta > 0$ such that if $0 < |x - a| < \delta$, then $|h(x) - l| < \varepsilon$ and $|f(x) - l| < \varepsilon$. Thus, if $0 < |x - a| < \delta$, then

$$l - \varepsilon < f(x) \le g(x) \le h(x) < l + \varepsilon,$$

so $|g(x) - l| < \varepsilon$.

14. (a) We ought to have

$$\lim_{x \to 0} \frac{f(bx)}{x} = \lim_{x \to 0} \frac{bf(bx)}{bx} = b \lim_{x \to 0} \frac{f(bx)}{bx} = b \lim_{y \to 0} \frac{f(y)}{y} = bl.$$

The next to last equality can be justified as follows. If $\varepsilon > 0$ there is a $\delta > 0$ such that if $0 < |y| < \delta$, then $|f(y)/y| < \varepsilon$. Then if $0 < |x| < \varepsilon/|b|$, we have $0 < |bx| < \varepsilon$, so $|f(bx)/bx| < \varepsilon$.

(b) In this case, $\lim\limits_{x \to 0} f(bx)/x = \lim\limits_{x \to 0} f(0)/x$ does not exist, unless $f(0) = 0$.

(c) Part (a) shows that $\lim\limits_{x \to 0} (\sin 2x)/x = 2 \lim\limits_{x \to 0} (\sin x)/x$. We can also use the following computation:

$$\lim_{x \to 0} \frac{\sin 2x}{x} = \lim_{x \to 0} \frac{2(\sin x)(\cos x)}{x} = 2 \lim_{x \to 0} \cos x \lim_{x \to 0} \frac{\sin x}{x} = 2 \lim_{x \to 0} \frac{\sin x}{x}.$$

(Of course this method won't work in general for $\lim\limits_{x \to 0} (\sin bx)/x$.)

15. (i)

$$\lim_{x \to 0} \frac{\sin 2x}{x} = 2 \lim_{x \to 0} \frac{\sin x}{x} = 2\alpha, \qquad \text{by Problem 14.}$$

(ii)

$$\lim_{x \to 0} \frac{\sin ax}{\sin bx} = \lim_{x \to 0} \frac{\sin ax}{x} \cdot \lim_{x \to 0} \frac{x}{\sin bx}$$

$$= a\alpha \cdot \frac{1}{b\alpha} = \frac{a}{b}.$$

(iii)

$$\lim_{x\to 0} \frac{\sin^2 2x}{x} = \lim_{x\to 0} \sin 2x \cdot \lim_{x\to 0} \frac{\sin 2x}{x} = 0 \cdot 2\alpha = 0.$$

(iv)

$$\lim_{x\to 0} \frac{\sin^2 2x}{x^2} = \left(\lim_{x\to 0} \frac{\sin 2x}{x}\right)^2 = 4\alpha^2.$$

(v)

$$\lim_{x\to 0} \frac{1-\cos x}{x^2} = \lim_{x\to 0} \frac{(1-\cos x)(1+\cos x)}{x^2(1+\cos x)} = \lim_{x\to 0} \frac{\sin^2 x}{x^2(1+\cos x)}$$

$$= \frac{\alpha^2}{2}.$$

(vi)

$$\lim_{x\to 0} \frac{\tan^2 x + 2x}{x + x^2} = \lim_{x\to 0} \frac{\dfrac{\sin^2 x}{x\cos^2 x} + 2}{1 + x}$$

$$= \lim_{x\to 0} \left(\frac{\sin x}{x} \cdot \frac{\sin x}{\cos^2 x} + 2\right) \bigg/ 1$$

$$= \alpha \cdot 0 + 2 = 2.$$

(vii)

$$\lim_{x\to 0} \frac{x\sin x}{1-\cos x} = \lim_{x\to 0} \frac{x\sin x(1+\cos x)}{(1-\cos x)(1+\cos x)} = \lim_{x\to 0} \frac{x\sin x(1+\cos x)}{\sin^2 x}$$

$$= \frac{2}{\alpha}.$$

(viii)

$$\lim_{h\to 0} \frac{\sin(x+h) - \sin x}{h} = \lim_{h\to 0} \frac{\sin x\cos h + \cos x\sin h - \sin x}{h}$$

$$= \lim_{h\to 0} \sin x \frac{(\cos h - 1)}{h} + \cos x \frac{\sin h}{h}$$

$$= \alpha\cos x \qquad [\text{we have } \lim_{h\to 0} \frac{\cos - 1}{h} = 0 \text{ by (v)}].$$

(ix)

$$\lim_{x\to 1} \frac{\sin(x^2 - 1)}{x - 1} = \lim_{x\to 1} \frac{(x+1)\sin(x^2-1)}{(x+1)(x-1)} = \lim_{x\to 1} \frac{(x+1)\sin(x^2-1)}{x^2 - 1}$$

$$= 2\lim_{x\to 1} \frac{\sin(x^2-1)}{x^2 - 1}$$

$$= 2\lim_{h\to 0} \frac{\sin h}{h} \qquad [\text{same reasoning as in Problem 14(a)}]$$

$$= 2\alpha.$$

(x)

$$\lim_{x\to 0} \frac{x^2(3+\sin x)}{(x+\sin x)^2} = \lim_{x\to 0} \frac{3+\sin x}{\left(1+\dfrac{\sin x}{x}\right)^2} = \frac{3}{(1+\alpha)^2}.$$

(xi)

$$\lim_{x\to 1}(x^2-1)^3 \sin\left(\frac{1}{x-1}\right)^3 = 0 \quad [\text{since } |\sin 1/(x-1)^3| \le 1 \text{ for all } x \ne 0].$$

16. (a) Intuitively, if $f(x)$ is close to l, then $|f(x)|$ is close to $|l|$. In fact, given $\varepsilon < 0$ there is a $\delta > 0$ such that if $0 < |x-a| < \delta$, then $|f(x)-l| < \varepsilon$. But $\big||f(x)|-|l|\big| \le |f(x)-l| < \varepsilon$ (by Problem 1-12(vi)).

(b) This follows from (a) and Theorem 2, since

$$\max(f,g) = \frac{f+g+|f-g|}{2},$$
$$\min(f,g) = \frac{f+g-|f-g|}{2}.$$

18. Pictorially, this means that f is bounded in any interval around a.

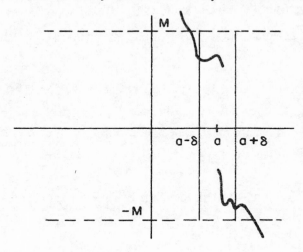

Choose $\delta > 0$ so that $|f(x)-l| < 1$ for $0 < |x-a| < \delta$ (we are picking $\varepsilon = 1$). Then $l-1 < f(x) < l+1$, so we can let $M = \max(|l+1|, |l-1|)$.

19. For any $\delta > 0$ we have $f(x) = 0$ for some x satisfying $0 < |x-a| < \delta$ (namely, irrational x with $0 < |x-a| < \delta$) and also $f(x) = 1$ for some x satisfying $0 < |x-a| < \delta$ (namely, rational x with $0 < |x-a| < \delta$). This means that we cannot have $|f(x)-l| < 1/2$ for all x no matter what l is. (There is a slight bit of cheating here; see Problem 8-5.)

20. Consider, for simplicity, the case $a > 0$. The basic idea is that since $f(x)$ is close to a for all rational x close to a, and close to $-a$ for all irrational x close to a, we cannot have $f(x)$ close to any fixed number. To make this idea work, we note that for any $\delta > 0$ there are x with $0 < |x - a| < \delta$ and $f(x) > a/2$ as well as x with $0 < |x - a| < \delta$ and $f(x) < -a/2$. Since the distance between $a/2$ and $-a/2$ is a, this means that we cannot have $|f(x) - l| < a$ for all such x, no matter what l is.

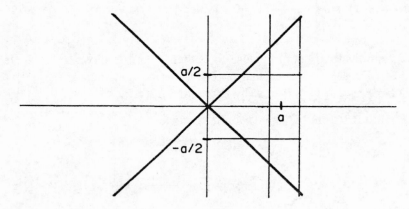

21. (a) Follows from (b), since $|\sin 1/x| \leq 1$ for all x ($\neq 0$).

(b) If $\delta > 0$ is such that $|g(x)| < \varepsilon/M$ for all x with $0 < |x| < \delta$, then $|g(x)h(x)| < \varepsilon$ for all such x.

22. If $\lim\limits_{x \to 0} f(x)$ does exist, then it is clear that $\lim\limits_{x \to 0}[f(x) + g(x)]$ does not exist whenever $\lim\limits_{x \to 0} g(x)$ does not exist [this was Problem 8(b) and (c)]. On the other hand, if $\lim\limits_{x \to 0} f(x)$ does not exist, choose $g = -f$; then $\lim\limits_{x \to 0} g(x)$ does not exist, but $\lim\limits_{x \to 0}[f(x) + g(x)]$ does exist.

23. (a) If $\lim\limits_{x \to 0} f(x)g(x)$ existed, then $\lim\limits_{x \to 0} g(x) = \lim\limits_{x \to 0} f(x)g(x)/f(x)$ would also exist.

(b) Clearly, if $\lim\limits_{x \to 0} f(x)g(x)$ exists, then $\lim\limits_{x \to 0} g(x) = 0$.

(c) In case (1) of the hint, we clearly cannot have $\lim\limits_{x \to 0} f(x) = 0$, so by assumption the limit does not exist at all. Let $g = 1/f$. Since it is *not* true that $\lim\limits_{x \to 0} |f(x)| = \infty$, it follows that if $\lim\limits_{x \to 0} g(x)$ exists, then $\lim\limits_{x \to 0} g(x) \neq 0$. But this would imply that $\lim\limits_{x \to 0} f(x)$ exists, so $\lim\limits_{x \to 0} g(x)$ does not exist. On the other hand, $\lim\limits_{x \to 0} f(x)g(x)$ clearly exists. In case (2), choose x_n as in the hint. Define $g(x) = 0$ for $x \neq x_n$, and $g(x) = 1$ for $x = x_n$. Then $\lim\limits_{x \to 0} g(x)$ does not exist, but $\lim\limits_{x \to 0} f(x)g(x) = 0$.

24. Given $\varepsilon > 0$, pick n with $1/n \leq \varepsilon$ and let δ be the minimum distance from a to all points in $A_1, \ldots, A_n$ (except a itself if a is one of these points). Then $0 < |x - a| < \delta$ implies that x is not in $A_1, \ldots, A_n$, so $f(x) = 0$ or $1/m$ for $m \geq n$, so $|f(x)| < \varepsilon$.

26. (a) Although $\lim\limits_{x \to 1} 1/x = 1$ is true, it is *not* true that for *all* $\delta > 0$ there is an $\varepsilon > 0$ with $|1/x - 1| < \varepsilon$ for $0 < |x - 1| < \delta$. In fact, if $\delta = 1$, there is no such ε, since $1/x$ can be arbitrarily large for $0 < |x - 1| < 1$.

Moreover, *any* bounded function f automatically satisfies the condition, whether $\lim\limits_{x \to a} f(x) = l$ is true or not.

(b) If f is a constant function, $f(x) = c$, this condition does not hold, since $|f(x) - c| < 1$ certainly does not imply that $0 < |x - a| < \delta$ for any δ.

Moreover, the function $f(x) = x$, for example, satisfies this condition no matter what a and l are.

27. (i), (ii), (iii), (iv) Both one-sided limits exist for all a.

(v) Both one-sided limits exist for $a \neq 0$ and neither exits for $a = 0$.

(vi) Both one-sided limits exist for all a with $|a| < 1$; moreover, $\lim\limits_{x \to 1^-} f(x)$ and $\lim\limits_{x \to 1^+} f(x)$ exist.

28. (a) (i), (ii) Both one-sided limits exist for all a.

(iii), (iv) Neither one-sided limit exists for any a.

(v) Both one-sided limits exists for all a.

(vi) Both one-sided limits exist for all a whose decimal expansion contains at least one 1; in addition, the right-hand limit exists for all a whose decimal expansion contains no 1's, but which end in 0999 . . . .

(b) The answers are the same as in part (a).

31.

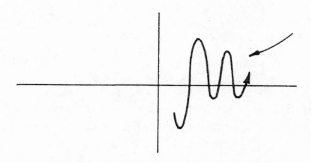

Let $l = \lim\limits_{x \to a^-} f(x)$ and $m = \lim\limits_{x \to a^+} f(x)$. Since $m - l > 0$, there is a $\delta > 0$ so that

$$|f(x) - l| < \frac{m-l}{2} \qquad \text{when} \qquad a - \delta < x < a,$$

$$|f(y) - m| < \frac{m-l}{2} \qquad \text{when} \qquad a < y < a + \delta.$$

This implies that

$$f(x) < l + \frac{m-l}{2} = m - \frac{m-l}{2} < f(y).$$

The converse is false, as shown by $f(t) = t$ and any a. It is only possible to conclude that $\lim\limits_{x \to a^-} f(x) \le \lim\limits_{x \to a^+} f(x)$.

32. Naturally we are assuming $a_n \ne 0$ and $b_m \ne 0$. If $x \ne 0$, then

$$\frac{a_n x^n + \cdots + a_0}{b_m x^m + \cdots + b_0} = \frac{a_n + \dfrac{a_n - 1}{x} + \cdots + \dfrac{a_0}{x^n}}{\dfrac{b_m}{x^{n-m}} + \cdots + \dfrac{b_0}{x^n}} = \frac{f(x)}{g(x)}.$$

If $m < n$, then $\lim\limits_{x \to \infty} f(x) = a_n$ but $\lim\limits_{x \to \infty} g(x) = 0$. This implies that $\lim\limits_{x \to \infty} f(x)/g(x)$ does not exist—otherwise we would have

$$\lim\limits_{x \to \infty} f(x) = [\lim\limits_{x \to \infty} f(x)/g(x)] \cdot [\lim\limits_{x \to \infty} g(x)] = 0.$$

If $m \ge n$, we write

$$\frac{a_n x^n + \cdots + a_0}{b_m x^m + \cdots + b_0} = \frac{\dfrac{a_n}{x^{m-n}} + \cdots + \dfrac{a_0}{x^m}}{b_m + \cdots + \dfrac{b_0}{x^m}} = \frac{f(x)}{g(x)}.$$

Then $\lim\limits_{x \to \infty} f(x) = 0$ if $m > n$, and a_n if $m = n$, while $\lim\limits_{x \to \infty} g(x) = b_m$. So $\lim\limits_{x \to \infty} f(x)/g(x) = 0$ if $m > n$, and a_n/b_m if $m = n$.

33. (i)

$$\lim\limits_{x \to \infty} \frac{x + \sin^3 x}{5x + 6} = \lim\limits_{x \to \infty} \frac{1 + \dfrac{\sin^3 x}{x}}{5 + \dfrac{6}{x}}$$

$$= \frac{1}{5}.$$

(ii)

$$\lim_{x \to \infty} \frac{x \sin x}{x^2 + 5} = \lim_{x \to \infty} \frac{1}{x + \dfrac{5}{x}} \cdot \sin x$$

$$= 0, \qquad \text{since } |\sin x| \le 1.$$

(iii)

$$\lim_{x \to \infty} \sqrt{x^2 + x} - x = \lim_{x \to \infty} \frac{\left(\sqrt{x^2 + x} - x\right)\left(\sqrt{x^2 + x} + x\right)}{\sqrt{x^2 + x} + x}$$

$$= \lim_{x \to \infty} \frac{x}{\sqrt{x^2 + x} + x}$$

$$= \lim_{x \to \infty} \frac{1}{\sqrt{1 + \dfrac{1}{x}} + 1}$$

$$= \frac{1}{2}.$$

(iv) The limit

$$\lim_{x \to \infty} \frac{x^2(1 + \sin^2 x)}{(x + \sin x)^2} = \lim_{x \to \infty} \frac{1 + \sin^2 x}{\left(1 + \dfrac{\sin x}{x}\right)^2}$$

does not exist, since $\left(1 + \dfrac{\sin x}{x}\right)^2 \to 1$ but $1 + \sin^2 x$ does not approach a limit as $x \to \infty$.

35. (i)

$$\lim_{x \to \infty} \frac{\sin x}{x} = 0,$$

since $|\sin x| \le 1$ for all x.

(ii)

$$\lim_{x \to \infty} x \sin \frac{1}{x} = \lim_{x \to \infty} \frac{\sin \dfrac{1}{x}}{\dfrac{1}{x}} = \lim_{x \to 0^+} \frac{\sin x}{x} \qquad \text{by Problem 34}$$

$$= \alpha.$$

36. $\displaystyle\lim_{x \to -\infty} f(x) = l$ means that for all $\varepsilon > 0$ there is some N such that $|f(x) - l| < \varepsilon$ for some $x < N$.

(a) The answer is the same as when $x \to \infty$ (Problem 32).

(b) If $l = \lim\limits_{x\to\infty} f(x)$, then for every $\varepsilon > 0$ there is some N such that $|f(x) - l| < \varepsilon$ for $x > N$. Now if $x < -N$, then $-x > N$, so $|f(-x) - l| < \varepsilon$. So $\lim\limits_{x\to-\infty} f(-x) = l$.

(c) If $l = \lim\limits_{x\to-\infty} f(x)$, then for every $\varepsilon > 0$ there is some N such that $|f(x) - l| < \varepsilon$ for $x < N$, and we can assume that $N < 0$. Now is $1/N < x < 0$, then $1/x < N$, so $|f(1/x) - l| < \varepsilon$.

37.

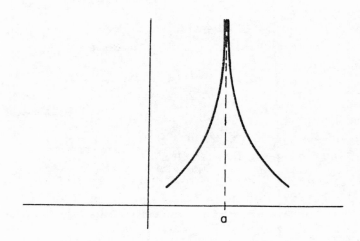

(a) Given $N > 0$, let $\delta = 1/\sqrt{N}$. Then $0 < |x - 3| < \delta$ implies that $(x-3)^2 < 1/N$, so $1/(x - 3)^2 > N$.

(b) Given $N > 0$, so that $1/N > 0$, choose $\delta > 0$ such that $|g(x)| < \varepsilon/N$ for $0 < |x - a| < \delta$. Then $0 < |x - a| < \delta$ implies that $|f(x)/g(x)| > \varepsilon \cdot (N/\varepsilon) = N$.

38. (a) $\lim\limits_{x\to a^+} f(x) = \infty$ means that for all N there is a $\delta > 0$ such that, for all x, if $a < x < a + \delta$, then $f(x) > N$.

$\lim\limits_{x\to a^-} f(x) = \infty$ means that for all N there is a $\delta > 0$ such that, for all x, if $a - \delta < x < a$, then $f(x) > N$.

$\lim\limits_{x\to\infty} f(x) = \infty$ means that for all N there is some M such that, for all x, if $x > M$, then $f(x) > N$.

It is also possible to define

$$\lim_{x \to -\infty} f(x) = \infty,$$

$$\lim_{x \to a} f(x) = -\infty,$$

$$\lim_{x \to a^+} f(x) = -\infty,$$

$$\lim_{x \to a^-} f(x) = -\infty,$$

$$\lim_{x \to \infty} f(x) = -\infty,$$

$$\lim_{x \to -\infty} f(x) = -\infty.$$

(b) Given $N > 0$, choose $\delta = 1/N$. If $0 < x < \delta$, then $1/x > N$.

(c) If $\lim_{x \to \infty} f(1/x) = \infty$, then for all N there is some M such that $f(1/x) < N$ for $x > M$. Choose $M > 0$. If $0 < x < 1/M$, then $x > M$, so $f(x) > N$. Thus $\lim_{x \to 0^+} f(x) = \infty$. The reverse direction is similar.

39. (i)

$$\lim_{x \to \infty} \frac{x^3 + 4x - 7}{7x^2 - x + 1} = \lim_{x \to \infty} \frac{x + \dfrac{4}{x} - \dfrac{7}{x^2}}{7 - \dfrac{1}{x} + \dfrac{1}{x^2}} = \infty.$$

(ii) $\lim_{x \to \infty} x(1 + \sin^2 x) = \infty$, since $1 \le 1 + \sin^2 x \le 2$ for all x.

(iii) $\lim_{x \to \infty} x \sin^2 x$ does not exist, since $\sin^2 x$ oscillates between 0 and 1.

(iv)

$$\lim_{x \to \infty} x^2 \sin \frac{1}{x} = \lim_{x \to \infty} x \cdot x \sin \frac{1}{x} = \infty,$$

since $\lim_{x \to \infty} x \sin 1/x = \alpha$, by Problem 35(ii).

(v)

$$\lim_{x \to \infty} \sqrt{x^2 + 2x} - x = \lim_{x \to \infty} \frac{\left(\sqrt{x^2 + 2x} - x\right)\left(\sqrt{x^2 + 2x} + x\right)}{\sqrt{x^2 + 2x} + x}$$

$$= \lim_{x \to \infty} \frac{2x}{\sqrt{x^2 + 2x} + x}$$

$$= \lim_{x \to \infty} \frac{2}{\sqrt{1 + \dfrac{2}{x}} + 1} = 1.$$

(vi)

$$\lim_{x\to\infty} x\left(\sqrt{x+2}-\sqrt{x}\right) = \lim_{x\to\infty}\frac{x\left(\sqrt{x+2}-\sqrt{x}\right)\left(\sqrt{x+2}+\sqrt{x}\right)}{\sqrt{x+2}+\sqrt{x}}$$

$$= \lim_{x\to\infty}\frac{2x}{\sqrt{x+2}+\sqrt{x}}$$

$$= \lim_{x\to\infty}\frac{2}{\sqrt{\dfrac{1}{x}+\dfrac{2}{x^2}}+\sqrt{\dfrac{1}{x}}} = \infty.$$

(vii)

$$\lim_{x\to\infty}\frac{\sqrt{|x|}}{x} = \lim_{x\to\infty}\frac{\sqrt{x}}{x} = \lim_{x\to\infty}\frac{1}{\sqrt{x}} = 0.$$

40. (a) The figure below shows one side of the n-gon, subtending an angle of $2\pi/n$. Angle BOC is thus π/n, so $BC = r\sin(\pi/n)$, and $AC = 2r\sin(\pi/n)$. So the whole perimeter is $2rn\sin(\pi/n)$.

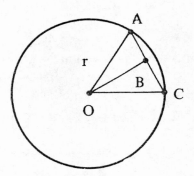

(b) As n becomes very large this approaches

$$\lim_{x\to\infty} 2rx\sin\left(\frac{\pi}{x}\right) = \lim_{x\to\infty}\pi 2r\frac{x}{\pi}\sin\left(\frac{\pi}{x}\right)$$

$$= 2\pi r\alpha,$$

where $\alpha = \lim_{x\to\infty}(\sin x)/x$, by Problem 35(ii). [Since you know that the perimeter should approach the circumference of the circle, which is $2\pi r$, you can guess that $\lim_{x\to 0}(\sin x)/x = 1$, when x is in radians.]

41. How do we know that $\sqrt{a^2-\varepsilon}$ and $\sqrt{a^2+\varepsilon}$ exist!? In Chapter 7 we prove (Theorem 8) that every positive number has a square root, but the proof of this theorem uses the fact that $f(x) = x^2$ is continuous, which is essentially what we are trying to prove. In fact, the existence of square roots is essentially equivalent to the continuity of f—compare Problem 8-8.

CHAPTER 6

1. (ii) No F, since $\lim\limits_{x \to 0} |x|/x$ does not exist.

(iv) No F, since $F(a)$ would have to be 0 for irrational a, and then F is not continuous at a is a is rational.

2. Problem 4-17:

(i), (ii), (iii) All points except integers.

(iv) All points.

(v) All points except 0 and $1/n$ for integers n.

(vi) All points in $(-1, 1)$ except 0 (where it is not defined) and $1/n$ for integers n.
Problem 4-19:

(i) All points not of the form $n + k/10$ for integers k and n.

(ii) All points not of the form $n + k/100$ for integers k and n.

(iii), (iv) No points.

(v) All points whose decimal expansion does not end $7999 \ldots$.

(vi) All points whose decimal expansion contains at least one 1.

3. (a) Clearly $\lim\limits_{h \to 0} f(h) = 0$, since $|h| < \delta$ implies that $|f(h) - f(0)| = |f(h)| < \delta$.

(b) Let $f(x) = 0$ for x irrational, and $f(x) = x$ for x rational.

(c) Notice that $|f(0)| \le |g(0)| = 0$, so $f(0) = 0$. Since g is continuous at 0, for every $\varepsilon > 0$ there is a $\delta > 0$ such that $|g(h) - g(0)| = |g(h)| < \varepsilon$ for $|h| < \delta$. Thus, if $|h| < \delta$, then $|f(h) - f(0)| = |f(h)| \le |g(h)| < \varepsilon$. So $\lim\limits_{h \to 0} f(h) = 0 = f(0)$.

4. Let $f(x) = 1$ for x rational, and $f(x) = -1$ for x irrational.

5. Let $f(x) = a$ for x irrational, and $f(x) = x$ for x rational.

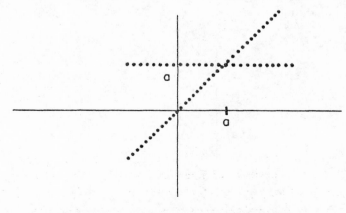

6. (a) Define f as follows (see the solution to Problem 4-17(vi)):

$$f(x) = \begin{cases} 0, & x \leq 0 \\ \dfrac{1}{\left[\dfrac{1}{x}\right]}, & 0 < x \leq 1 \\ 2, & x > 1. \end{cases}$$

(b) Let

$$f(x) = \begin{cases} -1, & x \leq 0 \\ \dfrac{1}{\left[\dfrac{1}{x}\right]}, & 0 < x \leq 1 \\ 2, & x > 1. \end{cases}$$

7. Note that $f(x + 0) = f(x) + f(0)$, so $f(0) = 0$. Now

$$\lim_{h \to 0} f(a + h) - f(a) = \lim_{h \to 0} f(a) + f(h) - f(a)$$
$$= \lim_{h \to 0} f(h)$$
$$= \lim_{h \to 0} f(h) - f(0) = 0,$$

since f is continuous at 0.

8. Since $(f + \alpha)(a) \neq 0$, Theorem 3 implies that $f + \alpha$ is non-zero in some open interval containing α.

9. (a) This is just a restatement of the definition: If the condition did not hold, then for every $\varepsilon > 0$ we would have $|f(x) - f(a)| \leq \varepsilon < 2\varepsilon$ for all x sufficiently close to a, i.e., for all x satisfying $|x - a| < \delta$ for some $\delta > 0$. If this were true for all ε, then f would be continuous at a.

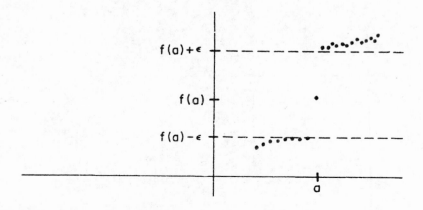

(b) If neither of these conditions held, then for every $\varepsilon > 0$ there would be $\delta_1, \delta_2 > 0$ such that $f(x) \geq f(a) - \varepsilon$ for $|x - a| < \delta_1$ and $f(x) \leq f(a) + \varepsilon$ for $|x - a| < \delta_2$. If $|x - a| < \delta = \min(\delta_1, \delta_2)$, then $f(a) - \varepsilon \leq f(x) \leq f(a) + \varepsilon$, so $|f(x) - f(a)| \leq \varepsilon$. Since this would be true for all $\varepsilon > 0$, it would follow that f is continuous at a.

10. (a)

$$\lim_{x \to a} |f|(x) = \left| \lim_{x \to a} f(x) \right| \qquad \text{by Problem 5-16}$$
$$= |f(a)| = |f|(a).$$

(b) The formulas for E and O in the solution to Problem 3-13 show that E and O are continuous if f is.

(c) This follows from part (a), since

$$\max(f, g) = \frac{f + g + |f - g|}{2},$$
$$\min(f, g) = \frac{f + g - |f - g|}{2}.$$

(d) Let $g = \max(f, 0)$ and $h = -\min(f, 0)$.

11. $1/g = f \circ g$ and f is continuous at $g(a)$ if $g(a) \neq 0$. So by Theorem 2, $1/g$ is continuous at a if $g(a) \neq 0$.

12. (a) Clearly G is continuous at a, since $G(a) = l = \lim_{x \to a} g(x) = \lim_{x \to a} G(x)$. So $f \circ G$ is continuous at a by Theorem 2. Thus

$$f(l) = f(G(a)) = (f \circ G)(a) = \lim_{x \to a} (f \circ G)(x) = \lim_{x \to 0} f(g(x)).$$

(b) Let $g(x) = l + x - a$ and

$$f(x) = \begin{cases} 0, & x \neq l \\ 1, & x = l. \end{cases}$$

Then $\lim_{x \to a} g(x) = l$ so $f\left(\lim_{x \to a} g(x) \right) = f(l) = 1$; but $g(x) \neq l$ for $x \neq a$, so $\lim_{x \to a} f(g(x)) = \lim_{x \to a} 0 = 0$.

13. (a) Since f is continuous on $[a, b]$ the limits $\lim_{t \to a^+} f(t)$ and $\lim_{t \to b^-} f(t)$ exist. Let

$$g(x) = \begin{cases} \lim_{t \to a^+} f(t), & x \leq a \\ f(x), & a < x < b \\ \lim_{t \to b^-} f(t), & b \leq x. \end{cases}$$

(b) Let $f(x) = 1/(x - a)$.

14. (a) The limit $\lim\limits_{t \to 0} f(a + t)$ exists, and equals $f(a) = g(a) = h(a)$, since

$$\lim_{t \to 0^+} f(a + t) = \lim_{t \to 0^+} g(a + t) = g(a),$$
$$\lim_{t \to 0^-} f(a + t) = \lim_{t \to 0^-} h(a + t) = h(a).$$

(b) f is continuous at c by (a), and at any $x \neq c$ in $[a, b]$, since f agrees with either g or h in some interval around x.

15. If f is continuous on $[a, b]$ and $f(a) > 0$, then there is some $\delta > 0$ such that for all x, if $a \leq x < x + \delta$, then $|f(x) - f(a)| < f(a)$. This last inequality implies that $f(x) > 0$. The proof for $f(b) > 0$ is similar.

16. (a) No in the first case; yes in the second.

(b) We have

$$\lim_{x \to a} g(x) = \lim_{x \to a} f(x) \qquad \text{since } g(x) = f(x) \text{ for } x \neq a.$$
$$= g(a) \qquad \text{by definition of } g(a).$$

(c) $g(x) = 0$ for all x.

(d) Since $g(a) = \lim\limits_{y \to a} f(y)$, by definition, it follows that for any $\varepsilon > 0$ there is a $\delta > 0$ such that $|f(y) - g(a)| < \varepsilon$ for $|y - a| < \delta$. This means that

$$g(a) - \varepsilon < f(y) < g(a) + \varepsilon$$

for $|y - a| < \delta$. So if $|x - a| < \delta$, we have

$$g(a) - \varepsilon \leq \lim_{y \to x} f(y) \leq g(a) + \varepsilon,$$

which shows that $|g(x) - g(a)| \leq \varepsilon$ for all x satisfying $|x - a| < \delta$. Thus g is continuous at a.

CHAPTER 7

1. (ii) Bonded above and below; no maximum or minimum value.

(iv) Bounded below but not above; minimum value 0.

(v) Bounded above and below. It is understood that $a > -1$ (so that $-a-1 < a+1$). If $-1 < a \leq 1/2$, then $a < -a - 1$, so $f(x) = a + 2$ for all x in $(-a - 1, a + 1)$, so $a + 2$ is the maximum and minimum value. If $-1/2 < a \leq 0$, then f has the minimum value a^2, and if $a \geq 0$, then f has the minimum value 0. Since $a + 2 > (a + 1)^2$ only for $[-1 - \sqrt{5}]/2 < a < [1 + \sqrt{5}]/2$, when $a \geq -1/2$ this function f has a maximum value only for $a \leq [1 + \sqrt{5}]/2$ (the maximum value being $a + 2$).

(vi) Bounded above and below. As in part (v), it is assumed that $a > -1$. If $a \leq -1/2$ then f has the minimum and maximum value $3/2$. If $a \geq 0$, then f has the minimum value 0, and the maximum value $\max(a^2, a + 2)$. If $-1/2 < a < 0$, then f has the maximum value $3/2$ and no minimum value.

(viii) Bounded above and below; maximum value 1; no minimum value.

(x) Bounded above and below; maximum value 0; the maximum value is a if a is rational, and there is no maximum value if a is irrational.

(xii) Bounded above and below; minimum value 0; maximum value $[a]$.

2. (ii) $n = -5$, since $f(-5) = 2(-5) + 1 < 0 < f(-4)$.

(iv) $n = 0$ since both roots of $f(x) = 0$ lie in $[0, 1]$.

3. (ii) If $f(x) = \sin x - x + 1$, then $f(0) > 0$ and $f(2) = (\sin 2) - 1 < 0$.

4. (a) Let $l = (n - k)/2$ and let

$$f(x) = (x^{2l} + 1)(x - 1)(x - 2) \cdots (x - k).$$

(b) If f has roots $a_1, \ldots, a_r$ with multiplicities $m_1, \ldots, m_r$, so that $k = m_1 + \cdots + m_r$, then

$$f(x) = (x - a_1)^{m_1} \cdots (x - a_r)^{m_r} g(x)$$

where g is a polynomial function of degree $n - (m_1 + \cdots + m_r) = n - k$ with no roots. It follows from Theorem 9 that $n - k$ is even.

6. If not, then f takes on both positive and negative values, so f would have the value 0 somewhere in $(-1, 1)$, which is impossible, since $\sqrt{1 - x^2} \neq 0$ for x in $(-1, 1)$.

8. If not, then $f(x) = g(x)$ for some x and $f(y) = -g(y)$ for some y. But f is either always positive or always negative, since $f(x) \neq 0$ for all x. So $g(x)$ and $g(y)$ have different signs. This implies that $g(z) = 0$ for some z, which is impossible, since $0 \neq f(z) = \pm g(z)$.

9. (a) $f(x) > 0$ for all $x \neq a$. For if $x_0 > a$ is the point with $f(x_0) > 0$, and if $f(x) < 0$ for some $x > a$, then $f(z) = 0$ for some z in the interval between x_0 and x; since $z \neq a$, this contradicts the hypothesis. The proof for $x < a$ is similar.

(b) $f(x) > 0$ for all $x > a$, and $f(x) < 0$ for all $x < a$ [the proof is essentially the same as for part (a)].

(c) For $y \neq 0$, let $f(x) = x^3 + x^2 y + x y^2 + y^3$ (to be very explicit we could write f_y instead of f). Since

$$f(x) = \frac{x^4 - y^4}{x - y} \qquad \text{for } x \neq y$$
$$f(y) = 4y^3 \neq 0$$

we have $f(x) = 0$ only when $x = -y$.

Say that $y > 0$. Then $f(y) = 4y^3 > 0$, while $f(-2y) = -5y^3 < 0$. It follows from part (b) that $f(x) > 0$ for $x > -y$ and $f(x) < 0$ for $x < -y$. Similarly, if $y < 0$, so that $y < 0 < -2y$, then $f(y) < 0$ while $f(-2y) > 0$, so again $f(x) > 0$ for $x > -y$ and $f(x) < 0$ for $x < -y$. In short, $x^3 + x^2 y + x y^2 + y^3 > 0$ for $x + y > 0$ and < 0 for $x + y < 0$.

12. (a) Use the proof in the solution to Problem 11, but applied to f and $-I$.

(b) Apply the same proof to f and g.

13. (a) No, f is not continuous on $[-1, 1]$. If $a < b$ are two points in $[-1, 1]$ with $a, b > 0$ or $a, b < 0$, then f takes every value between $f(a)$ and $f(b)$ on the interval $[a, b]$ since f is continuous on $[a, b]$. On the other hand, if $a < 0 < b$, then f takes on all values between -1 and 1 on $[a, b]$, so f certainly takes on all values between $f(a)$ and $f(b)$. The same argument works for $a = 0$ or $b = 0$ (because $f(0)$ was defined to be in $[-1, 1]$).

(b) If f were not continuous at a, then (by Problem 6-9(b)) for some $\varepsilon > 0$ there would be x arbitrarily close to a with $f(x) > f(a) + \varepsilon$ or $f(x) < f(a) - \varepsilon$, say the first. We can even assume that there are such x's arbitrarily close to a and $> a$, or else arbitrarily close to a and $< a$, say the first. Pick some $x > a$ with $f(x) > f(a) + \varepsilon$. By the Intermediate Value Theorem, there is x' between a and x with $f(x') < f(a) + \varepsilon$. But there is also y between a and x' with $f(y) > f(a) + \varepsilon$.

By the Intermediate Value Theorem, f takes on the value $f(a) + \varepsilon$ between x and x' and also between x' and y, contradicting the hypothesis.

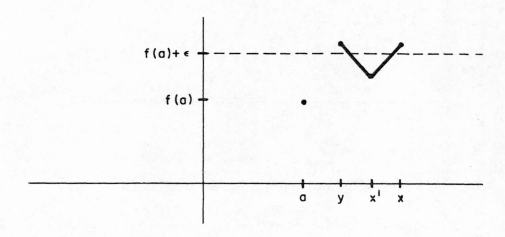

(c) As in (b), choose $x_1 > a$ with $f(x_1) > f(a) + \varepsilon$. Then choose x_1' between a and x_1 with $f(x_1') < f(a) + \varepsilon$. Then choose x_2 between a and x_1' with $f(x_2) > f(a) + \varepsilon$ and x_2' between a and x_2 with $f(x_2') < f(a) + \varepsilon$. Etc. Then f takes on the value $f(a) + \varepsilon$ on each interval $[x_n', x_n]$, contradicting the hypothesis.

14. (a) This is obvious since $|cf|(x) = |c| \cdot |f(x)|$ for all x in $[0, 1]$.

(b) We have

$$|f + g|(x) = |f(x) + g(x)| \le |f(x)| + |g(x)| \le |f|(x) + |g|(x).$$

If $|f + g|$ has its maximum value at x_0, then

$$\|f + g\| = |f + g|(x_0) \le |f|(x_0) + |g|(x_0) \le \|f\| + \|g\|.$$

If f and g are the two functions shown below, then

$$\|f\| = \|g\| = \|f + g\| = 1,$$

so $\|f + g\| \ne \|f\| + \|g\|$. (Notice that this happens even though we have $|f + g|(x) = |f|(x) + |g|(x)$ for all x.)

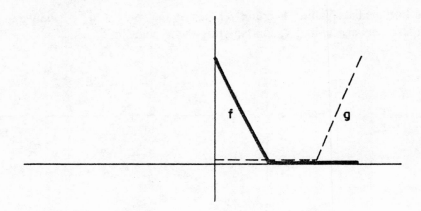

(c) Apply part (b) with f replaced by $h - g$ and g replaced by $g - f$.

15. (a) Choose $b > 0$ so that $|\phi(b)/b^2| < 1/2$. Then

$$b^n + \phi(b) = b^n \left(1 + \frac{\phi(b)}{b^n} \right) > \frac{b^n}{2} > 0.$$

Similarly, if $a < 0$ and $|\phi(a)/a^2| < 1/2$, then $a^n + \phi(a) < a^n/2 < 0$. So $x^n + \phi(x) = 0$ for some x in $[a, b]$.

(b) Choose $a > 0$ such that $a^n > 2\phi(0)$ and such that $|\phi(x)/x^n| < 1/2$ for $|x| > a$. Then for $|x| > a$ we have

$$x^n + \phi(x) = x^n \left(1 + \frac{\phi(x)}{x^n} \right) > \frac{x^n}{2} > \frac{a^n}{2} > \phi(0),$$

so the minimum of $x^n + \phi(x)$ for x in $[-a, a]$ is the minimum for all x.

16. If

$$f(x) = x^n + a_{n-1}x^{n-1} + \cdots + a_0,$$

let

$$M = \max(1, 2n|a_{n-1}|, \ldots, 2n|a_0|).$$

Then for all x with $|x| \geq M$ we have

$$\frac{1}{2} \leq 1 + \frac{a_{n-1}}{x} + \cdots + \frac{a_0}{x^n},$$

so

$$|f(x)| = \left| x^n \left(1 + \frac{a_{n-1}}{x} + \cdots + \frac{a_0}{x^n} \right) \right| \geq x^n/2.$$

If $b > M$ satisfies $|b^n| \geq 2f(0)$, then $|f(x)| \geq |f(0)|$ for $|x| \geq b$. So the minimum value of $|f(x)|$ on $[-b, b]$ is the minimum value on $\mathbf{R}$. (Naturally this problem can be generalized exactly as in Problem 15: If ϕ is continuous and $\lim\limits_{x \to \infty} \phi(x)/x^n =$

$0 = \lim\limits_{x \to -\infty} \phi(x)/x^n$, then there is some number y such that $|y^n + \phi(y)| \leq |x^n + \phi(x)|$ for all x.)

17. Pick $b > 0$ so that $f(x) < f(0)$ for $|x| > b$. Then the maximum of f on $[-b, b]$ is also the maximum on $\mathbf{R}$.

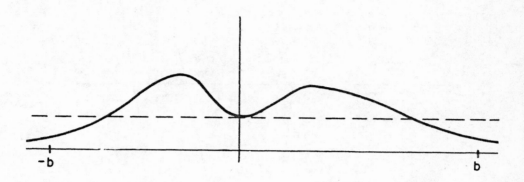

18. (a) Apply Theorem 3 to the (continuous) function

$$d(z) = \sqrt{(f(z))^2 + (z - x)^2}\,,$$

which gives the distance from $(x, 0)$ to $(z, f(z))$, for z in $[a, b]$.

(b) If $f(x) = x$ on (a, b), then no point of the graph is nearest to the point (a, a).

(c) Clearly the function d of part (a) satisfies $\lim\limits_{z \to \infty} d(z) = \infty = \lim\limits_{z \to -\infty} d(z)$, since $d(z) \geq |z - x|$. Choose $c > 0$ so that $d(z) > d(0)$ for $|z| > c$. Then the minimum of d on $[-c, c]$ will be the minimum of d on $\mathbf{R}$.

(d) By definition, $g(x) = \sqrt{(f(z))^2 + (z - x)^2}$ for some z in $[a, b]$. Now

$$\sqrt{(f(z))^2 + (z - y)^2} \leq \sqrt{(f(z))^2 + (z - x)^2} + |z - y| \qquad \text{for all } z.$$

So $g(y)$, the minimum of all $\sqrt{(f(z))^2 + (z - y)^2}$, is less than or equal to $|z - y| +$ the minimum of all $\sqrt{(f(z))^2 + (z - x)^2}$, which is $g(x) + |y - x|$. Since $|g(y) - g(x)| \leq |y - x|$ it follows that g is continuous (given $\varepsilon > 0$, let $\delta = \varepsilon$).

(e) Apply Theorem 3 to the continuous function g on $[a, b]$.

19. (a) If the continuous function g satisfied $g(x) \neq 0$ for all x, then either $g(x) > 0$ for all x or $g(x) < 0$ for all x, i.e., either $f(x) > f(x+1/n)$ or $f(x) < f(x+1/n)$ for all x. In the first case, for example, we would have

$$f(0) > f(1/n) > f(2/n) > \cdots > f(n/n) = f(1),$$

contradicting the hypothesis that $f(0) = f(1)$.

(b) The picture below illustrates such a function f when $1/4 < a < 1/3$.

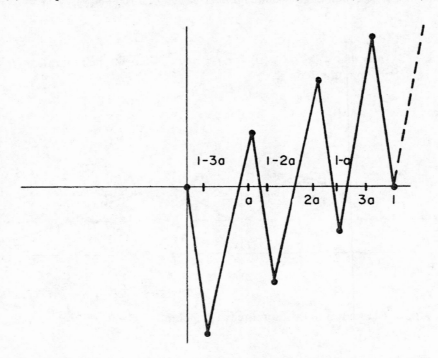

In general, if $1/(n + 1) < a < 1/n$, define f arbitrarily on $[0, a]$, subject only to the condition that $f(0) = 0$, $f(a) > 0$, and $f(1 - na) = -nf(a)$. Since $1/(n+1) < a < 1/n$, the numbers $0, 1-na$, and a are all distinct, so this is possible. Then define f on $[ka, (k + 1)a]$ by $f(ka + x) = f(x) + ka$. In particular, we have $f(1) = f(na + (1 - na)) = na + f(1 - na) = 0$, but $f(x + a) - f(a) = f(a) > 0$ for all x.

20. **(a)** If $f(a) = f(b)$ for $a < b$, then we cannot have $f(x_1) > f(a)$ and $f(x_2) < f(a)$ for some x_1, x_2 in $[a, b]$, since this would imply that $f(x) = f(a)$ for some x between x_1 and x_2, so that f would take on the value $f(a)$ three times. So either $f(x) > f(a)$ for all x in (a, b), or else $f(x) < f(a)$ for all x in (a, b), say the first. Pick any x_0 in (a, b).

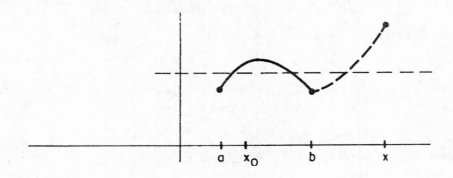

The Intermediate Value Theorem implies that f takes on all values between $f(a)$ and $f(x_0)$ in the interval $[a, x_0]$ and also in the interval $[x_0, b]$. So we cannot have $f(x) > f(a)$ for $x < a$ or $x > b$, since this would imply that f takes on these values yet a third time (on $[x, a]$ or $[b, x]$). So f is actually bounded above on **R** (since it is bounded on $[a, b]$), which means that f does not take on every value.

(b) Moreover, even if we allowed the situation where f did not take on all values, it would still be true that f actually has a maximum value M on **R** (the maximum on $[a, b]$ will be the maximum on **R**). Now f must take on this maximum value twice, say at x_0 and x_1. Pick $\alpha < x_0 < \beta < y_0 < \gamma$.

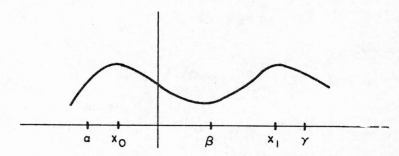

If m is the maximum of $f(\alpha)$, $f(\beta)$, $f(\gamma)$, then f takes on all values between m and M on each interval $[\alpha, x_0]$, $[x_0, \beta]$, $[\beta, x_1]$ and $[x_1, \gamma]$, which is impossible.

(c) The following picture, for $n = 5$, will indicate the general case.

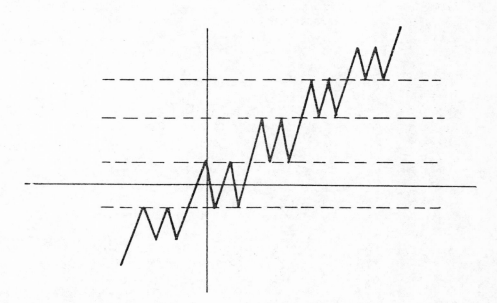

(d) Pick $x_1 < \cdots < x_n$ with $f(x_1) = \cdots = f(x_n) = a$. In each interval (x_i, x_{i+1}), either $f > a$ or $f < a$. Since n is even, there are an odd number, $n - 1$, of such intervals, so either $f > a$ in more than half of them, or $f < a$ in more than half of them. Thus $f > a$ in at least $n/2$ of them, or $f < a$ in at least $n/2$ of them, say the first. Then f takes on all values slightly larger than a at least twice in at least $n/2$ intervals. This shows that f cannot take on these values any where else, so f is bounded above. (Moreover, the same sort of argument as in part (c) shows that f would have to take on values slightly less that the maximum value at least $2n$ times.)

CHAPTER 8

1. (ii) 1 is the greatest element and -1 is the least element.

(iv) 0 is the least element, and the least upper bound is $\sqrt{2}$, which is not in the set.

(vi) Since $\{x : x^2 + x + 1 < 0\} = \left(\left[-1 - \sqrt{5}\right]/2, \left[-1 + \sqrt{5}\right]/2\right)$, the greatest lower bound is $\left[-1 - \sqrt{5}\right]/2$ and the least upper bound is $\left[-1 + \sqrt{5}\right]/2$; neither belongs to the set.

(viii) $1 - 1/2$ is the greatest element, and the greatest lower bound is -1, which is not in the set.

2. (b) Since A is bounded below, $B \neq \emptyset$. Since $A \neq \emptyset$, there is some x in A. Then any $y > x$ is not an upper bound for A, so no such y is in B, so B is bounded above. Let $\alpha = \sup B$. Then α is automatically $\geq$ any lower bound for A, so it suffices to prove that α *is* a lower bound for A. Now if α were not a lower bound for A, then there would be some x in A with $x < \alpha$. Since α is the *least* upper bound of B, this would mean that there is some y in B with $x < y < \alpha$. But this is impossible, since $x < y$ means that y is not a lower bound for A, so y would not be in B.

3. (a) No. For example, the functions f shown below have no second smallest x with $f(x) = 0$.

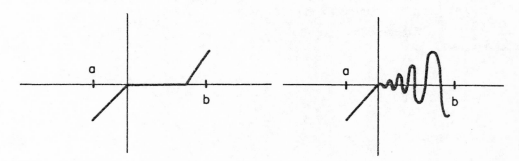

Since $b - a + x$ varies between b and a as x varies between a and b, the function $g(x) = f(b - a + x)$ satisfies $g(a) = f(b) > 0$ and $g(b) = f(a) < 0$. So there is a smallest y with $g(y) = 0$. Then $x = b - a + y$ is the largest x with $f(x) = 0$.

(b) Clearly $B \neq \emptyset$, since a is in B; in fact, there is some $\delta > 0$ such that B contains all points x satisfying $a \leq x < a + \delta$, by Problem 6-15, since f is continuous on $[a, b]$ and $f(x) < 0$. Similarly, b is an upper bound for B, and, in fact, there is a $\delta > 0$ such that all points x satisfying $b - \delta < x \leq b$ are upper bounds for A; this also follows from Problem 6-15, since f is continuous on $[a, b]$ and $f(b) > 0$.

Let $\alpha = \sup A$. Then $a < \alpha < b$. Suppose $f(\alpha) < 0$. By Theorem 6-3, there is a $\delta > 0$ such that $f(x) < 0$ for $\alpha - \delta < x < \alpha + \delta$. This would mean that $\alpha + \delta/2$ is in A, a contradiction. Similarly, suppose $f(\alpha) > 0$. Then there is a $\delta > 0$ such that $f(x) > 0$ for $\alpha - \delta < x < \alpha + \delta$. But then $\alpha - \delta/2$ would also be an upper bound for B, contradicting the fact that α is the least upper bound. So $f(\alpha) = 0$.

This α is the greatest x in $[a, b]$ with $f(x) = 0$. The sets A and B are different for the function shown below.

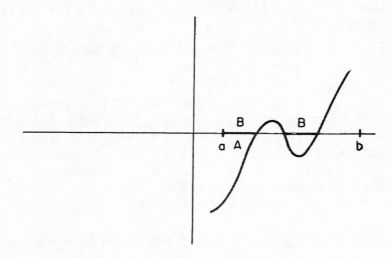

4. (a) Let c be the largest x in $[a, x_0]$ with $f(x) = 0$ and d the smallest x in $[x_0, b]$ with $f(x) = 0$.

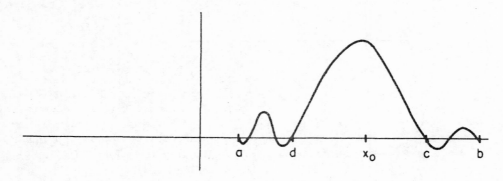

(b) Let c be the largest x in $[a, b]$ with $f(x) = f(a)$, and let d be the smallest x in $[c, b]$ with $f(x) = f(b)$.

6. (a) By definition of continuity, we have $f(a) = \lim\limits_{x \to a} f(x)$ for all a, so it suffices to prove that $\lim\limits_{x \to a} f(x) = 0$ (knowing that the limit l exists). Now given $\varepsilon > 0$, there is a $\delta > 0$ such that $|f(x) - l| < \varepsilon$ for all x satisfying $0 < |x - a| < \delta$. Since A is dense, there is a number x in A satisfying $0 < |x - a| < \delta$; so $|0 - l| < \varepsilon$. Since this is true for all $\varepsilon > 0$, it follows that $l = 0$.

(b) Apply part (a) to $f - g$.

(c) As in part (b), it obviously suffices to show that if f is continuous and $f(x) \geq 0$ for all numbers x in A, then $f(x) \geq 0$ for all x. Now there is a $\delta > 0$ such that, for all x, if $0 < |x - a| < \delta$, then $|f(x) - l| < |l|/2$. This implies that $f(x) < l + |l|/2$; if $l < 0$, it would follow that $f(x) < 0$, which would be false for those x in A which satisfy $0 < |x - a| < \delta$.

It is not possible to replace $\geq$ by $>$ throughout. For example, if $f(x) = |x|$, then $f(x) > 0$ for all x in the dense set $\{x : x \neq 0\}$, but it is not true that $f(x) > 0$ for all x.

7. According to Problem 3-16, we have $f(x) = cx$ for all rational x (where $c = f(1)$). Since f is continuous, it follows from Problem 6 that $f(x) = cx$ for all x (apply Problem 6 to f and $g(x) = cx$).

8. (a) The set $\{f(x) : x < a\}$ is bounded above (by $f(a)$); let $\alpha = \sup\{f(x) : x < a\}$. Then $\lim\limits_{x \to a^-} f(x) = \alpha$. Given any $\varepsilon > 0$, there is some $f(x)$ for $x < a$ with $f(x) > \alpha - \varepsilon$, since α is the least upper bound of $\{f(x) : x < a\}$. Let $\delta = a - x$. If $a - \delta < y < a$, then $x < y < a$, so $f(x) \leq f(y)$. This means that $\alpha \geq f(y) > \alpha - \varepsilon$, so surely $|f(y) - \alpha| < \varepsilon$.

The proof that $\lim\limits_{x \to a^+} = \inf\{f(x) : x > a\}$ is similar.

(b) It is clear from part (a) that

$$\lim_{x \to a^-} f(x) \leq f(a) \leq \lim_{x \to a^+} f(x).$$

If $\lim\limits_{x \to a} f(x)$ exists, it follows that

$$\lim_{x \to a} f(x) = \lim_{x \to a^-} f(x) \leq f(a) \leq \lim_{x \to a^+} f(x) = \lim_{x \to a} f(x),$$

so $\lim\limits_{x \to a} f(x) = f(a)$. Thus f is continuous at a, so f cannot have a removable discontinuity at a.

(c) If f is not continuous at some point a, then

$$\sup\{f(x) : x < a\} = \lim_{x \to a^-} f(x) < \lim_{x \to a^+} f(x) = \inf\{f(x) : x > a\}.$$

It follows that $f(x)$ cannot have any value between $\lim\limits_{x \to a^-} f(x)$ and $\lim\limits_{x \to a^+} f(x)$, except $f(a)$, so f cannot satisfy the Intermediate Value Theorem.

9. (a) is obvious for $\|\| \ \|\|$, since $|cf|(x) = |c| \cdot |f(x)|$ for all x in $[0, 1]$.

(b) We have $|f + g|(x) \leq |f|(x) + |g|(x)$ for all x in $[0, 1]$. Since $\||f + g|\|$ is $\sup\{|f + g|(x) : x \text{ in } [0, 1]\}$, there is some x_0 in $[0, 1]$ with

$$\||f + g|\| - |f + g|(x_0) < \varepsilon,$$

which implies that

$$\||f + g\|| - \left[|f|(x_0) + |g|(x_0)\right] < \varepsilon.$$

Since $|f|(x_0) \leq \||f\||$ and $|g|(x_0) \leq \||g\||$, it follows that

$$\||f + g\|| - \left[\||f\|| + \||g\||\right] < \varepsilon.$$

Since this is true for every $\varepsilon > 0$, it follows that $\||f + g\|| \leq \||f\|| + \||g\||$.

(c) follows from **(b)**, just as in Problem 7-14.

11. (a) We have $a_{n+1} \leq a_1/2^n < a_1/n$. Choose n so that $1/n < \varepsilon/a_1$. Then $a_{n+1} < \varepsilon$.

(b) Let R_i be the area of region number i in the following figure.

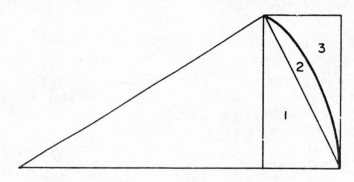

We must show that

$$R_2 < \tfrac{1}{2}(R_1 + R_2),$$

or

$$R_2 < R_1.$$

This is clear, since $R_2 < R_2 + R_3 = R_1$.

(c) Apply part (a) with $a_n = $ area of the circle minus the area of an inscribed regular polygon with 2^{n+1} sides; part (b) says that $a_{n+1} \leq a_n/2$.

(d) Let r_1 and r_2 be the radii of the two circles C_1 and C_2, and let A_i be the area of the region bounded by C_i. We know that there are numbers $\delta_1, \delta_2 > 0$ such that

$$\left| \frac{A_1}{A_2} - \frac{B_1}{B_2} \right| < \varepsilon$$

for any numbers B_1, B_2 with $|A_i - B_i| < \delta_i$. By part (c) there are numbers n_i such that the area of a regular polygon, with n_i sides, inscribed in C_i differs from A_i by less than δ_i. Let P_i be the area of a regular polygon inscribed in C_i with $\max(n_1, n_2)$ sides. Then

$$\left| \frac{A_1}{A_2} - \frac{P_1}{P_2} \right| < \varepsilon,$$

so

$$\left| \frac{A_1}{A_2} - \frac{{r_1}^2}{{r_2}^2} \right| < \varepsilon.$$

Since this is true for each $\varepsilon > 0$, it follows that $A_1/A_2 = {r_1}^2/{r_2}^2$.

14. (a) For each n and m we have $a_n \leq b_m$, because $a_n \leq a_{n+m} \leq b_{n+m} \leq b_m$. It follows from Problem 12 that $\sup\{a_n : n \text{ in } \mathbf{N}\} \leq \inf\{b_n : n \text{ in } \mathbf{N}\}$. Let x be any number between these two numbers. Then $a_n \leq x \leq b_n$ for all n, so x is in every I_n.

(b) Let $I_n = (0, 1/n)$.

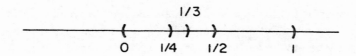

15. Let c be in each I_n. If $f(c) < 0$, then there is some $\delta > 0$ such that $f(x) < 0$ for all x in $[a, b]$ with $|x - c| < \delta$. Choose n with $1/2^n < \delta$. Since c is in I_n, which has total length $1/2^n$, it follows that all points x of I_n satisfy $|x - c| < \delta$. This contradicts the fact that f changes sign on I_n. Similarly, we cannot have $f(c) > 0$. So $f(c) = 0$.

16. Let c be in each I_n. Since f is continuous at c, there is a $\delta > 0$ such that f is bounded on the set of all points in $[0, 1]$ satisfying $|x - c| < \delta$. Choose n with $1/2^n < \delta$. Since c is in I_n, all points x of I_n satisfy $|x - c| < \delta$. This contradicts the fact that f is not bounded on I_n.

17. (a) (i) If x is in A then $x < \alpha$. So $y < x < \alpha$, so $y < \alpha$, so y is in A.

(ii) $\alpha - 1$ is in A.

(iii) $\alpha + 1$ is not in A.

(iv) If x is in A, then $x < \alpha$. Let $x' = (x + \alpha)/2$. Then $x < x' < \alpha$, so x' is in A.

(b) According to (iii) there is some y with y not in A. If $y < x$, then x cannot be in A, because (i) would imply that y is in A. Thus y is an upper bound for A, and $A \neq \emptyset$ by (ii), so $\sup A$ exists. Given x in A, choose x' in A with $x < x'$, by (iv). Then $x < x' \leq \sup A$, so $x < \sup A$. Conversely, if $x < \sup A$, then there is some y in A with $x < y$. Hence x is in A, by (i).

18. (a)

Almost upper bounds	*Almost lower bounds*
(i) All $\alpha > 0$.	All $\alpha \leq 0$.
(ii) All $\alpha > 0$.	All $\alpha < 0$.
(iii) All $\alpha > 0$.	All $\alpha \leq 0$.
(iv) All $\alpha > \sqrt{2}$.	All $\alpha \leq 0$.
(v) None.	None.
(vi) All $\alpha \geq [-1+\sqrt{5}]/2$.	All $\alpha \leq [-1+\sqrt{5}]/2$.
(vii) All $\alpha \geq 0$.	All $\alpha \leq [-1+\sqrt{5}]/2$.
(viii) All $\alpha > 1$.	All $\alpha \leq -1$.

(b) Every upper bound for A is surely an almost upper bound, so $B \neq \emptyset$. No lower bound for A can possibly be an almost lower bound (since A is infinite), so B is bounded below by any lower bound for A.

c. (i), (ii), (iii) 0.

(iv) $\sqrt{2}$.

(v) Does not exist.

(vi) $[-1+\sqrt{5}]/2$.

(vii) 0.

(viii) 1.

(d) $\underline{\lim} A = \sup C$, where C is the set of all almost lower bounds.

(i), (ii), (iii) (iv) 0.

(v) Does not exist.

(vi) $[-1+\sqrt{5}]/2$.

(vii) $[-1+\sqrt{5}]/2$.

(viii) -1.

19. (a) If x is an almost lower bound of A, and y is an almost upper bound, then there are only finitely many numbers in A which are $< x$ or $> y$. Since A is infinite, it follows that we must have $x \leq y$. Thus (Problem 12) $\underline{\lim} A \leq \overline{\lim} A$.

(b) This is clear, since $\overline{\lim} A \leq \alpha$ for any almost upper bound α, and $\alpha = \sup A$ is an almost upper bound.

(c) If $\overline{\lim} A < \sup A$, there is some almost upper bound x of A with $x < \sup A$. So there are only finitely many numbers of A which are greater than x (and there is at least one, since $x < \sup A$). The largest of these finitely many elements is the largest element of A.

(d) Reverse the inequalities in the arguments for parts (b) and (c).

20. (a) Notice that we must have $f(x) \leq f(\sup A)$, because f is continuous at $\sup A$ and there are points y arbitrarily close to $\sup A$ with $f(x) \leq f(y)$. (A simple ε-δ argument is being suppressed.) Now suppose that $\sup A < b$. Then $f(b) < f(x)$. Moreover, $\sup A$ is a shadow point, so there is some $z > \sup A$ with $f(z) > f(\sup A) \geq f(x)$. We cannot have $z \leq b$, for this would mean that z is in A. So $z > b$ and $f(b) < f(x) \leq f(z)$, contradicting the fact that b is not a shadow point.

(b) Since f is continuous at a, and $f(x) \leq f(b)$ for all x in (a, b), it follows that $f(x) \leq f(b)$ (either by a simple ε-δ argument, or using Problem 6, if you prefer).

(c) If $f(a) < f(b)$, then a would be a shadow point, so $f(a) = f(b)$.

CHAPTER 8, Appendix

1. (a) For $y > x$ we have, by the Mean Value Theorem,

$$y^\alpha - x^\alpha = (y - x)\alpha\xi^{\alpha-1} \qquad x < \xi < y.$$

So for $\alpha > 1$ we have

$$y^\alpha - x^\alpha \geq \alpha x^{\alpha-1}(y - x).$$

Since $x^{\alpha-1}$ is unbounded on $[0, \infty)$, we cannot make $y^\alpha - x^\alpha < \varepsilon$ simply by making $y - x$ less than any fixed δ. So f is not uniformly continuous on $[0, \infty)$ for $\alpha > 1$.

For $0 < \alpha < 1$ we have to be a little more careful. We have

$$y^\alpha - x^\alpha \leq \alpha y^{\alpha-1}(y - x)$$
$$\leq \alpha(y - x) \qquad \text{for } y \geq 1.$$

which at least shows that f is uniformly continuous on $[1, \infty)$. Since it is also uniformly continuous on $[0, 1]$ by the Theorem, it follows that it is uniformly continuous on $[0, \infty)$. (The argument for this is a simple corollary of the Lemma [with $c = \infty$].)

(b) $f(x) = \sin(1/x)$

(c) Just let f have portions with larger and larger slopes:

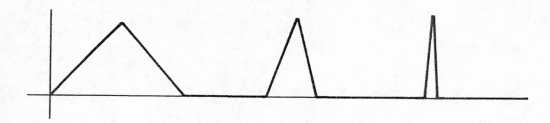

2. (a) Given $\varepsilon > 0$, choose $\delta > 0$ such that, for all x and y in A,

$$\text{if } |x - y| < \delta, \text{ then } |f(x) - f(y)|, \ |g(x) - g(y)| < \varepsilon/2.$$

Then also

$$|(f + g)(x) - (f + g)(y)| < \varepsilon.$$

(b) Choose $M > 0$ so that $|f(x)|, |g(x)| \leq M$ for all x in A. Given $\varepsilon > 0$, choose $\delta > 0$ such that, for all x and y in A,

$$\text{if } |x - y| < \delta, \text{ then } |f(x) - f(y)|, \ |g(x) - g(y)| < \frac{\varepsilon}{2M}.$$

Then also

$$|f(x)g(x) - f(y)g(y)| = |f(x)[g(x) - g(y)] + g(y)[f(x) - f(y)]|$$
$$\leq M \cdot \frac{\varepsilon}{2M} + M \cdot \frac{\varepsilon}{2M} = \varepsilon.$$

(c) Let $f(x) = x$ and $g(x) = \sin x$, both uniformly continuous on $[0, \infty)$. The product is not uniformly continuous on $[0, \infty)$, since there will be places where the graph is growing arbitrarily fast.

(d) Given $\varepsilon > 0$, choose $\varepsilon' > 0$ such that, for all α and β in B,

$$\text{if } |\alpha - \beta| < \varepsilon', \text{ then } |g(\alpha) - g(\beta)| < \varepsilon.$$

Then choose $\delta > 0$ such that, for all x and y in A,

$$\text{if } |x - y| < \delta, \text{ then } |f(x) - f(y)| < \varepsilon'.$$

It follows that

$$\text{if } |x - y| < \delta, \text{ then } |g(f(x)) - g(f(y))| < \varepsilon.$$

3. Given $\varepsilon > 0$, suppose f is not ε-good on $[a, b]$. Then, by the Lemma, either f is not ε-good on $[a, (a + b)/2]$ or f is not ε-good on $[(a + b)/2, b]$. Let I_1 be one of the halves on which f is not ε-good. Now bisect I_1, and let I_2 be a half on which f is not ε-good. Etc. Let x_0 be a point in all I_n. Choose $\delta > 0$ such that, if $|x - x_0| < \delta$, then $|f(x) - f(x_0)| < \varepsilon/2$. It follows that if $|x - x_0| < \delta$ and $|y - x_0| < \delta$, then $|f(x) - f(y)| < \varepsilon$, i.e., f is ε-good on $(x_0 - \varepsilon, x_0 + \varepsilon)$. But some I_n is contained in this interval, a contradiction.

4. Choose $\delta > 0$ such that, if x and y are in $[a, b]$ and $|y - x| \leq \delta$, then $|f(y) - f(x)| < \varepsilon$. Let $K = [(b - a)/\delta] + 1$. Then for any point x in $[a, b]$, there is a sequence

$$a = a_0, a_1, a_2, \ldots, a_k = x$$

with $k \leq K$ and $|a_{i+1} - a_i| < \delta$. It follows that

$$|f(a_1) - f(a)| < \varepsilon$$
$$|f(a_2) - f(a_1)| < \varepsilon$$
$$\cdot$$
$$\cdot$$
$$\cdot$$
$$|f(x) - f(a_{k-1})| < \varepsilon$$

which implies that

$$|f(x) - f(a)| < K\varepsilon$$

and hence

$$|f(x)| \leq f(a) + K\varepsilon$$

for all x in $[a, b]$.

CHAPTER 9

1. (b) The following figure illustrates the tangent lines to the graph of $f(x) = 1/x$.

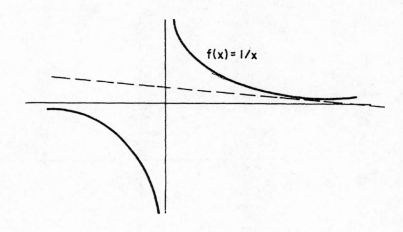

2. (b) The following figure illustrates the tangent lines to the graph of $f(x) = 1/x^2$.

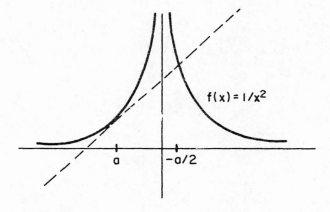

6. (a) The picture below indicates the relation between f' and $(f + c)'$.

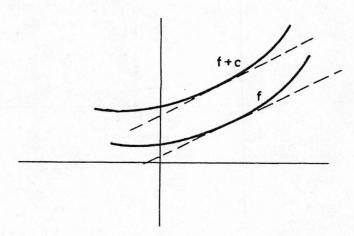

(b) The figure below indicates the relation between f' and $(cf)'$.

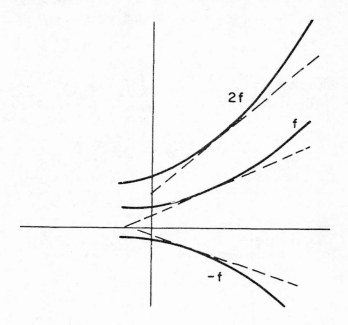

8. (a) The figure below indicates the relation between f' and g' if $g(x) = f(x+c)$.

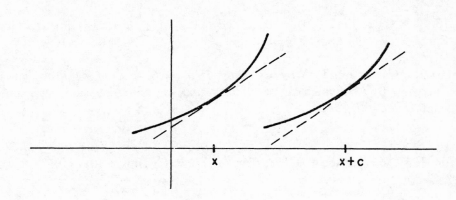

12. (a) $a'(t) = L(a(t))$ (the velocity at time t should be the velocity allowed at the point $a(t)$, where the car is located).

(b) The hypothesis means that $b(t) = a(t-1)$. Thus

$$b'(t) = a'(t-1) = L(a(t-1)) = L(b(t)).$$

(c) Suppose $b(t) = a(t) - c$. Then $b'(t) = a'(t) = L(a(t))$, whereas $b'(t)$ should be $L(b(t)) = L(a(t) - c)$. So B travels at the speed limit if the function L is periodic, with period c.

13. The limit

$$\lim_{t \to 0} \frac{h(a+t) - h(a)}{t}$$

exists, because

$$\lim_{t \to 0^+} \frac{h(a+t) - h(a)}{t} = \lim_{t \to 0^+} \frac{g(a+t) - g(a)}{t}$$
$$= \text{right-hand derivative of } g,$$

$$\lim_{t \to 0^-} \frac{h(a+t) - h(a)}{t} = \lim_{t \to 0^-} \frac{f(a+t) - f(a)}{t}$$
$$= \text{right-hand derivative of } f,$$

and these two limits are equal.

14.

$$f'(0) = \lim_{h \to 0} \frac{f(x) - f(0)}{h} = \lim_{h \to 0} \frac{f(h)}{h}.$$

Now

$$\frac{f(h)}{h} = \begin{cases} 0, & h \text{ irrational} \\ \dfrac{h^2}{h} = h, & h \text{ rational}, \end{cases}$$

so $\lim\limits_{h\to 0} f(h)/h = 0$.

15. (a) Notice that $f(0) = 0$. Since $|f(h)/h| \le h^2/|h| \le |h|$, it follows that $\lim\limits_{h\to 0} f(h)/h = 0$, i.e., $f'(0) = 0$.

(b) If $g(0) = 0$ and $g'(0) = 0$, then $f'(0) = 0$: For, $|f(h)/h| \le |g(h)/h| = |[g(h) - g(0)]/h|$, which can be made as small as desired, by choosing h sufficiently small, since $g'(0) = 0$.

16. Since $|f(0)| \le |0|^\alpha$, we have $f(0) = 0$. Now $|f(h)/h| \le |h|^{\alpha-1}$, and $\lim\limits_{h\to 0} |h|^{\alpha-1} = 0$, since $\alpha > 1$, so $\lim\limits_{h\to 0} f(h)/h = 0$. Thus $f'(0) = 0$.

17. $|f(h)/h| \ge |h|^{\beta-1}$; since $\beta - 1 < 0$, the number $|h|^{\beta-1}$ becomes large as h approaches 0, so $\lim\limits_{h\to 0} f(h)/h$ does not exist.

18. Since f is not continuous at a if a is rational, f is also not differentiable at rational a. If $a = m.a_1 a_2 a_3 \dots$ is irrational and h is rational, then $a + h$ is irrational, so $f(a + h) - f(a) = 0$. But if $h = -0.00\dots 0a_{n+1}a_{n+2}\dots$, then $a + h = m.a_1 a_2 \dots a_n 000\dots$, so $f(a + h) \ge 10^{-n}$, while $|h| < 10^{-n}$, so we have $|[f(a + h) - f(a)]/h| \ge 1$. Thus $[f(a + h) - f(a)]/h$ is 0 for arbitrarily small h and also has absolute value ≥ 1 for arbitrarily small h. It follows that $\lim\limits_{h\to 0}[f(a + h) - f(a)]/h$ cannot exist.

19. (a) For $t > 0$ we have
$$\frac{f(a+t) - f(a)}{t} \le \frac{g(a+t) - g(a)}{t} \le \frac{h(a+t) - h(a)}{t},$$
since $f(a) = g(a) = h(a)$. The left and right sides approach $f'(a) = h'(a)$ as $t \to 0^+$, so the middle term must also approach this limit. For $t < 0$ we have the inequalities reversed, which shows that as $t \to 0^-$ the middle term again approaches $f'(a) = h'(a)$.

(b) A counterexample without the condition $f(a) = g(a) = h(a)$ is shown below.

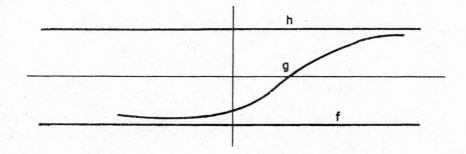

20. (a)

$$d = f(x) - f(a)(x-a) - f(a)$$
$$= x^4 - 4a^3(x-a) - a^4$$
$$= x^4 - 4a^3x + 3a^4$$
$$= (x-a)(x^3 + ax^2 + a^2x - 3a^3)$$
$$= (x-a)(x-a)(x^2 + 2ax + 3a^2).$$

(b) $f(x) - f(a)$ clearly has a as a root, so $f(x) - f(a)$ is divisible by $x - a$ by Problem 3-7. This means that $[f(x) - f(a)]/(x-a)$ is a polynomial function, so $d(x)/(x-a)$ is the polynomial function

$$h(x) = \frac{f(x) - f(a)}{x - a} - f'(a).$$

Then $\lim\limits_{x \to a} h(x) = 0$ by the definition of $f'(a)$. This implies that $h(a) = 0$, since the (polynomial) function h is continuous. So $d(x)/(x-a)$ has a as a root, so $d(x)/(x-a)$ is divisible by $(x-a)$, i.e., $d(x)$ is divisible by $(x-a)^2$.

22. (a)

$$f'(x) = \lim_{h \to 0} \frac{f(x+h) - f(x)}{h} = \lim_{h \to 0} \frac{f(x-h) - f(x)}{-h} = \lim_{h \to 0} \frac{f(x) - f(x-h)}{x}.$$

So

$$\lim_{h \to 0} \frac{f(x+h) - f(x-h)}{2h} = \frac{1}{2}\left[\lim_{h \to 0} \frac{f(x+h) - f(x)}{h} + \lim_{h \to 0} \frac{f(x) - f(x-h)}{h} \right]$$
$$= f'(x).$$

(b)

$$\frac{f(x+h) - f(x-k)}{h+k} = \frac{h}{h+k} \cdot \frac{f(x+h) - f(x)}{h} + \frac{k}{h+k} \cdot \frac{f(x) - f(x-k)}{k}.$$

Since $[f(x+h) - f(x)]/h$ and $[f(x) - f(x-k)]/k$ are close to $f'(x)$ when h and k are sufficiently small, this would seem to imply that

$$\frac{f(x+h) - f(x-k)}{h+k} \quad \text{is close to} \quad \left(\frac{h}{h+k} + \frac{k}{h+k}\right) f'(x) = f'(x).$$

However, some care is required to carry this argument out, for the following reason. If $h/(h+k)$ were very large, then

$$\frac{h}{h+k} \cdot \frac{f(x+h) - f(x)}{h}$$

could differ from $hf'(x)/(h+k)$ by a large amount, even if $[f(x+h) - f(x)]/h$ differed from $f'(x)$ by only a small amount. It will be essential to use the fact that both h and k are positive; otherwise $h/(h+k)$ could be made very large by choosing

k close to $-h$. In fact the result is false if h and k are allowed to have different signs, even when $h + k = 0$ is not allowed. The proper argument is as follows. If $\varepsilon > 0$ there is a $\delta > 0$ such that for $0 < h < \delta$ and $0 < k < \delta$ we have

$$-\varepsilon < \frac{f(x+h) - f(x)}{h} - f'(x) < \varepsilon,$$

$$-\varepsilon < \frac{f(x) - f(x-k)}{k} - f'(x) < \varepsilon.$$

Since $h, k > 0$, we can multiply these inequalities by $h/(h+k)$ and by $k/(h+k)$, respectively. Upon adding we obtain

$$-\varepsilon\left(\frac{h}{h+k} + \frac{k}{h+k}\right) < \frac{f(x+h) - f(x-k)}{h+k} - \left(\frac{h}{h+k} + \frac{k}{h+k}\right) f'(x)$$
$$< \varepsilon\left(\frac{h}{h+k} + \frac{k}{h+k}\right),$$

or

$$-\varepsilon < \frac{f(x+h) - f(x-k)}{h+k} - f'(x) < \varepsilon.$$

This proves the required limit.

23. If $g(x) = f(-x)$ then $g'(x) = -f'(-x)$, by Problem 8(b). But also $g(x) = f(x)$, so $g'(x) = f'(x)$, so $f'(x) = -f'(-x)$.

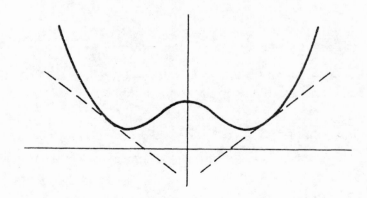

24. If $g(x) = f(-x)$, then $g'(x) = -f'(-x)$. But also $g(x) = -f(x)$, so $g'(x) = -f'(x)$, so $f'(x) = f'(-x)$.

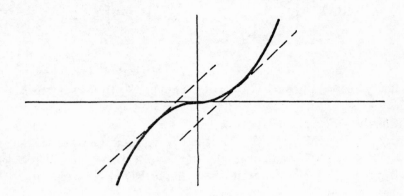

25. $f^{(k)}$ is even if k is even and f is even, or if k is odd and f is odd; $f^{(k)}$ is odd in the other two cases.

26. (ii) $f''(x) = 20x^3$.

(iv) $f''(x) = 20(x-3)^3$.

27. Proof by induction on k. The result is true for $k = 0$. If

$$S_n^{(k)}(x) = \frac{n!}{(n-k)!} x^{n-k},$$

then

$$S_n^{(k+1)}(x) = \frac{n!(n-k)}{(n-k)!} x^{n-k-1}$$

$$= \frac{n!}{[n-(k+1)]!} x^{n-(k+1)}.$$

28. (a) Since

$$f(x) = \begin{cases} x^3, & x > 0 \\ -x^3, & x < 0, \end{cases}$$

we have

$$f'(x) = \begin{cases} 3x^2, & x > 0 \\ -3x^2, & x < 0 \end{cases} \qquad f''(x) = \begin{cases} 6x, & x > 0 \\ -6x, & x < 0. \end{cases}$$

Moreover, $f'(0) = f''(0) = 0$. But $f'''(0)$ does not exist.

(b) The same sort of reasoning shows that

$$f'(x) = \begin{cases} 4x^3, & x > 0 \\ -4x^3, & x < 0 \end{cases} \qquad f''(x) = \begin{cases} 12x^2, & x > 0 \\ -12x^2, & x < 0 \end{cases}$$

$$f'''(x) = \begin{cases} 24x, & x > 0 \\ -24x, & x < 0 \end{cases}$$

and that $f'(0) = f''(0) = f'''(0) = 0$, but that $f^{(4)}(0)$ does not exist.

29. Clearly $f^{(k)}(x) = n!/(n-k)!x^{n-k}$ for $0 \le k \le n-1$ and $x > 0$, while $f^{(k)}(x) = 0$ for all k if $x < 0$. From these formulas it is easy to see that $f^{(k)}(0) = 0$ for $0 \le k \le n-1$. In particular, $f^{(n-1)}(x) = n!x$ for $x \ge 0$, and $f^{(n-1)}(x) = 0$ for $x \le 0$. So $f^{(n)}(0)$ does not exist, since $\lim\limits_{h \to 0^+} n!h/h = n!$, while $\lim\limits_{h \to 0^-} 0/h = 0$.

30. **(ii)** means that $f'(a) = -1/a^2$ if $f(x) = 1/x$.

(iv) means that $g'(a) = cf'(a)$ if $g(x) = cf(x)$.

(vi) means that $f'(a^2) = 3a^4$ if $f(x) = x^3$.

(viii) means that $g'(b) = cf'(cb)$ if $g(x) = f(cx)$.

(x) means that $f^{(k)}(a) = k!\binom{n}{k}a^{n-k}$ if $f(x) = x^n$.

CHAPTER 10

2. (ii) $\cos x + 2x \cos x^2$.

(iv) $\cos(\sin x) \cdot \cos x$.

(vi)

$$\frac{x \cos(\cos x)(-\sin x) - \sin(\cos x)}{x^2}.$$

(viii) $\cos(\cos(\sin x)) \cdot (-\sin(\sin x)) \cdot \cos x$.

2. (ii) $3\sin^2(x^2 + \sin x) \cdot \cos(x^2 + \sin x)x) \cdot (2x + \cos x)$.

(iv)

$$\cos\left(\frac{x^3}{\cos x^3}\right) \cdot \frac{(\cos x^3)3x^2 + x^3 \sin x^3 \cdot 3x^2}{\cos^2 x^3}.$$

(vi) $31^2(\cos x)^{31^2 - 1} \cdot (-\sin x)$.

(viii) $3\sin^2(\sin^2(\sin x)) \cdot \cos(\sin^2(\sin x)) \cdot 2\sin(\sin x) \cdot \cos(\sin x) \cdot \cos x$.

(x)

$$\cos(\sin(\sin(\sin(\sin x)))) \cdot \cos(\sin(\sin(\sin x))) \cdot \cos(\sin(\sin x)) \cdot \cos(\sin x) \cdot \cos x.$$

(xii)

$$5(((x^2 + x)^3 + x)^4 + x)^4 \cdot [1 + 4((x^2 + x)^3 + x)^3\{1 + 3(x^2 + x)^2[1 + 2x]\}].$$

(xiv)

$$\cos(6\cos(6\sin(6\cos 6x))) \cdot 6(-\sin(6\sin(6\cos 6x)) \cdot 6\cos(6\cos 6x) \cdot 6(-\sin 6x) \cdot 6$$

(xvi)

$$\frac{-\left[1 - \dfrac{-2(1 + \cos x)}{(x + \sin x)^2}\right]}{\left[x - \dfrac{2}{x + \sin x}\right]^2}.$$

(xviii)

$$\cos\left(\frac{x}{x - \sin\left(\dfrac{x}{x - \sin x}\right)}\right) \times$$

$$\times \frac{x - \sin\left(\dfrac{x}{x - \sin x}\right) - x\left[1 - \cos\left(\dfrac{x}{x - \sin x}\right)\dfrac{x - \sin x - x[1 - \cos x]}{(x - \sin x)^2}\right]}{\left[x - \sin\left(\dfrac{x}{x - \sin x}\right)\right]^2}.$$

3. See page 307 of the text.

4. (ii) $\cos(\sin x)$.

(iv) 0.

5. (ii) $(2x)^2$.

(iv) $17 \cdot 17$.

6. (ii) $f'(x) = g'(x \cdot g(a)) \cdot g(a)$.

(iv) $f'(x) = g'(x)(x - a) + g(x)$.

(vi) $f'(x) = g'((x - 3)^2) \cdot 2(x - 3)$.

8. If the two circles have radii $r_1(t) < r_2(t)$ at time t, with corresponding areas $A_i(t) = \pi r_i(t)^2$, then

$$\pi r_2(t)^2 - \pi r_1(t)^2 = 9\pi,$$
$$A_2'(t) = 10\pi.$$

Consequently,

$$10\pi - 2\pi r_1(t)r_1'(t) = 0.$$

Now the smaller circle has area 16π when $r_1(t) = 4$, so at this time $r_1'(t) = 5/4$. The circumference $C(t) = 2\pi r_1(t)$ thus satisfies $C'(t) = 2\pi r_1'(t) = 5\pi/2$ at this time.

9. Let $(a(t), 0)$ be the position of A at time t. Then at the time in question we have

$$a(t) = 5, \qquad a'(t) = 3.$$

If $\left(b(t), -\sqrt{3}\,b(t)\right)$ is the position of B at time t, then its distance from the origin is

$$\sqrt{b(t)^2 + 3b(t)^2} = -2b(t)$$

and its speed is $-2b'(t)$. At the time in question we have

$$b(t) = -3/2, \qquad b'(t) = -2.$$

The distance $d(t)$ between A and B satisfies

$$d(t)^2 = [a(t) - b(t)]^2 + 3b(t)^2,$$

so at the time in question

$$d(t) = \sqrt{\left(5 + \tfrac{3}{2}\right)^2 + 3\left(\tfrac{3}{2}\right)^2} = 7.$$

Moreover,

$$2d(t)d'(t) = 2[a(t) - b(t)] \cdot [a'(t) - b'(t)] + 6b'(t)b'(t).$$

Substituting the values found for $a(t)$, $b(t)$, $a'(t)$, $b'(t)$ and $d(t)$ at the time in question, we obtain $d'(t) = \frac{83}{14}$.

10. (ii) $(k \circ f)'(0) = k'(f(0)) \cdot f'(0) = 0$.

11. By definition

$$f'(0) = \lim_{x \to 0} \frac{f(x) - 0}{x}$$

$$= \lim_{x \to 0} \frac{g(x) \sin 1/x}{x}.$$

Now

$$\lim_{x \to 0} \frac{g(x)}{x} = \lim_{x \to 0} \frac{g(x) - g(0)}{x}$$

$$= g'(0) = 0.$$

Since $|\sin 1/x| \leq 1$, it follows that $f'(0) = 0$ (as in Problem 5-21).

13. (a) The Chain Rule and Problem 9-3 imply that

$$f'(x) = \frac{1}{2\sqrt{1 - x^2}} \cdot -2x$$

$$= -\frac{x}{\sqrt{1 - x^2}}.$$

(b) The tangent line through $\left(a, \sqrt{1 - a^2}\right)$ is the graph of

$$g(x) = -\frac{a}{\sqrt{1 - a^2}}(x - a) + \sqrt{1 - a^2}.$$

So if $f(x) = g(x)$, then

$$\sqrt{1 - x^2} = \frac{-a}{\sqrt{1 - a^2}}(x - a) + \sqrt{1 - a^2}.$$

Squaring yields

$$1 - x^2 = \frac{a^2(x - a)^2}{1 - a^2} - 2a(x - a) + 1 - a^2.$$

Multiplying through by $1 - a^2$, and multiplying out, everything reduces to

$$-x^2 - a^2 = -2ax,$$

i.e., $(x - a)^2 = 0$, so $x = a$. Notice that the same argument shows that g does not intersect the graph of $f(x) = -\sqrt{1 - x^2}$, which is the bottom half of the unit circle.

14. The graph of the function

$$f(x) = b\sqrt{1 - \frac{x^2}{a^2}}$$

is the top half of the ellipse with consists of all points (x, y) satisfying

$$\frac{x^2}{a^2} + \frac{y^2}{b^2} = 1.$$

Now

$$f'(x) = \frac{-bx}{a^2\sqrt{1 - \dfrac{x^2}{a^2}}}.$$

If the tangent line through $\left(c, b\sqrt{1 - c^2/a^2}\right)$ intersects the graph of f at x, then

$$b\sqrt{a - \frac{x^2}{a^2}} = \frac{-bc}{a^2\sqrt{1 - \dfrac{c^2}{a^2}}}(x - c) + b\sqrt{1 - \frac{c^2}{a^2}}.$$

The easiest way to solve this equation is to use the following trick. If we let $x' = x/a$ and $c' = c/a$, then the equation becomes

$$(1) \qquad b\sqrt{1 - (x')^2} = \frac{-b(c'a)}{a^2\sqrt{1 - (c')^2}}(x' - c') \cdot a + b\sqrt{1 - (c')^2},$$

or simply

$$\sqrt{1 - (x')^2} = \frac{-c'}{\sqrt{1 - (c')^2}}(x' - c') + \sqrt{1 - (c')^2}.$$

The solution to Problem 13 shows that $x' = c'$, so $x = c$.

For the hyperbola, we consider

$$f(x) = b\sqrt{\frac{x^2}{a^2} - 1}.$$

Then

$$f'(x) = \frac{bx}{a^2\sqrt{\dfrac{x^2}{a^2} - 1}},$$

so if the tangent line through $\left(c, b\sqrt{c^2/a^2 - 1}\right)$ intersects the graph at x, then

$$(2) \qquad b\sqrt{\frac{x^2}{a^2} - 1} = \frac{bc}{a^2\sqrt{\dfrac{c^2}{a^2} - 1}}(x - c) + b\sqrt{\frac{c^2}{a^2} - 1}.$$

Squaring equations (1) and (2) produces the same result, so the solutions of (2) are also $x = c$.

15. No. For example, g might be $-f$. If $f(a) \neq 0$ and $f \cdot g$ and f are differentiable at a, then g is differentiable at a.

16. (a) Since f is differentiable at a, it is continuous at a. Since $f(a) \neq 0$, it follows that $f(x) \neq 0$ for all x in an interval around a. So $f = |f|$ or $f = -|f|$ in this interval, so $|f|'(a) = f'(a)$ or $|f|'(a) = -f'(a)$. It is also possible to use the Chain Rule, and Problem 9-3: $|f| = \sqrt{f^2}$, so

$$|f|'(x) = \frac{1}{2\sqrt{f(x)^2}} \cdot 2f(x)f'(x)$$
$$= f'(x) \cdot \frac{f(x)}{|f(x)|}.$$

(b) Let $f(x) = x - a$.

(c) This follows from part (a), since $\max(f, g) = \left[f + g + |f - g|\right]/2$ and $\min(f, g) = \left[f + g - |f - g|\right]/2$.

(d) Use the same example as in part (b), choosing $g = 0$.

17. (a) We have

$$(f \circ g)'(x) = f'(g(x)) \cdot g'(x)$$
$$(f \circ g)''(x) = f''(g(x)) \cdot g'(x)^2 + f'(g(x)) \cdot g''(x)$$
$$(f \circ g)'''(x) = [f'''(g(x)) \cdot g'(x)^3 + 2f''(g(x)) \cdot g'(x)g''(x)]$$
$$+ [f''(g(x)) \cdot g'(x)g''(x) + f'(g(x)) \cdot g'''(x)]$$
$$= f'''(g(x)) \cdot g'(x)^3 + 3f''(g(x)) \cdot g'(x)g''(x) + f'(g(x))g'''(x).$$

So

$$\mathcal{D}(f \circ g) = \frac{(f \circ g)'''}{(f \circ g)'} - \frac{3}{2}\left(\frac{(f \circ g)''}{(f \circ g)'}\right)^2$$
$$= \frac{(f''' \circ g)g'^2}{f' \circ g} + \frac{3(f'' \circ g)g''}{f' \circ g} + \frac{g'''}{g'} - \frac{3}{2}\left(\frac{(f'' \circ g) \cdot g'}{f' \circ g} + \frac{g''}{g'}\right)^2$$
$$= \frac{(f''' \circ g)g'^2}{f' \circ g} + \frac{3(f'' \circ g)g''}{f' \circ g} + \frac{g'''}{g'} - \frac{3}{2}\left(\frac{(f'' \circ g) \cdot g'}{f' \circ g}\right)^2$$
$$- 3\frac{(f'' \circ g)g''}{f' \circ g} - \frac{3}{2}\left(\frac{g''}{g'}\right)^2$$
$$= \left[\frac{f'''}{f'} \circ g - \frac{3}{2}\frac{f'' \circ g}{f' \circ g}\right] \cdot g'^2 + \frac{g'''}{g'} - \frac{3}{2}\left(\frac{g''}{g'}\right)^2$$
$$= [\mathcal{D}f \circ g] \cdot g'^2 + \mathcal{D}g.$$

(b) We have

$$f'(x) = \frac{a(cx+d) - c(ax+b)}{(cx+d)^2} = \frac{ad - bc}{(cx+d)^2}$$

$$f''(x) = -\frac{2c(ad-bc)}{(cx+d)^3}$$

$$f'''(x) = \frac{6c^2(ad-bc)}{(cx+d)^4}.$$

So

$$\frac{f'''(x)}{f'(x)} - \frac{3}{2}\left(\frac{f'''(x)}{f'(x)}\right)^2 = \frac{6c^2}{(cx+d)^2} - \frac{3}{2}\left(\frac{-2c}{cx+d}\right)^2$$
$$= 0.$$

18. The proof is by induction on n. For $n = 1$, Leibnitz's formula is Theorem 4. Suppose that for a certain n, Leibnitz's formula is true for all numbers a such that $f^{(n)}(a)$ and $g^{(n)}(a)$ exist. Suppose that $f^{(n+1)}(a)$ and $g^{(n+1)}(a)$ exist. Then $f^{(n)}(x)$ and $g^{(n)}(x)$ must exist for all x in some interval around a. So Leibnitz's formula holds for all these x, that is,

$$(f \cdot g)^{(n)}(x) = \sum_{k=0}^{n}\binom{n}{k}f^{(k)}(x) \cdot g^{(n-k)}(x)$$

for all x in some interval around a. Differentiating, and using Theorem 4, we find that

$$(f \cdot g)^{(n+1)} = \sum_{k=0}^{n}\binom{n}{k}(f^{(k)} \cdot g^{(n-k)})'(a)$$

$$= \sum_{k=0}^{n}\binom{n}{k}\left[f^{(k+1)}(a)g^{(n-k)}(a) + f^{(k)}(a)g^{(n+1-k)}(a)\right]$$

$$= \sum_{k=1}^{n+1}\binom{n}{k-1}f^{(k)}(a)g^{(n+1-k)}(a)$$

$$\quad + \sum_{k=0}^{n}\binom{n}{k}f^{(k)}(a)g^{(n+1-k)}(a)$$

$$= \sum_{k=0}^{n+1}\binom{n+1}{k}f^{(k)}(a)g^{(n+1-k)}(a) \qquad \text{by Problem 2-3(a).}$$

19. The formulas

$$(f \circ g)'(x) = f'(g(x)) \cdot g'(x)$$
$$(f \circ g)''(x) = f''(g(x)) \cdot g'(x)^2 + f'(g(x)) \cdot g''(x)$$
$$(f \circ g)'''(x) = f'''(g(x)) \cdot g'(x)^3 + 3f''(g(x)) \cdot g'(x)g''(x) + f'(g(x))g'''(x),$$

lead to the following conjecture: If $f^{(n)}(g(a))$ and $g^{(n)}(a)$ exist, then also $(f \circ g)^{(n)}(a)$ exists and is a sum of terms of the form

$$c \cdot [g'(a)]^{m_1} \cdots [g^{(n)}(a)]^{m_n} \cdot f^{(k)}(g(a)),$$

for some number c, nonnegative integers $m_1, \ldots, m_n$, and a natural number $k \le n$. To prove this assertion by induction, note that it is true for $n = 1$ (with $a = m_1 = k = 1$). Now suppose that for a certain n, this assertion is true for all numbers a such that $f^{(n)}(g(a))$ and $g^{(n)}(a)$ exist. Suppose that $f^{(n+1)}(g(a))$ and $g^{(n+1)}(a)$ exist. Then $g^{(k)}(x)$ must exist for all $k \le n$ and all x in some interval around a, and $f^{(k)}(y)$ must exist for all $k \le n$ and all y in some interval around $g(a)$. Since g is continuous at a, this implies that $f^{(k)}(g(x))$ exists for all x in some interval around a. So the assertion is true for all these x, that is, $(f \circ g)^{(n)}$ is a sum of terms of the form

$$c \cdot [g'(x)]^{m_1} \cdots [g^{(n)}(x)]^{m_n} \cdot f^{(k)}(g(a)), \qquad m_1, \ldots, m_n \ge 0, \ 1 \le k \le n.$$

Consequently, $(f \circ g)^{(n+1)}(a)$ is a sum of terms of the form

$$c \cdot m_\alpha [g'(a)]^{m_1} \cdots [g^{(\alpha)}(a)]^{m_\alpha - 1} \cdots [g^{(n)}(a)]^{m_n} \cdot f^{(k)}(g(a)) \qquad m_\alpha > 0$$

or of the form

$$c \cdot [g'(a)]^{m_1 + 1} \cdots [g^{(n)}(a)]^{m_n} \cdot f^{(k+1)}(g(a)).$$

20. (a) We can choose

$$g(x) = \frac{a_n x^{n+1}}{n+1} + \frac{a_{n-1} x^n}{n} + \cdots + \frac{a_1 x^2}{2} + a_0 x + c$$

for any number c.

(b) Let

$$g(x) = \frac{b_2 x^{-1}}{-1} + \frac{b_3 x^{-2}}{-2} + \cdots + \frac{b_m x^{-m+1}}{-m+1}.$$

(c) No, the derivative of f is

$$f'(x) = n a_n x^{n-1} + \cdots + a_1 - \frac{b_1}{x^2} - \frac{2b_2}{x^3} - \cdots - \frac{m b_m}{x^{m+1}}.$$

21. (a) Let g be a polynomial function of degree $n - 1$ with precisely $n - 1$ roots (as in Problem 3-7(d)); then $g = f'$ for some polynomial function f of degree n (Problem 20).

(b) Proceed as in part (a), starting with a polynomial function g of degree $n - 1$ with no roots (notice that $n - 1$ is even).

(c) We can proceed as in part (a), or simply note that $f(x) = x^n$ has the desired property.

(d) Proceed as in part (a), starting with a polynomial function g of degree $n-1$ with k roots (this exists by Problem 7-4).

22. (a) If a is a double root of f, so that $f(x) = (x-a)^2 g(x)$, then $f'(x) = (x-a)^2 g'(x) + 2(x-a)g(a)$, so $f'(a) = 0$. Conversely, if $f(a) = 0$ and $f'(a) = 0$, then $f(x) = (x-a)g(x)$ for some g and $f'(x) = (x-a)g'(x) + g(x)$, so $0 = f'(a) = g(a)$; thus $g(x) = (x-a)h(x)$, so $f(x) = (x-a)^2 h(x)$.

(b) The only root of $0 = f'(x) = 2ax + b$ is $x = -b/2a$, so f has a double root if and only if

$$0 = f\left(-\frac{b}{2a}\right) = a\left(\frac{b^2}{4a^2}\right) + b\left(\frac{-b}{2a}\right) + c$$

$$= -\frac{b^2}{4a} + c,$$

or $b^2 - 4ac = 0$. Geometrically, this is precisely the condition that the graph of f touches the horizontal axis at the single point $-b/2a$ (compare with Figure 22 in Problem 9-20).

23. Since $d'(x) = f'(x) - f'(a)$, we have $d'(a) = 0$. So a is a double root of d.

24. (a) Clearly f will have to be of the form

$$f(x) = \prod_{\substack{j=1 \\ j \neq i}}^{n} (x - x_j)^2 (ax + b)$$

(because each x_j, $j \neq i$ is a double root, by Problem 22). It therefore suffices to show that a and b can be picked so that $f(x_i) = a_i$ and $f'(x_i) = b_i$. If we write f in the form $f(x) = g(x)(ax + b)$, then we must solve

$$[g(x_i)x_i] \cdot a + g(x_i) \cdot b = a_i$$
$$[g'(x_i)x_i + g(x_i)] \cdot a + g'(x_i) \cdot b = b_i.$$

These equations can always be solved because

$$[g(x_i)x_i] \cdot g'(x_i) - [g'(x_i)x_i + g(x_i)]g(x_i) = [g(x_i)]^2 \neq 0.$$

(b) Let f_i be the function constructed in part (a), and let $f = f_1 + \cdots + f_n$.

25. (a) If $g(a)$ and $g(b)$ had different signs, then $g(x)$ would be 0 for some x in (a, b), which implies that $f(x) = 0$, contradicting the fact that a and b are consecutive roots.

(b) We have

$$f'(x) = (x - b)g(x) + (x - a)g(x) + (x - a)(x - b)g'(x),$$

so

$$f'(a) = (a - b)g(a),$$
$$f'(b) = (b - a)g(b).$$

Since $g(a)$ and $g(b)$ have the same sign, $f'(a)$ and $f'(b)$ have different signs. So $f'(x) = 0$ for some x in (a, b), since f' is a continuous function.

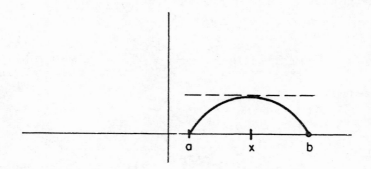

(c) Since

$$f'(x) = m(x - a)^{m-1}(x - b)^n g(x) + (x - a)^m n(x - b)^{n-1} g(x)$$
$$+ (x - a)^m (x - b)^n g'(x),$$

we have

$$h(a) = m(a - b)g(a),$$
$$h(b) = n(a - b)g(b),$$

so $h(a)$ and $h(b)$ have different signs, so $h(x) = 0$ for some x in (a, b), which implies that $f'(x) = 0$.

26.

$$f'(0) = \lim_{h \to 0} \frac{f(h) - f(0)}{h}$$
$$= \lim_{h \to 0} \frac{hg(h) - 0}{h}$$
$$= \lim_{h \to 0} g(h) = g(0), \qquad \text{since } g \text{ is continuous at } 0.$$

27. Let

$$g(x) = \begin{cases} \dfrac{f(x)}{x}, & x \neq 0 \\ f'(0), & x = 0. \end{cases}$$

Then $f(x) = xg(x)$ for all x, and

$$g(0) = f'(0) = \lim_{x \to 0} \frac{f(x) - f(0)}{x} = \lim_{x \to 0} g(x),$$

so g is continuous at 0.

28. The proof is by induction on k. For $k = 1$ we have

$$f'(x) = -nx^{-n-1}$$

$$= (-1)^1 \frac{(n + 1 - 1)!}{(n - 1)!} x^{-n-1} \qquad \text{for } x \neq 0.$$

Suppose that

$$f^{(k)}(x) = (-1)^k \frac{(n + k - 1)!}{(n - 1)!} x^{-n-k}$$

$$= (-1)^k k! \binom{n + k - 1}{n - 1} x^{-n-k} \qquad \text{for } x \neq 0.$$

Then

$$f^{(k+1)}(x) = (-1)^k \frac{(-n - k)(n + k - 1)!}{(n - 1)!} x^{-n-k-1}$$

$$= (-1)^{k+1} \frac{(n + k)!}{(n - 1)!} x^{-n-(k+1)} \qquad \text{for } x \neq 0.$$

29. If $x = f(x)g(x)$, then $1 = f'(x)g(x) + f(x)g'(x)$. In particular, $1 = f'(0)g(0) + f(0)g'(0) = 0$, a contradiction.

30. (a) Using Problem 28 and the Chain Rule, we obtain

$$f^{(k)}(x) = (-1)^k \frac{(n + k - 1)!}{(k - 1)!} (x - a)^{-n-k} \qquad \text{for } x \neq a.$$

(b) Since

$$f(x) = \frac{1}{x^2 - 1} = \frac{1}{2} \left(\frac{1}{x - 1} - \frac{1}{x + 1} \right),$$

we obtain, using part (a),

$$f^{(k)}(x) = \frac{(-1)^k (n + k - 1)!}{2(k - 1)!} [(x - 1)^{-n-k} - (x + 1)^{-n-k}].$$

31, 32. The formulas

$$f(x) = x^m \sin \frac{1}{x},$$

$$f'(x) = mx^{m-1} \sin \frac{1}{x} - x^{m-2} \cos \frac{1}{x},$$

$$f''(x) = m(m-1)x^{m-2} \sin \frac{1}{x} - mx^{m-3} \cos \frac{1}{x} - (m-2)x^{m-3} \cos \frac{1}{x}$$
$$- x^{m-4} \sin \frac{1}{x},$$

$$= m(m-1)x^{m-2} \sin \frac{1}{x} + (2-2m)x^{m-3} \cos \frac{1}{x} - x^{m-4} \sin \frac{1}{x},$$

$$f'''(x) = m(m-1)(m-2)x^{m-3} \sin \frac{1}{x} - m(m-1)x^{m-4} \cos \frac{1}{x}$$
$$+ (m-3)(2-2m)x^{m-4} \cos \frac{1}{x} + (2-2m)x^{m-5} \sin \frac{1}{x}$$
$$- (m-4)x^{m-5} \sin \frac{1}{x} + x^{m-6} \cos \frac{1}{x},$$

suggest the following conjecture: If $f(x) = x^m \sin 1/x$, for $x \neq 0$, then

$$f^{(k)}(x) = ax^{m-k} \sin \frac{1}{x}$$

$$+ \sum_{l=k+1}^{2k-1} \left(a_l x^{m-l} \sin \frac{1}{x} + b_l x^{m-l} \cos \frac{1}{x} \right) \pm \begin{cases} x^{m-2k} \sin \frac{1}{x}, & k \text{ even} \\ x^{m-2k} \cos \frac{1}{x}, & k \text{ odd} \end{cases}$$

for certain numbers a, a_l, b_l. Once this conjecture is made, it is easy to check it by induction. In fact, differentiating the first term yields

$$a(m-k)x^{m-(k+1)} \sin \frac{1}{x} - ax^{m-(k+2)} \cos \frac{1}{x},$$

and the second half of this expression can be incorporated in the sum $\sum_{l=k+2}^{2k+1}$ appearing in the desired expression for $f^{(k+1)}(x)$. Similarly, differentiating the last term yields

$$\begin{cases} \pm(m-2k)x^{m-(2k+1)} \sin \frac{1}{x} \mp x^{m-2(k+1)} \cos \frac{1}{x}, & k \text{ even } (k+1 \text{ odd}) \\ \pm(m-2k)x^{m-(2k+1)} \cos \frac{1}{x} \pm x^{m-2(k+1)} \sin \frac{1}{x}, & k \text{ odd } (k+1 \text{ even}) \end{cases}$$

and the first half of each expression can be incorporated in the sum $\displaystyle\sum_{l=k+2}^{2k+1}$. Finally,

each term appearing in the sum $\displaystyle\sum_{l=k+1}^{2k-1}$ yield upon differentiation two terms that can

be incorporated in the new sum $\displaystyle\sum_{l=k+2}^{2k+1}$.

It follows, in particular, that if $m = 2n$, then $f^{(k)}(x)$ always has a factor of at least x^2 for $k < n$ (while the remaining factor is bounded in an interval around 0). So if we define $f(0) = 0$, then

$$f'(0) = \lim_{h \to 0} \frac{f(h) - f(0)}{h}$$

$$= \lim_{h \to 0} \frac{f(h)}{h} = 0, \qquad \text{since } f(h) \text{ has a factor of at least } h^2;$$

consequently, if $2 \leq n$, then

$$f''(0) = \lim_{h \to 0} \frac{f'(h) - f'(0)}{h}$$

$$= \lim_{h \to 0} \frac{f'(h)}{h} = 0, \qquad \text{since } f'(h) \text{ has a factor of at least } h^2;$$

consequently, if $3 \leq n$, then $f'''(0) = 0$, etc. This argument (which is really another inductive argument) shows that $f'(0) = \cdots = f^{(n)}(0) = 0$. On the other hand, $f^{(n)}(x)$ is a sum of terms which do have a factor of at least x^2, together with $\pm \sin 1/x$ or $\cos 1/x$, so $f^{(n)}$ is *not* continuous at 0.

If $m = 2n + 1$, then $f^{(k)}$ always has a factor of at least x^2 for $k < n$, so $f'(0) = \cdots = f^{(n)} = 0$, but $f^{(n)}(x)$ is a sum of terms which do have a factor of at least x^2, together with $\pm x \cos 1/x$ or $\pm x \sin 1/x$. It follows that $f^{(n)}$ is continuous, but not differentiable, at 0.

33. (ii)

$$\frac{dz}{dx} = \frac{dz}{dy} \cdot \frac{dy}{dx} = (\cos y) \cdot (-\sin x) = \cos(\cos x) \cdot (-\sin x).$$

(iv)

$$\frac{dz}{dx} = \frac{dz}{dv} \cdot \frac{dv}{du} \cdot \frac{du}{dx}$$

$$= (\cos v)(-\sin u)(\cos x) = \cos(\cos(\sin x)) \cdot (-\sin(\sin x)) \cdot \cos x.$$

CHAPTER 11

1. (ii) $f'(x) = 5x^4 + 1 = 0$ for no x;

$f(-1) = -1, \ f(1) = 3$;

maximum $= 3$, minimum $= -1$.

(iv) $f'(x) = -\dfrac{(5x^4 + 1)}{(x^5 + x + 1)^2} = 0$ for no x;

$f(-1/2) = 32/15, \ f(1) = 1/3$;

maximum $= 32/15$, minimum $= 1/3$.

(Notice that $g(x) = x^5 + x + 1$ is increasing, since $g'(x) = 5x^4 + 1 > 0$ for all x; since $g(-1/2) = 15/32 > 0$, this shows that $g(x) \neq 0$ for all x in $[-1/2, 1]$, so f is differentiable on $[-1/2, 1]$.)

(vi) f is not bounded above or below on $[0, 5]$.

2. (i) $-4/3$ is a local maximum point, and 2 is a local minimum point.

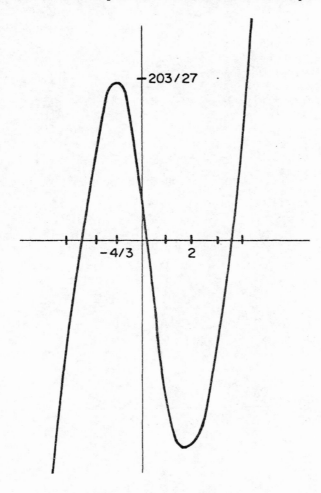

(ii) No local maximum or minimum points.

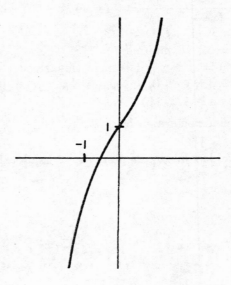

(iii) 0 is a local minimum point, and there are no local maximum points.

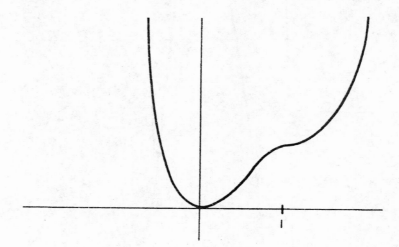

(iv) No local maximum or minimum points. In the figure below, a is the unique root of $x^5 + x + 1 = 0$.

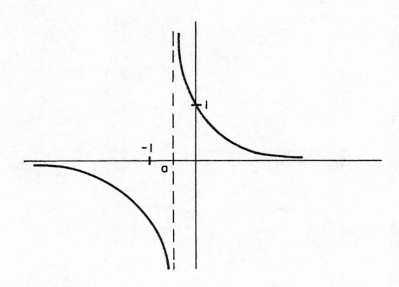

(v) $-1 + \sqrt{2}$ is a local maximum point, and $-1 - \sqrt{2}$ is a local minimum point.

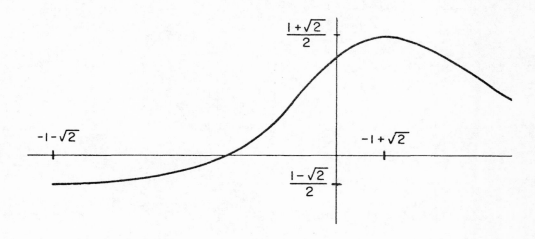

(vi) No local maximum or minimum points, since

$$f'(x) = -\frac{(1 + x^2)}{(x^2 - 1)^2} < 0 \qquad \text{for } x \neq \pm 1.$$

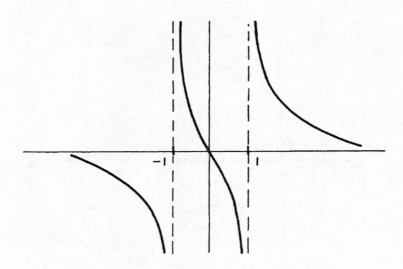

3. (i) f is odd;
$$f'(x) = 1 - \frac{1}{x^2} = \frac{x^2 - 1}{x^2};$$
$f'(x) = 0$ for $x \neq \pm 1$, $f'(x) > 0$ for $|x| > 1$;
$f(1) = 2$, $f(-1) = -2$.

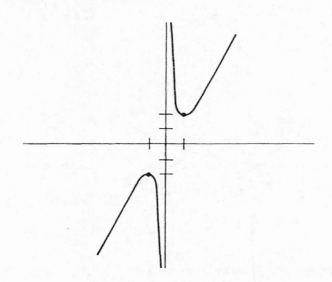

(ii) $f'(x) = 1 - \dfrac{6}{x^3} = \dfrac{x^3 - 6}{x^3}$;

$f'(x) = 0$ for $x = \sqrt[3]{6}$, $f'(x) > 0$ for $x > \sqrt[3]{6}$ and $x < 0$;

$f(x) = 0$ for $x = -\sqrt[3]{3}$.

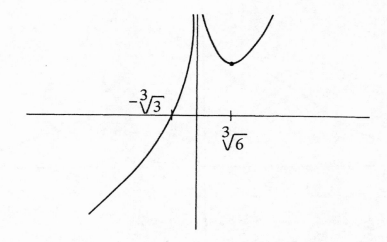

(iii) f is even;

$f'(x) = \dfrac{2x(x^2 - 1) - 2xx^2}{(x^2 - 1)^2} = \dfrac{-2x}{(x^2 - 1)^2}$;

$f'(x) = 0$ for $x = 0$, $f'(x) < 0$ for $x > 0$, $f'(x) > 0$ otherwise;

$f(0) = 0$.

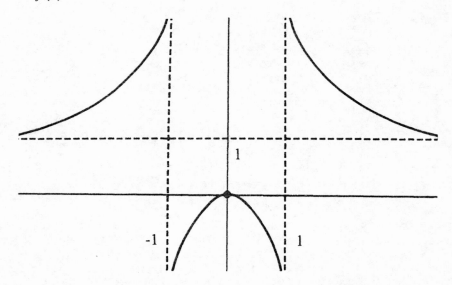

(iv) f is even, $f(x) > 0$ for all x;

$$f'(x) = \frac{-2x}{(1+x^2)^2};$$

$f'(x) = 0$ for $x = 0$, $f'(x) > 0$ for $x < 0$, $f'(x) < 0$ otherwise;

$f(0) = 1$.

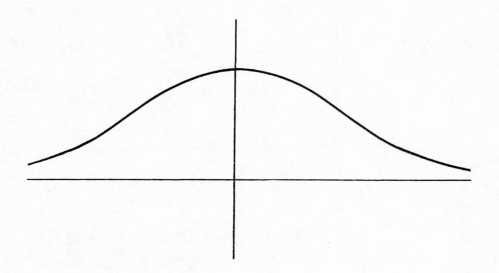

4. (b) Suppose x and y are points in $[a_{j-1}, a_j]$ and $[a_j, a_{j+1}]$, respectively, with $|x - a_j| = |y - a_j|$.

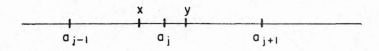

Then

$$|y - a_i| = |x - a_i| + |y - x| \qquad \text{for } i \leq j - 1,$$
$$|y - a_i| = |x - a_i| - |y - x| \qquad \text{for } i \geq j + 1.$$

So

$$f(y) = f(x) + |y - x| \cdot \{(j - 1) - (n - j)\}$$
$$= f(x) + |y - x| \cdot \{2j - n - 1\}.$$

This shows that f decreases until it reaches the "middlemost a_i" and then increases.

The minimum occurs at $a_{(n-1)/2}$ if n is odd and on the whole interval $[a_{n/2}, a_{n/2+1}]$ if n is even.

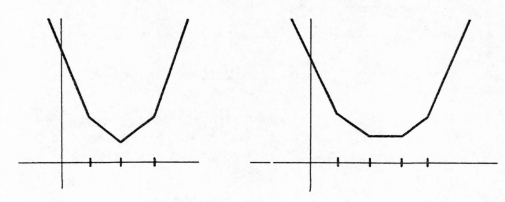

(c) We have

$$f(x) = \begin{cases} \dfrac{1}{1-x} + \dfrac{1}{1+a-x}, & x < 0 \\[2mm] \dfrac{1}{1+x} + \dfrac{1}{1+a-x}, & 0 < x < a \\[2mm] \dfrac{1}{1+x} + \dfrac{1}{1+x-a}, & a < x, \end{cases}$$

so

$$f'(x) = \begin{cases} \dfrac{1}{(1-x)^2} + \dfrac{1}{(1+a-x)^2}, & x < 0 \\[2mm] \dfrac{-1}{(1+x)^2} + \dfrac{1}{(1+a-x)^2}, & 0 < x < a \\[2mm] \dfrac{-1}{(1+x)^2} - \dfrac{1}{(1+x-a)^2}, & a < x. \end{cases}$$

Thus f is increasing on $(-\infty, 0]$ and decreasing on $[a, \infty)$, so the maximum of f on $[0, a]$ is the maximum on **R**. If $f'(x) = 0$ for x in $(0, a)$, then

$$(1 + x)^2 - (1 + a - x)^2 = 0,$$

whose only solution is $x = a/2$. Since

$$f\left(\frac{a}{2}\right) = \frac{4}{2+a} < \frac{2+a}{1+a} = f(0) = f(a),$$

the maximum value is $(2 + a)/(1 + a)$.

5. (ii) All irrational x are local minimum points, and all rational x are local maximum points.

(iv) All $1/n$ for n in **N** are local maximum points, and all other x are local minimum points.

6. (a) The distance $d(x)$ from (x_0, y_0) to $(x, f(x))$ satisfies

$$D(x) = [d(x)]^2 = (x - x_0)^2 + (mx + b - y_0)^2,$$

so the minimum occurs when

$$0 = D'(\bar{x}) = 2(\bar{x} - x_0) + 2m(m\bar{x} + b - y_0)$$

or

$$\bar{x} = \frac{x_0 + m(y_0 - b)}{1 + m^2}.$$

(b) The slope $(m\bar{x} + b - y_0)/(\bar{x} - x_0)$ of the line from (x_0, y_0) to $(\bar{x}, m\bar{x} + b)$ must satisfy

$$\frac{m\bar{x} + b - y_0}{\bar{x} - x_0} \cdot m = -1$$

which yields the same result.

(c) For $b = 0$ we have

$$\bar{x} = \frac{x_0 + my_0}{1 + m^2},$$

hence

$$\bar{x} - x_0 = \frac{m(y_0 - mx_0)}{1 + m^2},$$

$$m\bar{x} - y_0 = \frac{mx_0 - y_0}{1 + m^2}.$$

So the distance d from (x_0, y_0) to $(\bar{x}, m\bar{x})$ is

$$\frac{1}{1 + m^2}\sqrt{(y_0 - mx_0)^2(1 + m^2)} = \frac{|mx_0 - y_0|}{\sqrt{1 + m^2}}.$$

The general case can now be solved as in Problem 4-22, to give

$$\frac{|mx_0 - y_0 + b|}{\sqrt{1 + m^2}}.$$

(d) For $B \neq 0$, this line is the graph of $f(x) = (-A/B)x - C/B$, so the distance is

$$\frac{\left|\left(-\dfrac{A}{B}\right)x_0 - y_0 - \dfrac{C}{B}\right|}{\sqrt{1 + \dfrac{A^2}{B^2}}} = \frac{|Ax_0 + By_0 + C|}{\sqrt{A^2 + B^2}}.$$

For $B = 0$ we have the line parallel to the y-axis through $-C/A$. The distance to (x_0, y_0) is

$$\left|x_0 - \left(\frac{-C}{A}\right)\right| = \frac{|Ax_0 + C|}{|A|},$$

which is the same result for $B = 0$.

7. If $g(x) = f(x)^2$, then

$$g'(x) = 2f(x)f'(x),$$

so the critical points of g are those of f, together with the zeros of f (notice that g may be differentiable at points where $f(x) = 0$ even when f isn't, e.g., $f(x) = |x|$).

11. Let x be the height of the cone. The volume $V(x)$ is given by

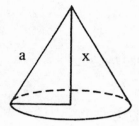

$$V(x) = \frac{1}{3}x\pi\left(\sqrt{a^2 - x^2}\right)^2 = \frac{\pi}{3}(a^2x - x^3).$$

So the volume is greatest when

$$0 = V'(x) = \frac{\pi}{3}[a^2 - 3x^2],$$

or $x = a/\sqrt{3}$. For this x we have

$$V(x) = \frac{\pi}{3}\left(\frac{a^3}{\sqrt{3}} - \frac{a^3}{3\sqrt{3}}\right)$$
$$= \frac{2\sqrt{3}a^3}{27}.$$

12. In the Figure below we have

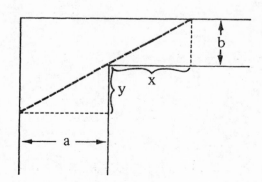

$$\frac{b}{x} = \frac{y}{a},$$

so the length of the dashed line is

$$\sqrt{b^2 + x^2} + \sqrt{a^2 + \frac{a^2b^2}{x^2}} = \sqrt{b^2 + x^2} + \frac{a}{x}\sqrt{x^2 + b^2} = \left(1 + \frac{a}{x}\right)\sqrt{x^2 + b^2}.$$

The maximum length of a ladder which can be carried horizontally around the corner is the *minimum* length of this dashed line. This occurs when

$$0 = -\frac{a}{x^2}\sqrt{x^2 + b^2} + \left(1 + \frac{a}{x}\right)\frac{x}{\sqrt{x^2 + b^2}}$$

$$= \left[-\frac{a}{x^2}(x^2 + b^2) + x + a\right] \cdot \frac{1}{\sqrt{x^2 + b^2}},$$

or

$$ax^2 + ab^2 = x^3 + ax^2,$$
$$x = a^{1/3}b^{2/3},$$

and the length is

$$\left(1 + \frac{a^{2/3}}{b^{2/3}}\right)\sqrt{a^{2/3}b^{4/3} + b^2} = (b^{2/3} + a^{2/3})\sqrt{\frac{a^{2/3}b^{4/3} + b^2}{b^{4/3}}}$$

$$= (b^{2/3} + a^{2/3})^{3/2}.$$

13. If $R(\theta)$ is the appropriate value of R for given θ, we have

$$\frac{\theta}{2} \cdot R(\theta)^2 = A.$$

The perimeter for this θ will have value

$$P(\theta) = \theta R(\theta) + 2R(\theta)$$
$$= \sqrt{2A}(\theta + 2) \cdot \theta^{-1/2}.$$

So the minimum occurs when

$$0 = P'(\theta) = \sqrt{2A}\left[\frac{1}{\theta^{1/2}} - \frac{\theta + 2}{2\theta^{3/2}}\right]$$

$$= \sqrt{2A} \cdot \frac{\theta - 2}{2\theta^{3/2}}$$

or $\theta = 2$ radians, and $R = \sqrt{A}$.

14. If

$$f(x) = x + \frac{1}{x} \qquad (x > 0)$$

then

$$f'(x) = 1 - \frac{1}{x^2},$$

which has the minimum value for $x = 1$, with $f(x) = 2$.

15. If x is the height of the trapezoid, then the area is

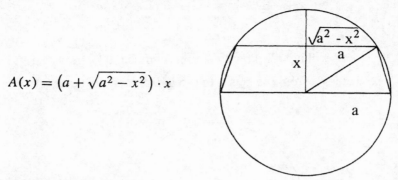

$$A(x) = \left(a + \sqrt{a^2 - x^2}\right) \cdot x$$

so the maximum occurs when

$$0 = A'(x) = a + \sqrt{a^2 - x^2} - \frac{x^2}{\sqrt{a^2 - x^2}}$$

$$= \frac{a\sqrt{a^2 - x^2} + a^2 - 2x^2}{\sqrt{a^2 - x^2}},$$

or

$$a^2(a^2 - x^2) = (2x^2 - a^2)^2 = 4x^4 - 4x^2a^2 + a^4,$$

so

$$4x^4 = 3x^2a^2,$$

$$x = \frac{\sqrt{3}a}{2}.$$

The area is

$$\left(a + \sqrt{a^2 - \frac{3a^2}{4}}\right)\frac{\sqrt{3}a}{2} = \left(a + \frac{a}{2}\right)\frac{\sqrt{3}a}{2} = \frac{3\sqrt{3}a^2}{4}.$$

16. The vertex of the right angle will obviously be to the left of the center of the circle. If x is the distance from the center to the vertex, then the length L of $A + B$ is

$$L(x) = a + x + \sqrt{a^2 - x^2}$$

so the maximum occurs when

$$0 = L'(x) = 1 - \frac{x}{\sqrt{a^2 - x^2}},$$

or

$$a^2 - x^2 = x^2,$$

$$x = a/\sqrt{2}.$$

The length is

$$a + \frac{a}{\sqrt{2}} + \sqrt{a^2 - \frac{a^2}{2}} = (1 + \sqrt{2})a.$$

17. **(i)** Obviously walking all the way around will be the longest path.

(ii) Let θ be the angle from the center to the point where he lands after rowing.

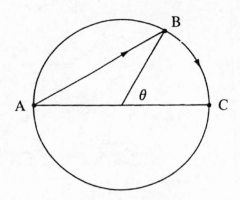

Then

$$\overset{\frown}{BC} = \theta, \qquad \overline{AB} = \sqrt{2 + 2\cos\theta} \qquad \text{(by the law of cosines)}.$$

The total time required is

$$T(\theta) = \frac{\sqrt{2 + 2\cos\theta}}{2} + \frac{\theta}{4}.$$

Now

$$0 = T'(\theta) = \frac{-\sin\theta}{2\sqrt{2 + 2\cos\theta}} + \frac{1}{4}$$

when

$$4\sin\theta = 2\sqrt{2 + 2\cos\theta} \implies 16\sin^2\theta = 8 + 8\cos\theta,$$

thus

$$2(1 - \cos^2\theta) = 1 + \cos\theta \implies 2\cos^2\theta + \cos\theta - 1 = 0,$$

hence

$$\cos\theta = \tfrac{1}{2} \text{ or } -1.$$

Here $\cos\theta = -1$ for $\theta = \pi$, one of the endpoints of the interval $[0, \pi]$ that we must consider for θ, while $\cos\theta = \frac{1}{2}$ for $\theta = \frac{\pi}{3}$. We have

$$T(0) = 1 \text{ hour, to row across}$$
$$T(\tfrac{\pi}{3}) = \tfrac{\sqrt{3}}{2} + \tfrac{\pi}{12} \text{ hours}$$
$$T(\pi) = \tfrac{\pi}{4} \text{ hours, to walk around.}$$

Since

$$\tfrac{\sqrt{3}}{2} + \tfrac{\pi}{12} > \tfrac{\pi}{4} \qquad (\text{i.e., } \tfrac{\sqrt{3}}{2} > \pi(\tfrac{1}{4} - \tfrac{1}{12}) = \tfrac{\pi}{6}),$$

walking around is fastest.

18. If $x = \overline{BC}$ and $y = \overline{AB}$

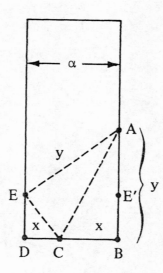

then we have

$$ED = \sqrt{x^2 - (\alpha - x)^2} = \sqrt{2\alpha x - \alpha^2} \qquad \text{from } \triangle EDC$$
$$\alpha^2 + (y - \overline{ED})^2 = y^2 \qquad \text{from } \triangle EE'A,$$

so

$$\alpha^2 + \left(y - \sqrt{2\alpha x - \alpha^2}\right)^2 = y^2$$
$$-y\sqrt{2\alpha x - \alpha^2} + \alpha x = 0$$
$$y^2(2\alpha x - \alpha^2) = \alpha^2 x^2$$
$$y^2 = \frac{\alpha^2 x^2}{2\alpha x - \alpha^2} = \frac{\alpha x^2}{2x - \alpha}.$$

The square of the length of the crease is

$$x^2 + y^2 = x^2 + \frac{\alpha x^2}{2x - \alpha} = \frac{2x^3}{2x - \alpha},$$

so the length is smallest when

$$0 = 6x^2(2x - \alpha) - 4x^3 = 8x^3 - 6x^2\alpha = x^2(8x - 6\alpha),$$

or $x = 3\alpha/4$. For this x the length is

$$\sqrt{\frac{2\left(\dfrac{3\alpha}{4}\right)^3}{\dfrac{3\alpha}{2} - \alpha}} = \frac{3\sqrt{3}\alpha}{4}.$$

20.

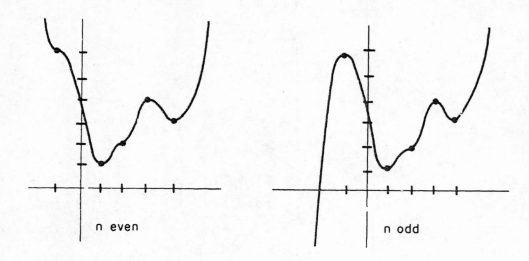

21. (a) 1 is a local minimum point, and 2 is a local maximum point. The nature of the critical points -1 and 3 can be determined by the behavior of $f(x)$ for large $|x|$:

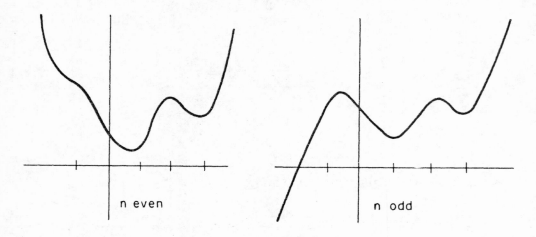

(b) No, for if 2 were the largest critical point, then f would have to be decreasing on $(3, \infty)$, since 2 is a local maximum point.

22. Let $f(x) = r(x)/s(x)$ for polynomial functions r and s. It is possible that r and s have a common root a, but in this case $r(x) = (x-a)r_1(x)$ and $s(x) = (x-a)s_1(x)$ for certain polynomial functions r_1 and s_1 (Problem 3-7). This means that $f(a)$ is undefined but that $f(x) = r_1(x)/s_1(x)$ for $x \neq a$ (and $s_1(x) \neq 0$). After factoring out all common linear factors of r and s, we find that the graph of f consists, except for a finite number of points, of the graph of

$$g(x) = \frac{a_n x^n + a_{n-1} x^{n-1} + \cdots + a_0}{b_m x^m + b_{m-1} x^{m-1} + \cdots + b_0} = \frac{p(x)}{q(x)},$$

where p and q have no common roots. The function g is defined at all points a except those with $q(a) = 0$ (of which there are at most m). Near such a point a the graph of g looks like (a), (b), (c), or (d), depending on the sign of $p(a)$ and whether a is a local maximum or minimum point for q or whether q is increasing or decreasing in an interval around a.

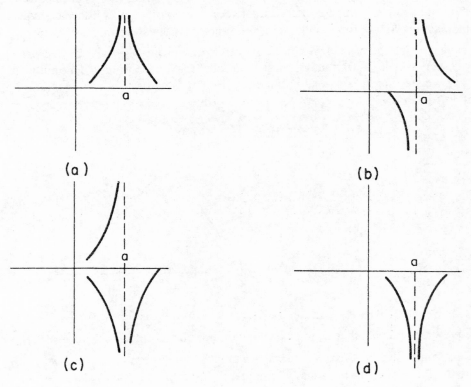

(a)

(b)

(c)

(d)

Since

$$g'(x) = \frac{q(x)p'(x) - p(x)q'(x)}{[q(x)]^2}$$

and $qp' - pq'$ is a polynomial function of degree at most $m + n$, there are at most $m + n$ local maximum and minimum points. On the intervals between these points and the points of discontinuity, g is either increasing or decreasing. The behavior of $g(x)$ for large x or large negative x has been discussed in Problems 5-32 and 5-36.

23. (a) This follows from Problem 3-7 and the fact that the difference of the two polynomial functions has degree at most $\max(m, n)$.

(b) If $m \geq n$, let f_1 be a polynomial function which has m roots, and let $f(x) = f_1(x) + x^n$ and $g(x) = x^n$.

24. (a) The polynomial function f', of degree $n - 1$, has k roots, and no multiple roots, since $f''(x) \neq 0$ when $f'(x) = 0$. It follows from Problem 7-4 that $n - 1 - k$ is even.

(b) Since $n - 1 - k$ is even, there is a polynomial function g of degree $n - 1$ with exactly k roots. Let f be a polynomial function of degree n with $f' = g$ (Problem 10-20).

(c) Let $l = k_1 + k_2$ and let $a_l < a_{l-1} < \cdots < a_1$ be all the local maximum and minimum points. On the intervals between these points f is either decreasing or increasing. Since $\lim\limits_{x \to \infty} f(x) = \infty$, the function f must be increasing on (a_1, ∞). Thus a_1 must be a local minimum point. Consequently, f must be decreasing on (a_2, a_1), which shows that a_2 must be a local maximum point. Continuing in this way we see that a_k is a local minimum point if k is odd and a local maximum point if k is even.

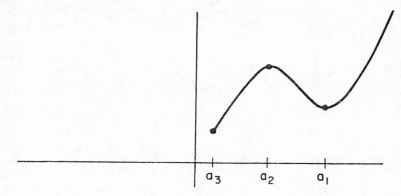

Now if n is even, then a_l must be a local minimum point, since $\lim\limits_{x \to -\infty} f(x) = \infty$. Thus l must be odd, so $a_1, a_3, \ldots, a_l$ are the local minimum points, and $a_2, \ldots, a_{l-1}$ are the local maximum points. Consequently $k_2 = k_1 + 1$. If n is odd, then a must be a local maximum point, since $\lim\limits_{x \to -\infty} f(x) = -\infty$. The same sort of reasoning then shows that $k_1 = k_2$.

(d) The hypotheses imply that $n - 1 - (k_1 + k_2)$ is even. Let $l = [n - 1 - (k_1 + k_2)]/2$, and choose a polynomial function f of degree n with f' as in the hint. Since $(1 + x^2)^l > 0$ for all x, it follows that $f'(x) > 0$ for $x > a_{k_1 + k_2}$ and that the sign of f' changes as we go from (a_{i-1}, a_i) to (a_{i-2}, a_{i-1}). Thus $a_{k_1 + k_2}, a_{k_1 + k_2 - 2}, \ldots$ are local minimum points and $a_{k_1 + k_2 - 1}, a_{k_1 + k_2 - 3}, \ldots$ are local maximum points.

26. Note that f is increasing. If $f()1/2) \geq 0$, then $f(3/4) \geq M/4$, so certainly $f \geq M/4$ on the interval $[3/4, 1]$. On the other hand, if $f(1/2) \leq 0$, then $f(1/4) \leq -M/4$, so $f \leq -M/4$ on the interval $[0, 1/4]$.

27. (a) Apply the Mean Value Theorem to $f - g$: If $x > a$, then

$$\frac{f(x) - g(x)}{x - a} = \frac{f(x) - g(x) - [f(a) - g(a)]}{x - a}$$
$$= f'(y) - g'(y) \qquad \text{for some } y \text{ in } (a, x)$$
$$> 0.$$

Since $x - a > 0$, it follows that $f(x) - g(x) > 0$. Similarly, if $x - a < 0$, then $f(x) < g(x)$.

(b) An example is shown below.

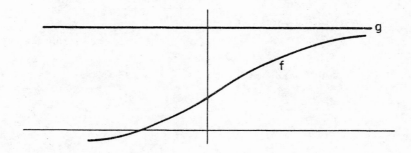

30. (a) The position at time t is

$$\big((v \cos \alpha)t, \ -16t^2 + (v \sin \alpha)t\big).$$

If $\cos \alpha = 0$, so that the cannon ball is shot straight up, then these points all lie on a straight line. If $\cos \alpha \neq 0$, then the set of all such points is equal to the set of all points

$$\left(t, \ -\frac{16t^2}{v \cos \alpha} + (\tan \alpha)t\right),$$

so the path of the cannon ball lies on the graph of

$$f(x) = \frac{-16x^2}{v \cos \alpha} + (\tan \alpha)x,$$

which is the graph of a parabola.

(b) The cannon ball hits the ground at time $t > 0$ when

$$0 = -16t^2 + (v \sin \alpha)t,$$

or $t = (v \sin \alpha)/16$ (of course we consider only $\alpha > 0$). It has then traveled a horizontal distance of

$$d(\alpha) = (v \cos \alpha) \cdot \frac{v \sin \alpha}{16}$$
$$= \frac{v^2 \sin \alpha \cos \alpha}{16}.$$

Now $d(\alpha)$ is a maximum at that α for which

$$0 = d'(\alpha) = \frac{v^2}{16}[\cos^2 \alpha - \sin^2 \alpha],$$

so $\tan \alpha = \pm 1$. Since only positive α are considered, α is a 45° angle.

31. (a) Such a function f is pictured below. As an explicit example we can take $f(x) + (\sin x^2)/x$. Then $\lim\limits_{x \to \infty} f(x) = 0$, but

$$f'(x) = \frac{2x^2 \sin x^2 - \sin x^2}{x^2}$$
$$= 2 \sin x^2 - \frac{\sin x^2}{x^2},$$

so $\lim\limits_{x \to \infty} f'(x)$ does not exist.

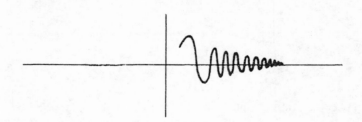

(b) Let $l = \lim\limits_{x \to \infty} f'(x)$. If $l < 0$, then there would be some N such that $|f'(x) - l| < |l|/2$ for $x > N$. This would imply that $f'(x) > |l|/2$. But that would imply, by the Mean Value Theorem, that

$$f(x) > f(N) + \frac{(x - N)|l|}{2} \qquad \text{for } x > N,$$

which would mean that $\lim\limits_{x \to \infty} f(x)$ does not exist. Similarly, $\lim\limits_{x \to \infty} f'(x)$ cannot be < 0.

(c) Let $l = \lim\limits_{x \to \infty} f''(x)$. If $l > 0$, then, as in part (a), we have $\lim\limits_{x \to \infty} f'(x) = \infty$. Another application of the Mean Value Theorem shows that $\lim\limits_{x \to \infty} f(x) = \infty$, contradicting the hypothesis. Similarly, $\lim\limits_{x \to \infty} f''(x)$ cannot be < 0.

32. If $g(x) \neq 0$ for all x in (a, b), then the function $h(x) = f(x)/g(x)$ is differentiable on (a, b), and by hypothesis

$$h'(x) = \frac{g(x)f'(x) - f(x)g'(x)}{[g(x)]^2} = 0.$$

This means that f/g is constant on (a, b), so $f = c \cdot g$ on (a, b) for some $c \neq 0$. Since f and g are continuous it follows that $f(a) = c \cdot g(a)$, so $g(a) = 0$, contradicting the hypothesis.

33. We have

$$f'(x) = \lim_{y \to x} \frac{f(y) - f(x)}{y - x}.$$

Now

$$\left| \frac{f(y) - f(x)}{y - x} \right| \leq |x - y|^{n-1},$$

and $\lim_{y \to x} |x - y|^{n-1} = 0$, since $n - 1 > 0$. Consequently $f'(x) = 0$ for all x, so f is constant.

34. (a) Since $|f(x) - f(x+h)| \leq C|h|^{\alpha}$, it follows that $\lim_{h \to 0} f(x+h) = f(x)$.

(b) Given $\varepsilon > 0$, choose $\delta = \varepsilon^{1/\alpha}/C$. Then for all x and y in the interval with $|x - y| < \delta$ we have

$$|f(x) - f(y)| \leq C|x - y|^{\alpha} < \frac{C(\varepsilon^{1/\alpha})^{\alpha}}{C} = \varepsilon.$$

(c) If f is differentiable at x, then

$$\lim_{y \to x} \frac{f(y) - f(x)}{y - x} = f'(x),$$

so for all y in some interval around x we have

$$\left| \frac{f(y) - f(x)}{y - x} - f'(x) \right| < 1,$$

hence

$$\left| \frac{f(y) - f(x)}{y - x} \right| < 1 + |f'(x)|,$$

$$|f(y) - f(x)| \leq (1 + |f'(x)|)|y - x|,$$

so we can choose $C = 1 + |f'(x)|$. (Actually, we can choose $C = \varepsilon + |f'(x)|$ for any $\varepsilon > 0$.) The converse is not true, e.g., $f(x) = |x|$.

(d) No, because the derivative f', and hence the required C, may not be bounded on $[a, b]$. For example, $f(x) = x^2 \sin 1/x^2$ on $[0, 1]$.

(e) Same proof as Problem 33.

35. Let

$$f(x) = a_0 x + \frac{a_1 x^2}{2} + \cdots + \frac{a_n x^{n+1}}{n+1}.$$

Then $f(0) = 0$ and $f(1) = 0$ by hypothesis. Rolle's Theorem implies that for some x in $(0, 1)$ we have

$$0 = f'(x) = a_0 + a_1 x + \cdots + a_n x^n.$$

36. If $f_m(x_0) = f_m(x_1) = 0$ for $x_0 < x_1$ in $[0, 1]$, then $f_m'(x) = 0$ for some x which is in (x_0, x_1), and hence satisfies $0 < x < 1$. But

$$f_m'(x) = 3x^2 - 3 = 3(x^2 - 1),$$

so $f_m'(x) = 0$ only for $x = \pm 1$.

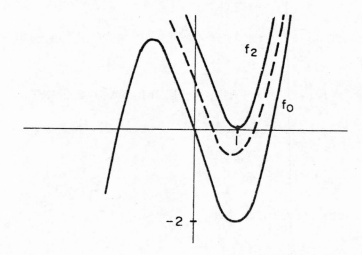

37. Problem 7-11 shows that there is at least one x. Suppose there were two, $x_0 < x_1$. The Mean Value Theorem, applied to $[x_0, x_1]$, would imply that

$$f'(x) = \frac{f(x_1) - f(x_0)}{x_1 - x_0} = \frac{x_1 - x_0}{x_1 - x_0} = 1$$

for some x in $[x_0, x_1]$, contradicting the hypothesis.

38. **(a)** Clearly f has at least two zeros, in fact at least two zeros in $[-1, 1]$, since $f(0) < 0$ while $f(\pm 1) > 0$. If f had more than two zeros, then f' would have at least two zeros. But

$$f'(x) = 2x + \sin x$$

and this is an increasing function, since

$$f''(x) = 2 + \cos x \geq 1 \qquad \text{for all } x.$$

(b) We have $f(0) < 0$, while $f(x)$ will be > 0 for large enough $|x|$, since $|x \sin x|$ is small compared to $2x^2$ and $|\cos^2 x| \leq 1$. In fact, writing

$$f(x) = x(2x - \sin x) - \cos^2 x,$$

and noting that $2x - \sin x > 1$ for $x > 1$, we see that $f(x) > 0$ for $x > 1$, and also for $x < -1$, since f is even. So f has at least two zeros in $[-1, 1]$, and no zeros outside of $[-1, 1]$. If f had more than two zeros, then f' would have two zeros in $[-1, 1]$. But

$$f'(x) = 4x - \sin x - x \cos x + 2 \cos x \sin x$$
$$= 4x - \sin x - x \cos x + \sin 2x$$

and this is increasing on $[-1, 1]$, since

$$f''(x) = 4 - 2 \cos x + x \sin x + 2 \cos 2x$$

which is ≥ 1 on $[-1, 1]$, since $x \sin x > 0$ on $[-1, 1]$, while $|\cos x|, |\cos 2x| \leq 1$.

39. (a) Suppose that $f''(x) < 4$ for all x in $[0, 1/2]$. Then, by the Mean Value Theorem, for all x in $[0, 1/2]$ we have

$$\frac{f'(x) - f'(0)}{x - 0} = f''(x') \qquad \text{for some } x' \text{ in } [0, x]$$
$$< 4,$$

so $f'(x) < 4x$. Applying the Mean Value Theorem again, we have

$$\frac{f(x) - f(0)}{x - 0} = f'(x') \qquad \text{for some } x' \text{ in } (0, x)$$
$$< 4x' < 4x,$$

so $f(x) < 4x^2$. Consequently $f(1/2) < 1/2$.

The same sort of analysis can be applied to f on $[1/2, 1]$ if $f''(x) < -4$ for all x in $[0, 1/2]$. It is a little more convenient to introduce the function $g(x) = 1 - f(1 - x)$, which satisfies $g(0) = 0$ and $g''(x) = -f''(1 - x) < 4$ for x in $[0, 1/2]$. As we have just shown,

$$1/2 > g(1/2) = 1 - f(1/2),$$

so $f(1/2) > 1/2$, contradicting the result found before.

(b) Note first that we cannot have $f''(x) = 4$ for $0 < x < 1/2$ and also $f''(x) = -4$ for $1/2 < x \leq 1$, since this would imply that $f'(x) = 4x$ for $0 \leq x \leq 1/2$ and $f''(x) = -4x$ for $1/2 \leq x \leq 1$, in which case $f''(1/2)$ would not exist. On the other hand, if we have $f''(x) \leq 4$ for all x in $(0, 1/2)$ but $f''(x) < 4$ for at least one x, then we have $f'(x) < 4x$ for at least one x, and consequently for all larger x in $(0, 1/2)$, and therefore $f(x) < 4x^2$ for these x, so that $f(1/2) < 1/2$; if we also had $f''(x) \geq -4$ for all x in $(1/2, 1)$, then $f(1/2) \geq 1/2$, a contradiction.

40. If $g(x) = f(xy)$, then

$$g'(x) = y \cdot f'(xy)$$

$$= y \cdot \frac{1}{xy} = \frac{1}{x} = f'(x).$$

So there is a number c such that $g(x) = f(x) + c$ for all $x > 0$. Now

$$f(y) = g(1) = f(1) + c = c,$$

so $g(x) = f(x) + f(y)$.

41. Suppose $f(a) = f(b) = 0$. If x is a local maximum point of f on $[a, b]$, then $f'(x) = 0$ and $f''(x) \leq 0$; from the equation

$$f''(x) + f'(x)g(x) - f(x) = 0$$

we can conclude that $f(x) \leq 0$. Similarly, f cannot have a negative local minimum on (a, b).

42. If $f(x_i) = 0$ for $x_1 < x_2 < \cdots < x_{n+1}$, then $f'(x) = 0$ for some x in each of the n intervals (x_i, x_{i+1}). Consequently $f''(x) = 0$ for $n - 1$ numbers x, etc. (In other words, we are all set up for a proof by induction.)

43. If x is one of the x_i, then $f(x) - P(x) = 0 = Q(x)$, so we can choose any c. Otherwise, let

$$F(t) = Q(x)[f(t) - P(t)] - Q(t)[f(x) - P(x)].$$

Then for $i = 1, \ldots, n + 1$ we have

$$F(x_i) = 0, \qquad \text{since } f(x_i) - P_i = 0 \text{ and } Q(x_i) = 0$$

and also

$$F(x) = 0.$$

By Problem 42, we have $F^{(n+1)}(c) = 0$ for some c in (a, b). That is,

$$0 = F^{(n+1)}(c) = Q(x)[f^{(n+1)}(c) - 0] - (n + 1)![f(x) - P(x)].$$

45. This is a trivial consequence of the Mean Value Theorem because if we define

$$g(x) = \begin{cases} \lim_{y \to a^+} f(y), & x = a \\ f(x), & a < x < b \\ \lim_{y \to b^-} f(y), & x = b, \end{cases}$$

then g is continuous on $[a, b]$ and differentiable on (a, b) and $g'(x) = f'(x)$ for x in (a, b), so there is some x in (a, b) with

$$f'(x) = g'(x) = \frac{g(b) - g(a)}{b - a}.$$

46. We have

$$[f(b) - f(a)]g'(x) = f'(x)[g(b) - g(a)].$$

If $g'(x) = 0$, then $f'(x)[g(b) - g(a)] = 0$. But this contradicts the assumption that $g(b) \neq g(a)$, and the fact that $f'(x) \neq 0$ (since $g'(x) = 0$).

47. Let

$$h(x) = f(x)g(b) + g(x)f(a) - f(x)g(x).$$

Then

$$h(a) = h(b) = f(a)g(b),$$

so by Rolle's Theorem there is some x in (a, b) with

$$0 = h'(x) = f'(x)g(b) + g'(x)f(a) - f'(x)g(x) - f(x)g'(x),$$

or

$$f'(x)[g(b) - g(x)] = g'(x)[f(x) - f(a)].$$

Since $g'(x) \neq 0$ for all x in (a, b), we also have $g(b) \neq g(x)$ for x in (a, b) (otherwise Rolle's Theorem, applied to the interval $[x, b]$, would imply that $g'(x) = 0$ for some x' in (x, b).)

50. Since $g(0) = 0$, and g is continuous at 0, we have $\lim_{x \to 0} g(x) = 0$. Therefore, by l'Hôpital's Rule

$$f'(0) = \lim_{x \to 0} \frac{f(x)}{x} = \lim_{x \to 0} \frac{g(x)}{x^2}$$

$$= \lim_{x \to 0} \frac{g'(x)}{2x} = \lim_{x \to 0} \frac{g(x) - g'(0)}{2x} = \frac{1}{2}g''(0) = \frac{17}{2}.$$

(The limit $\lim_{x \to 0} g'(x)/2x$ could also be found by l'Hôpital's Rule.)

51. (a) Use exactly the same proof as for l'Hôpital's Rule, but consider only x in $(a, a + \delta)$ or in $(a - \delta, a)$, respectively.

(b) Again the proof of l'Hôpital's Rule will work, almost verbatim. (It is tempting to apply l'Hôpital's Rule to g/f: Since $\lim_{x \to 0} g'(x)/f'(x) = 0$, it follows that $\lim_{x \to a} g(x)/f(x) = 0$. Unfortunately, this implies only that $\lim_{x \to a} |f(x)/g(x)| = \infty$.)

(c) Since $\lim\limits_{x\to 0^+} f(1/x) = \lim\limits_{x\to\infty} f(x) = 0$ and $\lim\limits_{x\to 0^+} g(1/x) = \lim\limits_{x\to\infty} g(x) = 0$, part (a) implies that

$$\lim_{x\to\infty} \frac{f(x)}{g(x)} = \lim_{x\to 0^+} \frac{f(1/x)}{g(1/x)} = \lim_{x\to 0^+} \frac{-(1/x^2)f'(1/x)}{-(1/x^2)g'(1/x)}$$

$$= \lim_{x\to\infty} \frac{f'(x)}{g'(x)} = l.$$

(d) Similar to part (c), using the case $x \to a^+$ of part (b) instead of part (a).

52. (a) For any $\varepsilon > 0$ there is some a such that

$$\left| \frac{f'(x)}{g'(x)} - l \right| < \varepsilon \qquad \text{for } x \geq a.$$

This means, in particular, that $g'(x) \neq 0$ for $x > a$; it follows that $g(x) - g(a) \neq 0$ for $x > a$ (by Rolle's Theorem). Therefore the Cauchy Mean Value Theorem can be written in the form

$$\frac{f(x) - f(a)}{g(x) - g(a)} = \frac{f'(x')}{g'(x')} \qquad \text{for some } x' \text{ in } (a, x).$$

Since $x' > a$, the desired inequality follows.

(b) We have

$$\frac{f(x)}{g(x)} = \frac{f(x) - f(a)}{g(x) - g(a)} \cdot \frac{f(x)}{f(x) - f(a)} \cdot \frac{g(x) - g(a)}{g(x)},$$

where $f(x) - f(a) \neq 0$, $g(x) \neq 0$ for large enough x, since $\lim\limits_{x\to\infty} f(x) = \lim\limits_{x\to\infty} g(x) = \infty$. These limits also imply that

$$\lim_{x\to\infty} \frac{f(x)}{f(x) - f(a)} = \lim_{x\to\infty} \frac{g(x) - g(a)}{g(x)} = 1.$$

It follows that $f(x)/g(x)$ can be made as close to $[f(x) - f(a)]/[g(x) - g(a)]$ as desired by choosing x large enough. Together with part (a), this shows that

$$\left| \frac{f(x)}{g(x)} - l \right| < 2\varepsilon \qquad \text{for sufficiently large } x.$$

53. One other form of l'Hôpital's Rule will be used in later problems: If $\lim\limits_{x\to\infty} f(x) = \lim\limits_{x\to\infty} g(x) = \infty$ and $\lim\limits_{x\to\infty} f'(x)/g'(x) = \infty$, then $\lim\limits_{x\to\infty} f(x)/g(x) = \infty$. To prove this, apply Problem 52 to g/f: Since $\lim\limits_{x\to\infty} g'(x)/f'(x) = 0$, we have $\lim\limits_{x\to\infty} g(x)/f(x) = 0$. This implies (as we remarked in the solution to Problem 51) that $\lim\limits_{x\to\infty} |f(x)/g(x)| = \infty$. Since $\lim\limits_{x\to\infty} f(x) = \lim\limits_{x\to\infty} g(x) = \infty$, we can conclude that $\lim\limits_{x\to\infty} f(x)/g(x) = \infty$.

54. (a) Since a is a minimum point for f on $[a, b]$, for all sufficiently small $h > 0$ we have

$$\frac{f(a+h) - f(a)}{h} \geq 0;$$

this implies that $f'(a) \geq 0$. The proof that $f'(b) \leq 0$ is similar.

(b) Part (a) shows that we cannot have the minimum of f at a or at b, since we are assuming that $f'(x) < 0$ and $f'(b) > 0$. So the minimum occurs at some point x in (a, b). Then $f'(x) = 0$.

(c) Let $g(x) = f(x) - cx$. Then $g'(a) = f'(a) - c < 0$ and $g'(b) = f'(b) - c > 0$. So by part (b), $0 = g'(x) = f'(x) - c$ for some x in (a, b).

55. (a) A simple modification of the proof of Theorem 7 shows that if $\lim\limits_{x \to a^+} f'(x)$ exists, then

$$\lim_{x \to a^+} f'(x) = \lim_{h \to 0^+} \frac{f(a+h) - f(a)}{h} = f'(a).$$

Similarly, if $\lim\limits_{x \to a^-} f'(x)$ exists, then $\lim\limits_{x \to a^-} f'(x) = f'(a)$. So if both one-sided limits existed, f' would be continuous at a.

(b) Suppose, for example, that $\lim\limits_{x \to a^+} f'(x) = \infty$. This means that $f'(x) \geq f'(a) + 1$ for all $x > a$ sufficiently close to a. But by Darboux's Theorem, if x_0 is such an x, then f' takes on all values between $f'(a)$ and $f'(x_0)$ on the interval (a, x_0), a contradiction.

56. If $f(a) \neq 0$, then continuity of f implies that $f = |f|$ or $f = -|f|$ in some interval around a, so f is differentiable at a. If $f(a) = 0$, then a is a minimum point for $|f|$, so $|f|'(a) = 0$. This means that

$$0 = \lim_{h \to 0} \frac{|f(a+h)| - |f(a)|}{h}$$

$$= \lim_{h \to 0} \frac{|f(a+h)|}{h}.$$

This equation also says that $f'(a) = 0$.

57. (a) Let $f(x) = x^n + y^n - (x + y)^n$. If $f(x_0) = 0$ for some $x_0 \neq 0$, then Rolle's Theorem would imply that

$$0 = f'(x) = nx^{n-1} - n(x + y)^{n-1} \qquad \text{for some } x \text{ in } (0, x_0) \text{ or } (x_0, 0).$$

But this means that $x^{n-1} = (x + y)^{n-1}$ for $y \neq 0$, which is impossible, since $g(x) = x^{n-1}$ is increasing ($n - 1$ is odd).

(b) Now we have $f(0) = f(-y) = 0$. If f were zero at three points $a < b < c$, then Rolle's Theorem could be applied to $[a, b]$ and $[b, c]$ to prove that there are two numbers x with

$$0 = f'(x) = nx^{n-1} - n(x + y)^{n-1};$$

but this equation holds only for $x = -(x + y)$ (Problem 1-6).

58. The tangent line through (a, a^n) is the graph of

$$g(x) = na^{n-1}(x - a) + a^n$$
$$= na^{n-1}x + (1 - n)a^n.$$

If $g(x_0) = f(x_0)$ for some $x_0 \neq a$, then Rolle's Theorem may be applied to $g - f$ on the interval $[a, x_0]$, or $[x_0, a]$:

$$0 = g'(x) - f'(x) = na^{n-1} - nx^{n-1} \qquad \text{for some } x \text{ in } (a, x_0) \text{ or } (x_0, a).$$

This is impossible, since $x \neq a$ and $n - 1$ is odd, so $a^{n-1} \neq x^{n-1}$.

59. The tangent line through $(a, f(a))$ is the graph of

$$g(x) = f'(a)(x - a) + f(a)$$
$$= f'(x)x + f(a) - af'(a).$$

If $g(x_0) = f(x_0)$ for some $x_0 \neq a$, then

$$0 = g'(x) - f'(x) = f'(a) - f'(x) \qquad \text{for some } x \text{ in } (a, x_0) \text{ or } (x_0, a).$$

This is impossible, since f' is increasing.

60. Since

$$g'(x) = \frac{xf'(x) - f(x)}{x^2},$$

it suffices to show that

$$xf'(x) - f(x) > 0,$$

or

$$f'(x) > \frac{f(x)}{x} \qquad \text{for } x > 0.$$

Now the Mean Value Theorem, applied to f on $[0, x]$, shows that

$$\frac{f(x)}{x} = \frac{f(x) - f(0)}{x - 0} = f'(x') \qquad \text{for some } x' \text{ in } [0, x].$$
$$< f'(x), \qquad \text{since } f' \text{ is increasing.}$$

61. Let $g(x) = (1 + x)^n - (1 + nx)$. Then $g(0) = 0$, but

$$g'(x) = n(1 + x)^{n-1} - n.$$

Since $n - 1 \geq 0$ this means that

$$g'(x) < 0 \qquad \text{for } -1 < x < 0,$$
$$> 0 \qquad \text{for } x > 0.$$

Thus $g(x) > 0$ for $-1 < x < 0$ and $0 < x$.

62. (a) 0 is actually a minimum on all of **R**, since $f(0) = 0$ and $f(x) \geq 0$ for all x.

(b)

$$f'(0) = \lim_{h \to 0} \frac{h^4 \sin^2(1/h)}{h} = 0,$$

and

$$f'(h) = 4h^3 \sin^2(1/h) = 2h^2 \sin(1/h)\cos(1/h) \qquad \text{for } h \neq 0.$$

So

$$f'(0) = \lim_{h \to 0} \frac{4h^3 \sin^2(1/h) - 2h^2 \sin(1/h)\cos(1/h)}{h}$$
$$= 0.$$

63. (a) Since f' is continuous, $f'(x) > 0$ for all x in some interval around a, so f is increasing in this interval.

(b) We have

$$g'(x) = 2x \sin \frac{1}{x} - \cos \frac{1}{x}.$$

So $g'(x) = 1$ when $\cos 1/x = 1$ (and consequently $\sin 1/x = 0$), and $g'(x) = -1$ when $\cos 1/x = -1$.

(c) We have $f'(x) = \alpha + g'(x)$, so $f'(x) > 0$ when $g'(x) = 1$, and $f'(x) < 0$ when $g'(x) = -1$.

64. (a) We have

$$g(y) = \frac{2\sin y}{y} - \cos y,$$

so

$$g'(y) = \frac{2y\cos y - 2\sin y}{y^2} + \sin y.$$

So if $g'(y) = 0$, then

$$0 = 2y\cos y - 2\sin y + y^2 \sin y,$$

or

(1) $$\cos y = \frac{2\sin y - y^2 \sin y}{2y} = (\sin y)\left(\frac{2 - y^2}{2y}\right).$$

Hence

(2)
$$g(y) = \frac{2\sin y}{y} - (\sin y)\left(\frac{2-y^2}{2y}\right)$$

$$= (\sin y)\left(\frac{2}{y} - \frac{2-y^2}{2y}\right)$$

$$= (\sin y)\left(\frac{2+y^2}{2y}\right).$$

(b) Moreover, from (1) we have

$$1 - \sin^2 y = \cos^2 y = (\sin^2 y)\left(\frac{2-y^2}{2y}\right)^2,$$

so

$$\sin^2 y = \frac{1}{1 + \left(\dfrac{2-y^2}{2y}\right)^2} = \frac{4y^2}{4+y^4},$$

so, by (2),

$$|g(y)| = |\sin y| \cdot \left|\frac{2+y^2}{2y}\right|$$

$$= \frac{|2y|}{\sqrt{4+y^4}} \cdot \frac{2+y^2}{|2y|} = \frac{2+y^2}{\sqrt{4+y^4}}.$$

(c) We have

$$f'(x) = 1 + g(1/x).$$

Now we clearly have $g(y) < 0$ for arbitrarily large y (since $g(y)$ is practically $-\cos y$ for large y), so for arbitrarily large y we have

$$g(y) < -\frac{2+y^2}{\sqrt{4+y^4}} < -1$$

by part (b). Thus $f'(x) < 0$ for arbitrarily small x, while we also have $f'(x) > 0$ for arbitrarily small x.

(d) We have

$$f'(x) = \alpha + g(1/x).$$

For sufficiently large y we have $g(y) > -\alpha$. So for sufficiently small x we have $f'(x) > 0$.

65. (a) If the minimum of f on $[b, 1]$ occurred at some c with $b < c \le 1$, then f would clearly not be increasing at c, since we would have $f(x) \ge f(c)$ for all $x < c$ sufficiently close to c. Now if $0 \le a < b \le 1$, then the minimum of f on $[a, 1]$ is a, so $f(a) \le f(b)$. To obtain the strict inequality $f(a) < f(b)$, pick some

a' with $a < a' < b$ such that $f(a') > f(a)$ (this is possible since f is increasing at a); then $f(a) < f(a') \leq f(b)$.

(b) Let $\alpha = \sup S_b$. If $b \leq y < \alpha$, then there is some x in S_b with $y < x$. Therefore $f(y) \geq f(b)$. Moreover, since f is increasing at α, we have $f(\alpha) > f(x)$ for $x < \alpha$ sufficiently close to α, so $f(\alpha) > f(b)$. This shows that α is actually in $\sup S_b$. Now if $\alpha < 1$ there would be a $\delta > 0$ such that $f(x) > f(\alpha)$ for $\alpha < x < \alpha + \delta$. This shows that all such x are in S_b, contradicting the fact that $\alpha = \sup S_b$. So $\alpha = \sup S_b = 1$. So $f(y) \geq f(b)$ for all $y \geq b$.

(c) For sufficiently small h we have

$$f(a+h) > f(a) \qquad \text{if } h > 0,$$
$$f(a+h) < f(a) \qquad \text{if } h < 0.$$

This implies that

$$\frac{f(a+h) - f(a)}{h} > 0,$$

which implies that

$$f'(a) = \lim_{h \to 0} \frac{f(a+h) - f(a)}{h} \geq 0.$$

(d) Since $f'(a) > 0$, for sufficiently small h we have

$$\frac{f(a+h) - f(a)}{h} > 0.$$

This implies that $f(a+h) > f(a)$ for $h > 0$ and $f(a+h) < f(a)$ for $h < 0$.

(e) Part (d) implies that f is increasing at a for all a in $[0, 1]$, so part (a) implies that f is increasing on $[0, 1]$.

(f) If $\varepsilon > 0$, then $g'(a) = f'(a) + \varepsilon = \varepsilon > 0$ for all a in $[0, 1]$, so g is increasing on $[0, 1]$ by part (e), so $f(1) + \varepsilon > f(0)$, or $f(1) - f(0) > -\varepsilon$. Similarly, h is increasing on $[0, 1]$, so $\varepsilon - f(1) > -f(0)$, or $f(1) - f(0) < \varepsilon$. Thus $|f(1) - f(0)| < \varepsilon$. Since this is true for all $\varepsilon > 0$, it follows that $f(1) = f(0)$. (Of course, the same argument, applied to $[a, b]$, for $0 \leq a < b \leq 1$, shows that $f(a) = f(b)$.)

66. (a) Suppose f is not constant, so that $f(a') \neq f(b')$ for some $a' < b'$ in $[a, b]$. To be specific, say $f(a') < f(b')$. By Problem 8-4(b), there are $a' \leq c < d \leq b'$ with $f(c) = f(a') < f(b') = f(d)$ and $f(c) < f(x) < f(d)$ for all x in (c, d). But then a' is not a local maximum for f.

(b) We can assume $f(a_0) < f(x) < f(b_0)$ by Problem 8-4(b) [renaming c to be a_0 and d to be b_0]. By Theorem 1 of the Appendix to Chapter 8 there is some $k \geq 2$ such that

$$|f(x) - f(y)| < \frac{f(b_0) - f(a_0)}{2} \qquad \text{for } |x - y| \leq \delta = \frac{b_0 - a_0}{k}.$$

Let $c_i = a_0 + i\delta$. Since

$$f(c_1) - f(a_0) < \frac{f(b_0) - f(a_0)}{2}$$

$$f(b_0) - f(c_{k-1}) < \frac{f(b_0) - f(a_0)}{2}$$

we must have

$$f(c_1) < f(c_{k-1}).$$

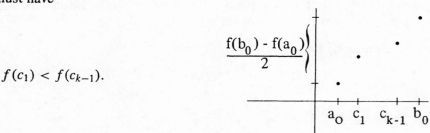

Consequently there is some i with $1 \le i < k - 1$ such that $f(c_i) < f(c_{i+1})$. Let $c_i = a_1$ and $c_{i+1} = b_1$. Then $a_0 < a_1 < b_1 < b_0$ and $f(a_1) < f(b_1)$. Moreover, we can assume that $f(a_1) < f(x) < f(b_1)$ for all $a_1 < x < b_1$ [use Problem 8-4(b) again].

Continuing in this way, we find intervals $[a_n, b_n]$ with $a_n < a_{n+1} < b_{n+1} < b_n$, and $f(a_n) < x < f(b_n)$ for $a_n < x < b_n$; moreover, we can assume that $b_n - a_n < 1/n$. Now let x be in all $[a_n, b_n]$. Then every interval around x contains some $[a_k, b_k]$, with $f(x_k) < f(x) < f(b_k)$; hence x is not a local maximum or minimum.

67. (a) The local strict maximum points are the rational numbers.

(b) Let x be a point in all intervals $I_n = [a_n, b_n]$. Since the points x_n are chosen to be distinct, $x = x_n$ for at most one n. Since x is a local strict maximum point, there is a $\delta > 0$ such that x is a strict maximum point for f on $(x - \delta, x + \delta)$. But I_n is contained in $(-x - \delta, x + \delta)$ for all sufficiently large n; choose such an n for which $x \ne x_n$. Then $f(x) > f(x_n)$, since I_n is contained in $(x - \delta, x + \delta)$, while $f(x_n) > f(x)$, since x is in I_n.

1. (i) $f''(x) = 6x - 2 > 0$ for $x > 1/3$.

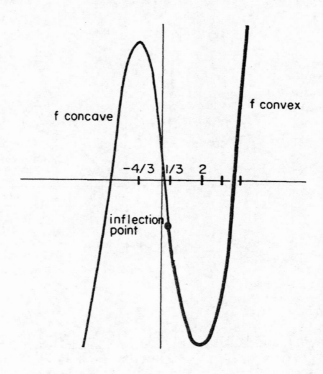

(ii) $f''(x) = 20x^3 + 1 > 0$ for $x > -1/\sqrt[3]{20}$.

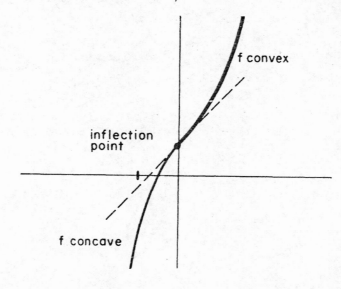

(iii) $f''(x) = 36x^2 - 48x + 12 = 12(3x^2 - 4x + 1) = 12(3x - 1)(x - 1) > 0$ for $x < 1/3$ or $x > 1$.

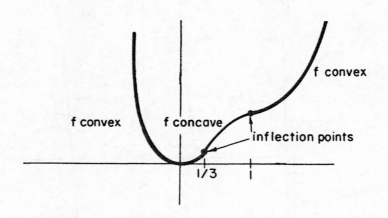

(iv) We have

$$f''(x) = \frac{-(x^5 + x + 2)^2 20x^3 + (5x^4 + 1)2(x^5 + x + 1)(5x^4 + 1)}{(x^5 + x + 1)^4}$$

$$= \frac{2(x^5 + x + 1)[(5x^4 + 1)^2 - 10x^3(x^5 + x + 1)]}{(x^5 + x + 1)^4}$$

$$= \frac{2}{(x^5 + x + 1)^3}[15x^8 - 10x^3 + 1].$$

To determine the sign of $f''(x)$ it suffices to determine the sign of

$$g(x) = 15x^8 - 10x^3 + 1.$$

Now

$$g'(x) = 120x^7 - 30x^2 = 30x^2(4x^5 - 1).$$

So $g'(x) = 0$ for $x = 0$ or $x = \sqrt[5]{1/4}$. We have $g(0) = 1$ and

$$g(\sqrt[5]{1/4}) = (\sqrt[5]{1/4})^3 \left[15 \cdot \frac{1}{4} - 10\right] + 1$$

$$= (\sqrt[5]{1/4})^3 \left(\frac{-25}{4}\right) + 1$$

$$< 0,$$

since $\sqrt[5]{1/4} > 4/25$. So g attains it (negative) minimum value at $\sqrt[5]{1/4}$. Moreover, since $h(x) = 4x^5 - 1$ is increasing, $g'(x) > 0$ for $x > \sqrt[5]{1/4}$ and $g'(x) < 0$ for

$x < \sqrt[5]{1/4}$. So g is decreasing on $(-\infty, \sqrt[5]{1/4}\,]$ and increasing on $[\sqrt[5]{1/4}, \infty)$.

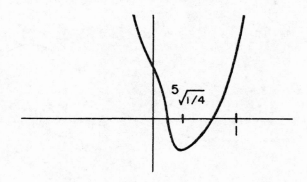

Consequently, g has two zeros, both in $[0, 1]$, since $g(1) > 0$. It follows that if a is the unique root of $x^5 + x + 1 = 0$, then $f''(x) < 0$ for $x < a$, but $f''(x) > 0$ for all $x > a$ except those x in a certain interval contained in $(0, 1)$. Thus the graph of f is convex on (a, ∞), except for a bump lying over some interval contained in $(0, 1)$.

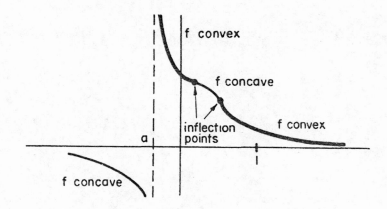

(v) We have

$$f''(x) = \frac{(x^2+1)^2(-2-2x) - (1-2x-x^2)2(x^2+1)2x}{(x^2+1)^4}$$

$$= \frac{2}{(x^2+1)^3}[x^3 + 3x^2 - 3x - 1] = \frac{2}{(x^2+1)^3}(x-1)(x^2+4x+1)$$

$$= \frac{2}{(x^2+1)^3}(x-1)\big(x - [-2+\sqrt{3}\,]\big)\big(x - [-2-\sqrt{3}\,]\big),$$

so $f''(x) > 0$ for $-2 - \sqrt{3} < x < -2 + \sqrt{3}$ and $x > 1$.

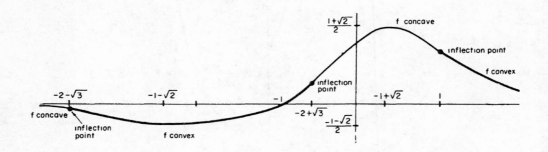

(vi)

$$f''(x) = \frac{(x^2 - 1)^2(-2x) + (1 + x^2)2(x^2 - 1)2x}{(x^2 - 1)^4}$$

$$= \frac{2x}{(x^2 - 1)^3}[x^2 + 3],$$

so $f''(x) > 0$ for $x > 1$ and $-1 < x < 0$.

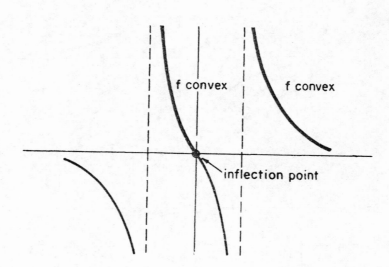

2. If $f(0) = 0$, the graph looks like the following.

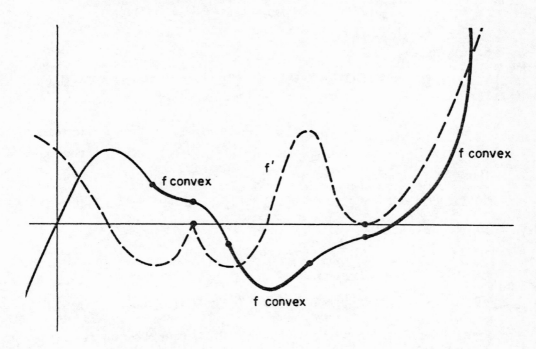

3. Two such functions are shown below.

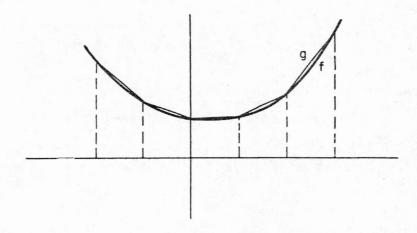

4. According to Problem 4-2, the points in (x, y) are precisely those of the form $tx + (1 - t)y$ for $0 < t < 1$. Definition 2 thus shows that f is convex if and only if

$$\frac{f(tx + (1 - t)y) - f(x)}{tx + (1 - t)y - x} < \frac{f(y) - f(x)}{y - x},$$

which is equivalent to

$$f(tx + (1-t)y) < tf(x) + (1-t)f(y).$$

5. (a) We have

$$g(tx + (1-t)y) < tg(x) + (1-t)g(y) \qquad \text{since } g \text{ is convex,}$$

so

$$f(g(tx + (1-t)y) < f(tg(x) + (1-t)g(y)) \qquad \text{since } f \text{ is increasing}$$
$$< tf(g(x)) + (1-t)f(g(y)) \qquad \text{sicne } f \text{ is convex.}$$

Thus, $f \circ g$ is convex.

(b) Let $f(x) = 1 + x^2$, $x > 0$ and $g(x) = 1/x$, $x < 0$.

(c) We have

$$(f \circ g)' = (f' \circ g)g'$$
$$(f \circ g)'' = (f'' \circ g)g'^2 + (f' \circ g)g''.$$

Since f'', g'', $g'^2 \geq 0$ it follows that $(f \circ g)'' > 0$ if $f' > 0$.

6. (a) Since f is convex, f' is increasing. If f' isn't either always negative or always positive, let $c = \sup\{x : f'(x) < 0\}$. Then $f' < 0$ to the left of c and $f' > 0$ to the right of c. [Actually, f' will be continuous, see Problem 10; so c can be described more simply as the zero of f'.]

(b) For $x < y$ consider

$$(f \circ g)'(x) = f'(g(x)) \cdot g'(x),$$
$$(f \circ g)'(y) = f'(g(y)) \cdot g'(y).$$

Suppose first that g is increasing. Then

$$0 \leq g'(x) < g'(y) \qquad \text{since } g \text{ is increasing and convex.}$$

Moreover, $g(x) < g(y)$ implies that

$$0 \leq f'(g(x)) < f'(g(y)) \qquad \text{since } f \text{ is increasing and convex.}$$

It follows that

$$f'(g(x)) \cdot g'(x) < f'(g(y)) \cdot g'(y).$$

Next suppose that g is decreasing. Then

$$g'(x) < g'(y) \leq 0,$$

and $g(x) > g(y)$ implies that

$$f'(g(x)) > f'(g(y)) \geq 0.$$

It again follows that $f'(g(x)) \cdot g'(x) < f'(g(y)) \cdot g'(y).$

Finally, suppose that g is decreasing to the left of c and increasing to the right of c. If $x < y \leq c$ or $c \leq x < y$, then we have already shown that $f'(g(x)) \cdot g'(x) < f'(g(y)) \cdot g'(y)$. If $x < c < y$, then

$$f'(g(x)) \cdot g'(x) < f'(g(c)) \cdot g'(c) < f'(g(y)) \cdot g'(y),$$

so we still have $f'(g(x)) \cdot g'(x) < f'(g(y)) \cdot g'(y)$. Thus, $(f \circ g)'$ is increasing.

(c) *Lemma.* Suppose f is convex on an interval and $a < b$ are points in this interval. If $f(a) < f(b)$, then f is increasing to the right of b; and if $f(a) > f(b)$, then f is decreasing to the left of b.

Proof. Consider the case $f(a) < f(b)$ (the proof in the other case is similar or one can apply this first case to $g(x) = f(-x)$).

If $b < d$, then the definition of convexity shows immediately that we cannot have $f(d) \leq f(b)$. Moreover, if $b < d_1 < d_2$, then the same argument shows (since we now know that $f(b) < f(d_1)$) that $f(d_1) < f(d_2)$. Thus f is increasing on $[b, \infty)$.

With the aid of this lemma we can now prove the theorem. Since f is not constant, there is some $a < b$ with $f(a) \neq f(b)$. We consider only the case $f(a) < f(b)$. We already know from the lemma that f is increasing to the right of b. Suppose now that the minimum of f on $[a, b]$ occurs at some c in (a, b). Then f is decreasing to the left of a by the lemma. Moreover, if a' is any number with $a < a' < c$, then we must have $f(a') > f(c)$ (if we had $f(a') = f(c)$, then $f(x) < f(c)$ for x in (a', c), contradicting the fact that c is the minimum point). So the lemma also implies that f is decreasing to the left of a' for all such a'. This shows that f is decreasing to the left of c. Similarly, f is increasing to the right of c.

On the other hand, suppose that the minimum of f on $[a, b]$ occurs at a. The same sort of reasoning as before shows that f is increasing to the right of a. There are then two possibilities:

It may happen that $f(d) > f(a)$ for some $d < a$. In this case, the minimum of f on $[d, a]$ occurs at some c with $d < c \leq a$. The same reasoning as before shows that f is decreasing to the left of c and increasing to the right of c.

It may also happen that $f(d) < f(a)$ for all $d < a$. Then we may apply the results already proved (for $a < b$) to $d < a$: If the minimum of f ever occurs at a point c in (d, a), then f is decreasing to the left of c and increasing to the right of c, but if the minimum is always at d, then f is increasing to the right of d for all d, so f is increasing.

7. Choose $x > 0$ so that $f(x) < f(0)$. The Mean Value Theorem implies that there is some x_0 in $(0, x)$ with $f'(x_0) < 0$. If we had $f'(y) \leq f'(x_0)$ for all $y \geq x_0$, then for all $x > x_0$ we would have

$$f(x) - f(x_0) \leq f'(x_0)(x - x_0),$$

which would imply that $f(x)$ is eventually negative (since $f'(x_0) < 0$). Therefore $f'(x_1) > f'(x_0)$ for some $x_1 > x_0$. This implies that the minimum of f' on $[0, x_1]$ occurs at some x in $(0, x_1)$. Then $f''(x) = 0$.

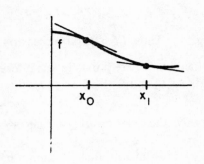
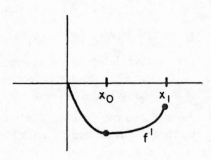

8. (a) This follows from Problem 4 with $t = 1/2$.

(b) The assertion is true for $n = 1$, i.e., $k = 1/2$. Suppose that for some n it is true for all x and y. If $k = m/2^{n+1}$ is in lowest terms, then k is odd. Consequently $k_1 = (m - 1)/2^{n+1}$ and $k_2 = (m + 1)/2^{n+1}$ can be expressed in the form $a/2^n$, so the assertion is true for k_1 and k_2. Notice also that $k = (k_1 + k_2)/2$. From the result for k_1 and k_2, and the assertion for $n = 1$ applied to $x' = k_1 x + (1 - k_1)y$ and $y' = k_2 x + (1 - k_2)y$ we obtain

$$f(kx + (1 - k)y) = f\left(\frac{x' + y+}{2}\right) < \frac{f(x')}{2} + \frac{f(y')}{2}$$
$$< \frac{k_1 f(x) + (1 - k_1)f(y)}{2} + \frac{k_2 f(y) + (1 - k_2)f(y)}{2}$$
$$= kf(x) + (1 - k)f(y).$$

(c) Let $0 < t < 1$. For any $\varepsilon > 0$ there is a number k of the form $m/2^n$ which is so close to t that

$$|f(kx + (1 - k)y) - f(tx + (1 - t)y)| < \varepsilon,$$
$$\left|[kf(x) + (1 - k)f(y)] - [tf(x) + (1 - t)f(y))]\right| < \varepsilon.$$

Then

$$f(tx + (1 - t)y) < f(kx + (1 - k)y) + \varepsilon$$
$$< kf(x) + (1 - k)f(y) + \varepsilon$$
$$< tf(x) + (1 - t(f(y) + 2\varepsilon.$$

Thus $f(tx + (1-t)y) \leq tf(x) + (1-t)f(y)$. The following diagram shows that if strict inequality holds for even one t, then it holds for all t (by applying the weak inequality to x and $tx + (1-t)y$ or to $tx + (1-t)y$ and y). But we have strict inequality for t of the form $m/2^n$, so we must have strict inequality for all t.

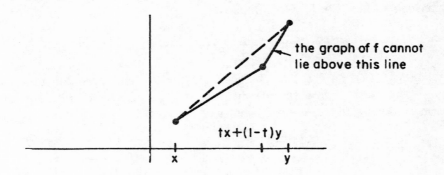

9. (a) Let x_α and x_β be the smallest and largest of the x_i. Then

$$x_a = \sum_{i=1}^{n} p_i x_\alpha \leq \sum_{i=1}^{n} p_i x_i \leq \sum_{i=1}^{n} p_i x_\beta = x_\beta.$$

(b) Part (a), applied to $p_1/t, \ldots, p_{n-1}/t$, shows that $(1/t)\sum_{i=1}^{n} p_i x_i$ lies between the smallest and largest of $x_1, \ldots, x_{n-1}$, so it certainly lies between the smallest and largest of $x_1, \ldots, x_n$.

(c) Jensen's inequality is true for $n = 1$. Suppose it is true for $n - 1$. Then by Problem 4 we have, since $p_n = 1 - t$,

$$f\left(\sum_{i=1}^{n} p_i x_i\right) = f\left(t \cdot (1/t)\sum_{i=1}^{n} p_i x_i + (1-t)x_n\right)$$
$$\leq tf\left(\sum_{i=1}^{n-1}(p_i/t)x_i\right) + (1-t)f(x_n)$$
$$\leq t\sum_{i=1}^{n-1}\frac{p_i}{t}f(x_i) + p_n f(x_n)$$
$$= \sum_{i=1}^{n} p_i f(x_i).$$

(The same sort of proof shows that strict inequality holds for $n > 1$ (begin by checking that strict inequality holds for $n = 2$).)

10. (a) As the proof of Theorem 1 shows, $[f(a+h) - f(a)]/h$ is decreasing as $h \to 0^+$, so

$$f'_+(a) = \lim_{h \to 0^+} \frac{f(a+h) - f(a)}{h} = \inf\left\{ \frac{f(a+h) - f(a)}{h} : h > 0 \right\}.$$

This inf exists because each quotient $[f(a+h) - f(a)]/h$ for $h > 0$ is greater than any one such quotient for $h < 0$. Similarly,

$$f'_-(a) = \sup\left\{ \frac{f(a+h) - f(a)}{h} : h < 0 \right\}.$$

The relation $f'_-(a) \leq f'_+(a)$ is obvious from the previous considerations. The functions f'_+ and f'_- are increasing, because if $a < b$, then (as in the proof of Theorem 1; see Figure 6 of the text) we have

$$f'_-(a) \leq f'_+(a) < \frac{f(a + (b-a)) - f(a)}{b - a} = \frac{f(b + (a-b)) - f(b)}{a - b}$$
$$< f'_-(b) \leq f'_+(b).$$

(b) If $b < a$, then as in part (a) (with a and b interchanged) we have

$$f'_+(b) < f'_-(a) \leq f'_+(a).$$

If f'_+ is continuous at a, then $\lim_{b \to a} f'_+(b) = f'_+(a)$, so we must have $f'_-(a) = f'_+(a)$.

To prove the converse, we first show that f'_+ will always be continuous on the right, i.e.,

$$\lim_{b \to a^+} f'_+(b) = f'_+(a).$$

In fact for any $\varepsilon > 0$ we can choose $c > a$ so that

$$\frac{f(c) - f(a)}{c - a} < f'_+(a) + \varepsilon.$$

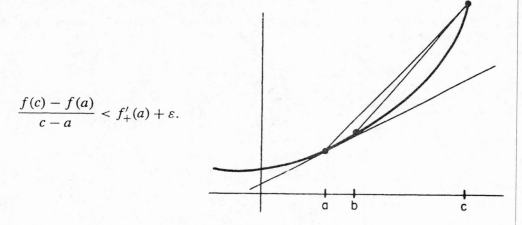

Since $f'_+(a)$ exists, f satisfies $\lim_{b \to a^+} f(b) = f(a)$ (as a matter of fact, f is continuous at a even if $f'_+(a)$ does not exist; see Problem 11). So we can choose $b > a$

close to a with $f(b)$ as close to $f(a)$ as desired. Thus we can choose $b > a$ so that

$$\frac{f(c) - f(b)}{c - b} < \frac{f(c) - f(a)}{c - a} + \varepsilon.$$

Therefore

$$f'_+(a) < f'_+(b) < \frac{f(c) - f(b)}{c - b}$$
$$< \frac{f(c) - f(a)}{c - a} + \varepsilon$$
$$< f'_+(a) + 2\varepsilon.$$

This shows that f'_+ is continuous on the right.

It remains to show that if $f'_+(a) = f'_-(a)$, then f'_+ is continuous on the left at a. Given $\varepsilon > 0$, choose $c < a$ so that

$$f'_+(a) - \varepsilon = f'_-(a) - \varepsilon < \frac{f(a) - f(c)}{c - a}.$$

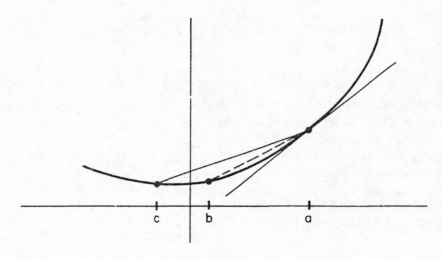

Then if $c < b < a$, the secant line through $(b, f(b))$ and $(a, f(a))$ lies between the tangent line at a and the secant line through $(c, f(c))$ and $(a, f(a))$, i.e.,

$$f'_+(a) - \varepsilon < \frac{f(a) - f(c)}{c - a} < \frac{f(a) - f(b)}{a - b} < f'_+(b) < f'_+(a).$$

This shows that $\lim\limits_{b \to a^-} f'_+(b) = f'_+(a)$.

11. (a) Let a be a point of the interval. Let $\varepsilon > 0$. Pick some $x_0 > a$. Notice that no matter what value $f(x_0)$ may have, the line segment between $(a, f(a))$ and $(x_0, f(x_0))$ eventually lies below the horizontal line at height $f(a) + \varepsilon$. Since the graph of f must lie below this line on (a, x_0), this shows that $f(x) < f(a) + \varepsilon$ for all $x > a$ sufficiently close to a. A similar argument works for all $x < a$ sufficiently close to a.

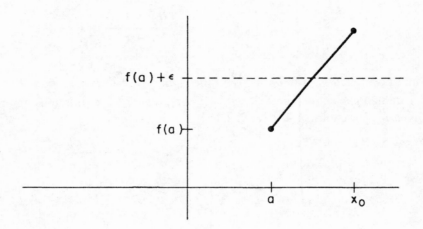

It remains to show that $f(x) > f(a) - \varepsilon$ for all x sufficiently close to a. If $f(x) \geq f(a)$ for all x there is nothing to prove, so suppose that $f(x_0) < f(a)$ for some x_0 with $x_0 > a$, say. Then we must have $f(y) > f(a)$ for all $y < a$, because of convexity, so all $y < a$ certainly satisfy $f(y) > f(a) - \varepsilon$. Moreover, if we pick some $y_0 < a$, then the line segment between $(y_0, f(y_0))$ and $(a, f(a))$ lies above the horizontal line at height $f(a) - \varepsilon$ in some interval to the right of a. Since the graph of f must lie *above* this line to the right of a, it follows that $f(x) > f(a) - \varepsilon$ for all $x > a$ sufficiently close to a.

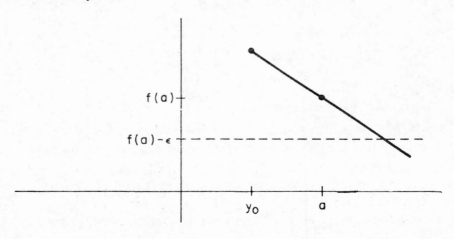

(b) The following Figure shows the possible kinds of discontinuities on a closed interval.

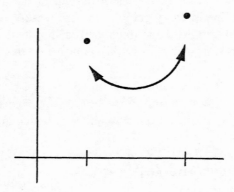

12. (a) Clearly f is weakly convex on an interval if and only if for all a and b in the interval, the line segment joining $(a, f(a))$ and $(b, f(b))$ lies *above or on* the graph of f. If f is actually convex, then it clearly contains no straight line segments. Conversely, suppose that f is weakly convex and its graph contains no straight line segments. To prove f convex we have to show that the line segment joining $(a, f(a))$ and $(b, f(b))$ cannot contain even one point $(x, f(x))$ for $a < x < b$.

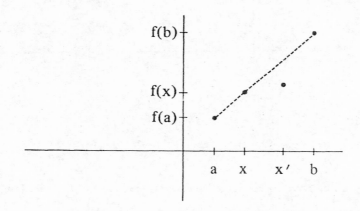

Suppose it did. Since the graph of f does not contain the entire line segment from $(x, f(x))$ to $(b, f(b))$, there must be some x' in (x, b) such that the point $(x', f(x'))$ lies below this line segment. But then we easily see that $(x, f(x))$ lies above the line segment from $(a, f(a))$ to $(x', f(x'))$, contradicting the fact that f is weakly convex.

(b) *Theorem* $1'$. If f is weakly convex and differentiable at a, then the graph of f lies above or on the tangent line through $(a, f(a))$ at all points. If $a < b$ and f is differentiable at a and b, then $f'(a) \le f'(b)$. *Lemma.* Suppose f is differentiable and f' is nondecreasing. If $a < b$ and $f(a) = f(b)$, then $f(x) \le f(a) = f(b)$

for $a < x < b$. *Theorem* $2'$. If f is differentiable and f' is nondecreasing, then f is weakly convex. *Theorem* $3'$. If f is differentiable and the graph of f lies above or on each tangent line at every point, then f is weakly convex. *Theorem* $4'$. If f is differentiable on an interval and intersects each of its tangent lines in an interval, then f is either weakly convex or weakly concave on that interval.

13. Suppose first that A_f is convex. Then for $x_1 < x_2$, the points $(x_1, f(x_1))$ and $(x_2, f(x_2))$ are in A_f, so all points of the line segment between them are in A_f. But this just means that all of these points lie above or on the graph of f, so f is weakly convex.

Conversely, suppose that f is weakly convex, and let (x_1, y_1) and (x_2, y_2) be two points of A_f, so that we have $f(x_i) \leq y_i$.

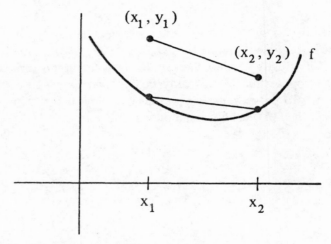

Modifying Problem 4 in the obvious way, we have

$$(*) \qquad \begin{aligned} f(tx_1 + (1-t)x_2) &\leq f(x_1) + (1-t)f(x_2) \\ &\leq ty_1 + (1-t)y_2 \end{aligned} \qquad \text{for } 0 \leq t \leq 1.$$

But every point of the line segment between (x_1, y_1) and (x_2, y_2) is of the form

$$(tx_1 + (1-t)x_2, ty_1 + (1-t)y_2),$$

and $(*)$ shows that these points are in A_f.

CHAPTER 12

1. (ii) $f^{-1}(x) = x^{1/3} + 1$. (If $y = f^{-1}(x)$, then $x = f(y) = (y-1)^3$, so $y = 1 + x^{1/3}$.)

(iv)

$$f^{-1}(x) = \begin{cases} (-x)^{1/2}, & x \le 0 \\ (1-x)^{1/3}, & x > 1. \end{cases}$$

(If $y = f^{-1}(x)$, then

$$x = f(y) = \begin{cases} -y^2, & y \ge 0 \\ 1 - y^3, & y < 0. \end{cases}$$

Since $-y^2 \le 0$ if $y \ge 0$ and $1 - y^3 > 1$ if $y < 0$, we have $y = (-x)^{1/2}$ for $x \le 0$ and $y = (1-x)^{1/3}$ for $x > 1$.)

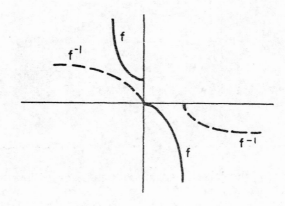

(vi) $f^{-1}(x) = x - [x/2]$ for $[x]$ even. (If $y = f^{-1}(x)$, then

$$x = f(y) = y + [y]$$
$$= y + n \qquad \text{for } n \le y < n + 1.$$

Thus

$$2n \le x < 2n + 1,$$

and

$$y = x - n = x - [x/2].)$$

165

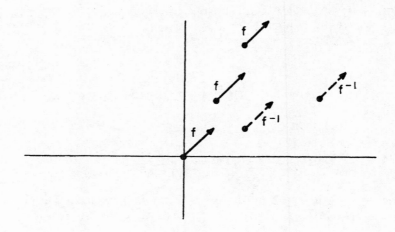

(viii)

$$f^{-1}(x) = \begin{cases} \dfrac{-1 + \sqrt{1 + 4x^2}}{2x}, & x \neq 0 \\ 0, & x = 0. \end{cases}$$

(If $y = f^{-1}(x)$, then $x = f(y) = y/(1 - y^2)$. So $xy^2 + y - x = 0$. If $x = 0$, then $y = 0$. If $x \neq 0$, then

$$y = \frac{-1 + \sqrt{1 + 4x^2}}{2x} \qquad \text{or} \qquad y = -1 - \sqrt{\frac{1 + 4x^2}{2x}}.$$

The first possibility is the correct one, since x and y must have the same sign.)

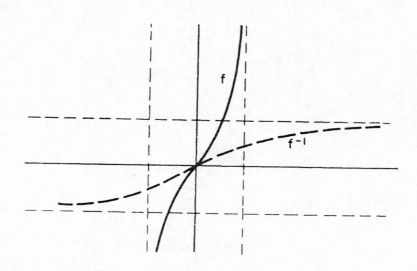

2. (i) f^{-1} is increasing and $f^{-1}(x)$ is not defined for $x \leq 0$.

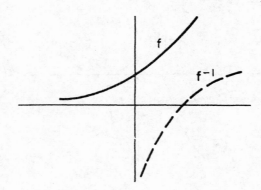

(ii) f^{-1} is increasing and $f^{-1}(x)$ is not defined for $x \geq 0$.

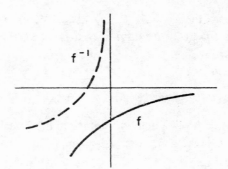

(iii) f^{-1} is decreasing and $f^{-1}(x)$ is not defined for $x \leq 0$.

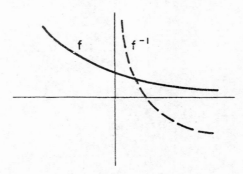

(iv) f^{-1} is decreasing and $f^{-1}(x)$ is not defined for $x \geq 0$.

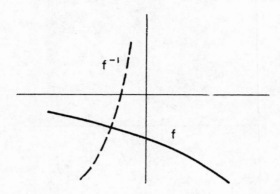

5. (b) If $h(x) = 1+x$, then $g = h \circ f$, so $g^{-1} = f^{-1} \circ h^{-1}$, so $g^{-1}(x) = f^{-1}(x-1)$. It is also possible to find g^{-1} directly: if $y = g^{-1}(x)$, then $x = g(y) = 1 + f(y)$, so $y = f^{-1}(x-1)$.)

7. (ii) Any interval $[a, b]$, since f is increasing.

(iv) Those intervals $[a, b]$ which are contained in the interval $(-\infty, -1-\sqrt{2}\,]$ or in $[-1 + \sqrt{2}, \infty)$ or in $[-1 + \sqrt{2}, \infty)$, since these are the intervals on which f is increasing or decreasing.

8. We have

$$g'(x) = (f^{-1})'(x) = \frac{1}{f'(f^{-1}(x))}$$
$$= \{1 + [f^{-1}(x)]^3\}^{1/2}$$
$$g''(x) = \frac{3}{2} \frac{[f^{-1}(x)]^2 (f^{-1})'(x)}{\{1 + [f^{-1}(x)]^3\}^{1/2}}$$
$$= \frac{3}{2} [f^{-1}(x)]^2$$
$$= \frac{3}{2} g(x)^2.$$

9. Apply Theorem 5 to f^{-1}.

10. (a) For $\mathcal{D}f(y)$ to exist $f'(y)$, $f''(y)$, and $f'''(y)$ must exist, with $f'(y) \neq 0$. Then

$$(f^{-1})'(x) = \frac{1}{f'(f^{-1}(x))}$$

$$(f^{-1})''(x) = \frac{-f''(f^{-1}(x)) \cdot (f^{-1})'(x)}{[f'(f^{-1}(x))]^2}$$

$$= \frac{-f''(f^{-1}(x))}{[f'(f^{-1}(x))]^3}$$

$$(f^{-1})'''(x) = \frac{1}{[f'(f^{-1}(x))]^6} \times \left\{ -[f'(f^{-1}(x))]^3 f'''(f^{-1}(x)) \cdot (f^{-1})'(x) \right.$$

$$\left. + f''(f^{-1}(x)) \cdot 3[f'(f^{-1}(x))^2 (f^{-1})'(x) \right\}$$

$$= \frac{-[f'(f^{-1}(x))]^3 f'''(f^{-1}(x)) + 3f''(f^{-1}(x)) \cdot [f'(f^{-1}(x))]^2}{[f'(f^{-1}(x))]^7}$$

all exist (compare Problem 21), with $(f^{-1})'(x) \neq 0$.

(b) Since we know that $\mathcal{D}f^{-1}(x)$ exists, we can use Problem 10-17(a) to write

$$1 = \mathcal{D}(f \circ f^{-1})(x) = [\mathcal{D}f(f^{-1}(x))] \cdot [(f^{-1})'(x)]^2 + \mathcal{D}f^{-1}(x)$$

$$= \frac{\mathcal{D}f(f^{-1}(x))}{[f'(f^{-1}(x))]^2} + \mathcal{D}f^{-1}(x),$$

or

$$\mathcal{D}f^{-1}(x) = 1 - \frac{\mathcal{D}f(f^{-1}(x))}{[f'(f^{-1}(x))]^2}.$$

11. (a) Let $f = g^{-1}$, where $g(x) = -x^5 - x$. Notice that g is one-one, since $g'(x) = -5x^4 - 1 < 0$, and that g takes on all values. So f is defined on **R**, and for all x we have

$$x = g(f(x)) = -[f(x)]^5 - f(x).$$

Moreover, f is differentiable, since $g'(x) \neq 0$ for all x.

(b)

$$f'(x) = (g^{-1})'(x) = \frac{1}{g'(g^{-1}(x))} \qquad \text{by Theorem 5}$$

$$= \frac{1}{g'(f(x))} = \frac{1}{-5[f(x)]^4 - 1}.$$

(c) Differentiating both sides of

$$[f(x)]^5 + f(x) + x = 0$$

yields

$$5[f(x)]^4 \cdot f'(x) + f'(x) + 1 = 0,$$

so

$$f'(x) = \frac{-1}{1 + 5[f(x)]^4}.$$

12. (a) $f(x) = \sqrt{1 - x^2}$ and $f(x) = -\sqrt{1 - x^2}$.

(b) There are no functions with this property.

(c) Let

$$\left.\begin{array}{c} g_1(x) \\ g_2(x) \\ g_3(x) \end{array}\right\} = \quad g(x) \text{ for } \quad \left\{\begin{array}{l} x < -1 \\ -1 < x < 1 \\ x > 1. \end{array}\right.$$

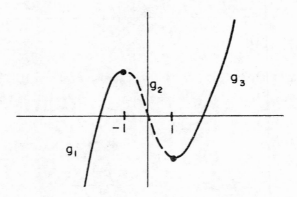

Then each g_i is one-one. If $f_i = g_i^{-1}$, then each f_i satisfies $[f_i(x)]^3 - 3f_i(x) = x$. The domain of

$$\left.\begin{array}{c} f_1 \\ f_2 \\ f_3 \end{array}\right\} \quad \text{is} \quad \left\{\begin{array}{l} (-\infty, 2) \\ (-2, 2) \\ (2, \infty). \end{array}\right.$$

(To find $y = f_i(x) = g_i^{-1}(x)$ explicitly we would have to solve the equation $x = g(y) = y^3 - 3y$. This can be done, but only with great difficulty; see Chapter 25.)

It is not hard to see that any continuous function f satisfying $[f(x)]^3 - 3f(x) = x$, and defined on an interval, must be (part of) some f_i. For such a function f satisfies $g(f(x)) = x$; this equation implies that f is one-one (Problem 3-23) and that f^{-1} coincides with g on the domain of f^{-1}. But the domain of f^{-1} is an interval, and the only intervals on which g is one-one are contained in $(-\infty, -1)$ or $(-1, 1)$, or $(1, \infty)$.

13. (a) Differentiating both sides of $[f(x)]^2 + x^2 = 1$ yields

$$2f(x)f'(x) + 2x = 0,$$

or

$$f'(x) = -\frac{x}{f(x)}.$$

(b) This equation is true for

$$f(x) = \sqrt{1 - x^2}, \quad \text{in which case} \quad f'(x) = \frac{-x}{\sqrt{1 - x^2}} = \frac{-x}{f(x)},$$

and

$$f(x) = -\sqrt{1 - x^2}, \quad \text{in which case} \quad f'(x) = \frac{x}{\sqrt{1 - x^2}} = \frac{-x}{f(x)}.$$

(c) We have

$$3[f(x)]^2 f'(x) - 3f'(x) = 1,$$

so

$$f'(x) = \frac{1}{3([f(x)]^2 - 1)}.$$

14. **(a)** Differentiating both sides of $x^3 + [f(x)]^3 = 7$ yields

$$3x^2 + 3[f(x)]^2 f'(x) = 0,$$
$$6x + 6f(x)[f'(x)]^2 + 3[f(x)]^2 f''(x) = 0,$$

or

$$f'(x) = \frac{-x^2}{[f(x)]^2},$$

$$f''(x) = \frac{-2x - 2f(x)\left[\dfrac{-x^2}{[f(x)]^2}\right]^2}{[f(x)]^2}$$

$$= \frac{-2x[f(x)]^4 - 2x^4 f(x)}{[f(x)]^6}.$$

(b) For this f we have

$$f'(-1) = -\frac{1}{4}$$

$$f''(-1) = \frac{2 \cdot 2^4 - 2 \cdot 2}{2^6}$$

$$= \frac{7}{16}.$$

15. Differentiating both sides of $3x^3 + 4x^2 f(x) - x[f(x)]^2 + 2[f(x)]^3 = 4$ yields

$$9x^2 + 8xf(x) + 4x^2 f'(x) - [f(x)]^2 - 2xf(x)f'(x) + 6[f(x)]^2 f'(x) = 0.$$

At the point $(-1, 1)$ we have

$$9 - 8 + 4f'(-1) - 1 + 2f'(-1) + 6f'(-1) = 0$$

or

$$f'(-1) = 0.$$

So the equation of the tangent line is $y = 1$.

16. Consider a differentiable function f which satisfies

$$[f(x)]^4 + [f(x)]^3 + xf(x) = 1;$$

then

$$4[f(x)]^3 f'(x) + 3[f(x)]^2 f'(x) + f(x) + xf'(x) = 0,$$

$$f'(x) = \frac{-f(x)}{4[f(x)]^3 + 3[f(x)]^2 + x}.$$

19. (ii) $\beta^{-1}(3) = -1$, since $\beta(-1) = h(0) = 3$. So

$$(\beta^{-1})'(3) = \frac{1}{\beta'(\beta^{-1}(3))} = \frac{1}{\beta'(-1)}$$

$$= \frac{1}{h'(0)}$$

$$= \frac{1}{\sin^2(\sin 1)}.$$

(The answer is not surprising, since the equation $\beta(x) = h(x + 1)$ implies that $\beta^{-1} = h^{-1} - 1$.)

21. As in Problems 10-19 and 10-31, the main difficulty is in formulating a reasonable conjecture for the form of $(f^{-1})^{(k)}(x)$. It is not hard to prove the following assertion by induction on k: If $f^{(k)}(f^{-1}(x))$ exists, and $f'(f^{-1}(x))$ is non-zero, then

$$(f^{-1})^{(k)}(x) = \frac{A(x)}{[f'(f^{-1}(x))]^m}$$

for some integer m, where $A(x)$ is a sum of terms of the form

$$[f'(f^{-1}(x))]^{m_1} \cdots [f^{(l)}(f^{-1}(x))]^{m_l}.$$

22. (a) Suppose f is increasing and g is decreasing, and $f(a) = g(a)$. If $a < b$, then

$$g(b) < g(a) = f(a) < f(b),$$

and similarly if $b < a$.

(b) Appropriate functions f and g are shown below (to be explicit we can take $g(x) = x$ and $f(x) = [x] + \sqrt{x - [x]}$ (Problem 4-17)).

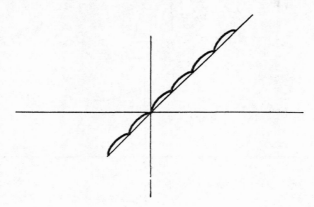

(c) Appropriate functions f and g are shown below. (Using the exponential function from Chapter 18, we can define $f(x) = e^x$ and $g(x) = -e^x$, but at the moment explicit definitions would be awkward.)

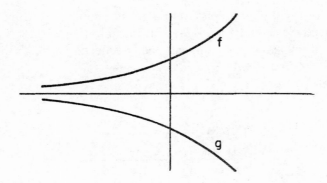

23. (a) The geometric idea behind the proof is indicated below: If $f(a) > a$, then $f(f(a)) = a < f(a)$. Since $f(a) > a$, and $f(b) < b$ for some b (namely, $f(a)$), it follows that $f(x) = x$ for some x in $[a, b]$.

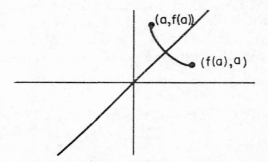

(b) Let f be any decreasing function on $(-\infty, a]$ which takes on all values $\geq a$, and define

$$g(x) = \begin{cases} f(x), & x \leq a \\ f^{-1}(x), & x \geq a. \end{cases}$$

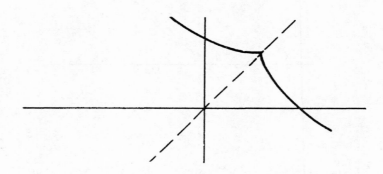

(c) If $f(x) < x$, then $x = f^{-1}(f(x)) < f^{-1}(x) = f(x)$, a contradiction. Similarly, we cannot have $f(x) > x$.

24. The functions with this property are precisely the one-one functions, because reflecting through the antidiagonal is the same as reflecting through the vertical axis, then reflecting through the diagonal, and finally reflecting through the vertical axis again.

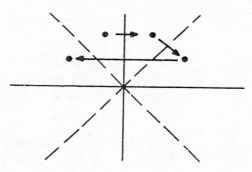

If a more analytic proof is desired, notice that the reflection of (a, b) through the antidiagonal is $(-b, -a)$. Thus if $(a, f(a))$ and $(b, f(b))$ are two points on the graph of f, we require that $(-f(a), -a)$ and $(-f(b), -b)$ should not have the same first coordinate if $a \neq b$. In other words $f(a)$ and $f(b)$ must be different. So f must be one-one.

25. (a) Since f is not increasing, there is some $x < y$ with $f(y) \leq f(x)$. Since f is nondecreasing, if $x \leq z \leq y$, then $f(x) \leq f(z) \leq f(y) \leq f(x)$. So $f(x) = f(z) = f(y)$.

(b) $f(x+h) \geq f(x)$ for $h > 0$ and $f(x+h) \leq f(x)$ for $h < 0$, so

$$\frac{f(x+h) - f(x)}{h} \geq 0$$

for all $h \neq 0$, so $f'(x) \geq 0$.

(c) If $y > x$, then

$$\frac{f(y) - f(x)}{y - x} = f'(z) \qquad \text{for some } z \text{ in } (x, y),$$

$$\geq 0$$

so $f(y) \geq f(x)$. Similarly, if $y < x$, then $f(y) \leq f(x)$.

26. (a) The idea behind the proof is indicated in the figure below. On the interval $[n, n+1]$, let g be the linear function with $g(n) = f(n+1)$ and $g(n+1) = f(n+2)$.

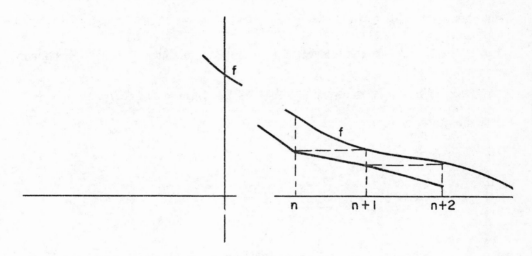

(b) On the interval $[n, n+1]$ let g be the linear function with $g(n) = f(n+1)/(n+1)$ and $g(n+2) = f(n+2)/(n+2)$.

1. (a) Set $s = x - a$, so that $x = a + s$.

(b) The tangent line according to our new definition consists of all points

$$c(a) + s \cdot c'(a) = \big(a, f(a)\big) + s \cdot \big(1, f'(a)\big)$$
$$= \big(a, f(a)\big) + \big(s, sf'(a)\big) = \big(a + s, f(a) + sf'(a)\big),$$

the same set of points obtained in part (a).

2. We have

$$c(t) = (f(t), t^2) = \begin{cases} (t^2, t^2) & t \geq 0 \\ (-t^2, t^2) & t \leq 0 \end{cases}$$

and these points are all on the graph of $h(x) = |x|$, since $\left|-t^2\right| = t^2$. On the other hand, if $S(t) = t^2$, then our straightforward definition of c' gives

$$c'(0) = (f'(0), S'(0)) = (0, 0),$$

since we have $f'(0) = S'(0) = 0$.

3. (a) Since $u' \neq 0$ on the interval, u is one-one on the interval, so u^{-1} exists, and each point

$$\big(u(t), v(t)\big) = \big(u(t), v(u^{-1}(u(t)))\big) = \big(u(t), v \circ u^{-1}(u(t))\big)$$

is on the graph of $v \circ u^{-1}$.

(b) We have

$$f' = (v \circ u^{-1})' = (v' \circ u^{-1}) \cdot (u^{-1})'$$
$$= \frac{v' \circ u^{-1}}{u' \circ u^{-1}},$$

so

$$f'(x) = f'(u(t)) = \frac{v'(t)}{u'(t)}.$$

(c) Then

$$f'' = \frac{(u' \circ u^{-1})(v' \circ u^{-1})' - (v' \circ u^{-1})(u' \circ u^{-1})'}{(u' \circ u^{-1})^2}$$

$$= \frac{\dfrac{(u' \circ u^{-1})(v'' \circ u^{-1})}{(u' \circ u^{-1})} - \dfrac{(v' \circ u^{-1})(u'' \circ u^{-1})}{(u' \circ u^{-1})}}{(u \circ u^{-1})^2}$$

$$= \frac{(u' \circ u^{-1})(v'' \circ u^{-1}) - (v' \circ u^{-1})(u'' \circ u^{-1})}{(u' \circ u^{-1})^3}$$

so

$$f''(x) = \frac{u'(t)v''(t) - v'(t)u''(t)}{(u'(t))^3}.$$

4. (i) Differentiating the equation $x^{2/3} + f(x)^{2/3} = 1$ yields

$$\frac{2}{3x^{1/3}} + \frac{2f'(x)}{3f(x)^{1/3}} = 0,$$

$$f'(x) = -\frac{f(x)^{1/3}}{x^{1/3}}.$$

(ii) Problem 3(b) gives

$$f'(x) = \frac{3\sin^2 t \cos t}{3\cos^2 t(-\sin t)} = -\frac{\sin t}{\cos t}$$

for $x = \cos^3 t$. Substituting this value of x into the equation $x^{2/3} + y^{2/3} = 1$ gives

$$\cos^2 t + f(x)^{2/3} = 1,$$

$$f(x)^{1/3} = \sqrt{1 - \cos^2 t} = \sin t,$$

so

$$f'(x) = -\frac{f(x)^{1/3}}{x^{1/3}}.$$

5. The square of the distance from P to $(u(t), v(t))$ is

$$(x_0 - u(t))^2 + (y_0 - v(t))^2,$$

which has it minimum at $\bar{t}$ when

(*) $$0 = 2[x_0 - u(\bar{t})] \cdot [-u'(\bar{t})] + 2[y_0 - v(\bar{t})] \cdot [-v'(\bar{t})].$$

If $u'(\bar{t}) \neq 0$, this can be written as

$$\frac{v'(\bar{t})}{u'(\bar{t})} \cdot \frac{y_0 - v(\bar{t})}{x_0 - u(\bar{t})} = -1;$$

thus the tangent line, with slope $v'(\bar{t})/u'(\bar{t})$, is perpendicular to the line from P to Q, with slope $[y_0 - v(\bar{t})]/[x_0 - u(\bar{t})]$.

If $u'(\bar{t}) = 0$, so that the tangent line is parallel to the first axis, then [since we assumed that $u'(\bar{t})$ and $v'(\bar{t})$ are not both 0], (*) implies that $y_0 - v(\bar{t}) = 0$, i.e., that the line from P to Q is parallel to the second axis, and thus perpendicular to the tangent line.

6. (a) Letting $u(\theta) = f(\theta)\cos\theta$, $v(\theta) = f(\theta)\sin\theta$, the slope of the point with polar coordinates $(f(\theta), \theta)$ is

$$\frac{v'(\theta)}{u'(\theta)} = \frac{f(\theta)\cos\theta + f'(\theta)\sin\theta}{-f(\theta)\sin\theta + f'(\theta)\cos\theta}.$$

(b) When $f(\theta) = 0$, this formula shows that the slope of the tangent line through the point with polar coordinates $(0, \theta)$ [i.e., the origin] is $\tan\theta$, which is just the slope of the line making an angle of θ with the positive horizontal axis.

For the Archimedian spiral $r = f(\theta) = \theta$ we have $f(0) = 0$ so the tangent line through the origin is the horizontal axis.

For the graph of $r = \cos 2\theta$ (Problem 3(iii) of that Appendix) we have

$$\cos 2\theta = 0 \qquad \text{for } \theta = 45°, 135°, \ldots$$

so the lines through the origin making angles with the horizontal axis of $45°$ and $-45°$ are tangent lines.

For $r = |\cos 2\theta|$ (Problem 3(v)) we clearly have the same tangent lines through the origin.

Similarly, for the graph of $r = \cos 3\theta$ (Problem 3(iv)) the lines through the origin making angles of $30°$, $60°$, $-30°$ and $-60°$ are tangent lines, and the same is true for $r = |\cos 3\theta|$ (Problem 3(vi)).

For the graph of the lemniscate $r^2 = 2a^2 \cos 2\theta$ (Problem 10 of that Appendix) we again have $r = 0$ for $\cos 2\theta = 0$, so the lines through the origin making angles of $45°$ and $-45°$ are tangent lines.

(c) We must have $f'(\theta) = 0$, since $f(\theta)$ is the distance from the origin to the point with polar coordinates $(f(\theta), \theta)$. According to part (a), the slope of the tangent line is then $-\cot\theta = -1/\tan\theta$. Since $\tan\theta$ is the slope of the line from the origin to the point with polar coordinates $(f(\theta), \theta)$, this shows, in agreement with Problem 5, that this line is perpendicular to the tangent line.

(d) By part (a) we have

$$\tan\alpha = \frac{f(\theta)\cos\theta + f'(\theta)\sin\theta}{-f(\theta)\sin\theta + f'(\theta)\cos\theta}$$

so

$$\tan(\alpha - \theta) = \frac{\tan\alpha - \tan\theta}{1 + \tan\alpha \tan\theta}$$

$$= \frac{\dfrac{f(\theta)\cos\theta + f'(\theta)\sin\theta}{-f(\theta)\sin\theta + f'(\theta)\cos\theta} - \dfrac{\sin\theta}{\cos\theta}}{1 + \dfrac{f(\theta)\cos\theta + f'(\theta)\sin\theta}{-f(\theta)\sin\theta + f'(\theta)\cos\theta} \cdot \dfrac{\sin\theta}{\cos\theta}}$$

$$= \frac{f(\theta)\cos^2\theta + f'(\theta)\sin\theta\cos\theta + f(\theta)\sin^2\theta - f'(\theta)\sin\theta\cos\theta}{-f(\theta)\sin\theta\cos\theta + f'(\theta)\cos^2\theta + f(\theta)\cos\theta\sin\theta + f'(\theta)\sin^2\theta}$$

$$= \frac{f(\theta)}{f'(\theta)}.$$

7. (a) (i) If

$$\left(x^2 + f(x)^2 + f(x)\right)^2 = x^2 + f(x)^2,$$

then

$$\left(x^2 + f(x)^2 + f(x)\right)\left[2x + 2f(x)f'(x) + f'(x)\right] = x + f(x)f'(x),$$

so

$$f'(x)\left\{\left[1 + 2f(x)\right]\left(x^2 + f(x)^2 + f(x)\right) - f(x)\right\} = x\left[1 - 2\left(x^2 + f(x)^2 + f(x)\right)\right]$$

so

$$f'(x) = \frac{x\left[1 - 2\left(x^2 + f(x)^2 + f(x)\right)\right]}{\left[1 + 2f(x)\right]\left(x^2 + f(x)^2 + f(x)\right) - f(x)}$$

$$= \frac{x\left[1 - 2\sqrt{x^2 + f(x)^2}\right]}{\left[1 + 2f(x)\right]\sqrt{x^2 + f(x)^2} - f(x)}.$$

(ii) At the point with polar coordinates $(x, \theta) = (1 - \sin\theta, \theta)$ the slope of the tangent line is

$$\frac{(1 - \sin\theta)\cos\theta - \cos\theta\sin\theta}{(-1 + \sin\theta)\sin\theta - \cos\theta\cos\theta} = \frac{\cos\theta(1 - 2\sin\theta)}{\sin^2\theta - \cos^2\theta - \sin\theta}$$

$$= \frac{\cos\theta(1 - 2\sin\theta)}{1 - 2\cos^2\theta - \sin\theta}.$$

(b) We have $r = 0$ for $\theta = 90°$, so the line through the origin making an angle of $90°$ with the horizontal axis is a tangent line. [More precisely, there is no tangent line at this point, but there are appropriate left- and right-hand derivatives of ∞ and $-\infty$.]

8. (a) From the Figure, the distance from P to the radius passing through Q is $a\sin t$. Since the distance from O to Q is at, the first coordinate of P is the difference, $at - a\sin t$.

Similarly, the second coordinate of P is a minus the distance from P to the center of the circle, and thus $a - a\cos t$.

(b) We have

$$u'(t) = a(1 - \cos t) \geq 0,$$

since $\cos t \leq 1$; in fact, $u'(t) > 0$ except at isolated points. So u is increasing.

(c) We have

$$v(t) = a - a\cos t$$

$$\cos t = \frac{a - v(t)}{a},$$

and thus
$$t = \pm \arccos \frac{a - v(t)}{a},$$

with the + sign for $t \geq 0$, and the − sign for $t < 0$, since arccos is always positive. Moreover,

$$\sin t = \pm\sqrt{1 - \cos^2 t}$$

$$= \pm\sqrt{1 - \left(\frac{a - v(t)}{a}\right)^2}$$

$$= \pm\frac{1}{a}\sqrt{[2a - v(t)]v(t)},$$

with the sign being the same as that of $\sin t$ Hence

$$u(t) = at - a\sin t$$

$$= \pm a \arccos \frac{a - v(t)}{a} \pm \sqrt{[2a - v(t)]v(t)},$$

where the first $\pm$ is the same as the sign of t and the second is the opposite of the sign of $\sin t$.

(d) For the first half of the first arch of the cycloid we have $t > 0$ and $\sin t > 0$, so

$$u(t) = a \arccos \frac{a - v(t)}{a} - \sqrt{[2a - v(t)]v(t)}.$$

This means that this curve consists of points

$$(u(t), v(t)) = \left(a \arccos \frac{a - v(t)}{a} - \sqrt{[2a - v(t)]v(t)},\ v(t)\right), \qquad 0 \leq t \leq 1$$

or of points

$$\left(a \arccos \frac{a - y}{a} - \sqrt{[2a - y]y},\ y\right) \qquad 0 \leq y \leq 2a,$$

which is indeed the graph of g^{-1}.

9. By the Cauchy Mean Value Theorem (Theorem 11-8), there is a number x in (a, b) with

$(*)$ $[u(b) - u(a)]v'(x) = [v(b) - v(a)]u'(x).$

If we write this as

$$\frac{v'(x)}{u'(x)} = \frac{v(b) - v(a)}{u(b) - u(a)}$$

then the right-hand side is just the slope of the line from P to Q, while Problem 3 shows that the left-hand side is the slope of the tangent line of the curve (since $u'(x) \neq 0$, u is one-one in an interval containing x, so part of the curve is the graph of a function $f = v \circ u^{-1}$).

Of course, this assumes that the denominators are not 0, so we really have to exercise more care. To begin with, note that either $u(b) \neq u(a)$ or $v(b) \neq v(a)$, since otherwise $P = Q$ and there is nothing to prove. We might as well assume $u(b) \neq u(a)$, since the whole argument can be made with u and v interchanged. Then the only problem is that we might have $u'(x) = 0$ in (*). Since $u(b) \neq u(a)$, this means that we must also have $v'(x) = 0$. This possibility can actually arise, as mentioned in Problem 2, and really should be eliminated by hypothesis.

10. (a) Obviously

$$|u(t) - l_1| \leq \sqrt{|u(t) - l_1|^2 + |v(t) - l_2|^2} = \|c(t) - l\|,$$

and similarly for $|v(t) - l_2|$.

Now suppose that $\lim_{t \to a} c(t) = l$ by the above definition. Given $\varepsilon > 0$, let $\delta > 0$ be the one given by the definition. Then for $0 < |t - a| < \delta$ we have

$$|u(t) - l_1| \leq \|c(t) - l\| < \varepsilon.$$

Thus $\lim_{t \to a} u(t) = l_1$. Similarly for v.

(b) Conversely, suppose that $\lim_{t \to a} c(t) = l$ according to the definition in terms of component functions, so that $\lim_{t \to a} u(t) = l_1$ and $\lim_{t \to a} v(t) = l_2$. Suppose we are given $\varepsilon > 0$. Choose $\delta_1, \delta_2 > 0$ so that

$$\text{if } 0 < |t - a| < \delta_1, \text{ then } |u(t) - l_1| < \frac{\varepsilon}{\sqrt{2}}$$

$$\text{if } 0 < |t - a| < \delta_2, \text{ then } |v(t) - l_2| < \frac{\varepsilon}{\sqrt{2}},$$

and let $\delta = \min(\delta_1, \delta_2)$. Then if $0 < |t - a| < \delta$ we have

$$|u(t) - l_1|^2 < \frac{\varepsilon^2}{2}$$

$$|v(t) - l_2|^2 < \frac{\varepsilon^2}{2},$$

and thus

$$\|c(t) - l\|^2 = |u(t) - l_1|^2 + |v(t) - l_2|^2 < \frac{\varepsilon^2}{2} + \frac{\varepsilon^2}{2} = \varepsilon^2,$$

so that

$$\|c(t) - l\| < \varepsilon.$$

CHAPTER 13

3. (a) Problem 2-7 shows that

$$\frac{\sum_{k=1}^{n} k^p}{n^{p+1}} = \frac{1}{p+1} + \frac{A}{n} + \frac{B}{n^2} + \cdots,$$

which can clearly be made as close to $1/(p+1)$ as desired by choosing n large enough.

(b) We have

$$L(f, P_n) = \frac{b^{p+1}}{n^{p+1}} \left[\sum_{k=0}^{n-1} k^p \right],$$

$$U(f, P_n) = \frac{b^{p+1}}{n^{p+1}} \left[\sum_{k=1}^{n} k^p \right].$$

Part (a) shows that $L(f, P_n)$ and $U(f, P_n)$ can be made as close to the number $b^{p+1}/(p+1)$ as desired by choosing n sufficiently large. As in Problem 1, this implies that $\int_0^b x^p \, dx = b^{p+1}/(p+1)$.

4. (a) We have

$$\frac{b}{a} = \frac{t_n}{t_0} = \frac{t_n}{t_{n+1}} \cdot \frac{t_{n-1}}{t_{n-2}} \cdots \frac{t_1}{t_0} = r^n,$$

so $r = (b/a)^{1/n} = c^{1/n}$. Similarly,

$$\frac{t_i}{a} = r^i,$$

so $t_i = ar^i = a \cdot c^{i/n}$.

(b) We have

$$U(f, P) = \sum_{i=1}^{n} [a \cdot c^{i/n}]^p \cdot [a \cdot c^{i/n} - a \cdot c^{(i-1)/n}]$$

$$= a^{p+1} (1 - c^{-1/n}) \sum_{i=1}^{n} \left(c^{(p+1)/n} \right)^i$$

$$= a^{p+1} (1 - c^{-1/n}) c^{(p+1)/n} \sum_{i=0}^{n-1} \left(c^{(p+1)/n} \right)^i$$

$$= a^{p+1} (1 - c^{-1/n}) c^{(p+1)/n} \frac{1 - c^{p+1}}{1 - c^{(p+1)/n}} \qquad \text{by Problem 2-5}$$

182

$$= a^{p+1}(1 - c^{p+1})c^{(p+1)/n}\frac{1 - c^{-1/n}}{1 - c^{(p+1)/n}}$$

$$= (a^{p+1} - b^{p+1})c^{(p+1)/n}\frac{1 - c^{-1/n}}{1 - c^{(p+1)/n}}$$

$$= (a^{p+1} - b^{p+1})c^{p/n}\frac{c^{1/n} - 1}{1 - c^{(p+1)/n}}.$$

Problem 2-5 also gives

$$1 + c^{1/n} + \cdots + c^{p/n} = \frac{1 - c^{(p+1)/n}}{1 - c^{1/n}}.$$

So

$$U(f, P) = (b^{p+1} - a^{p+1})c^{p/n} \cdot \frac{1}{1 + c^{1/n} + \cdots + c^{p/n}}.$$

Similarly,

$$L(f, P) = c^{-p/n}U(f, P) = (b^{p+1} - a^{p+1}) \cdot \frac{1}{1 + c^{1/n} + \cdots + c^{p/n}}.$$

(c) By making n large enough, we can make $c^{1/n}$ as close as we like to 1 (see Problem 22-10 for a rigorous proof). The same holds, of course, for each of the p numbers $c^{1/n}, \ldots, c^{p/n}$. So $U(f, P)$ and $L(f, P)$ can both be made as close as desired to

$$\frac{b^{p+1} - a^{p+1}}{1 + \underbrace{1 + \cdots + 1}_{p \text{ times}}} = \frac{b^{p+1} - a^{p+1}}{p + 1}.$$

5. (i) The integral is 0, since the part from -1 to 0 is the negative of the part from 0 to 1.

(ii) By the same reasoning the integral is

$$\int_{-1}^{1} 3\sqrt{1 - x^2}\,dx = 3\frac{\pi}{2},$$

since $f(x) = 3\sqrt{1 - x^2}$ is a semi-circle of radius $\sqrt{3}$ on $[-1, 1]$.

6. Since $\sin t > 0$ on $[0, \pi/2]$ (using radians) we clearly have

$$\int_{0}^{x} \frac{\sin t}{t + 1}\,dt > 0 \qquad \text{for } 0 < x \leq \pi/2.$$

Moreover, the integral $\int_{\pi/2}^{\pi} \sin t\,dt$ is exactly the negative of $\int_{0}^{\pi/2} \sin t\,dt$, while $1/(t + 1)$ is smaller on $[\pi/2, \pi]$ than on $[0, \pi/2]$, so the entire integral

$$\int_{\pi/2}^{\pi} \frac{\sin t}{t + 1}\,dt$$

is smaller in absolute value that the same integral on $[0, \pi/2]$. The same is certainly true of

$$\int_{\pi/2}^{x} \frac{\sin t}{t} \, dt$$

for all $\pi/2 \leq x \leq \pi$. This shows that

$$\int_{0}^{x} \frac{\sin t}{t+1} \, dt > 0 \qquad \text{also for } \pi/2 \leq x \leq \pi.$$

Etc.

7. (ii) $\int_{0}^{2} f = 0.$

(iv) f is not integrable.

(vi) f is integrable; a rigorous proof can be given in several ways, using various problems in this chapter, for example Problem 20. (Presumably, the integral of f is

$$\frac{1}{2} + \frac{1}{2}\left(\frac{1}{2} - \frac{1}{3}\right) + \frac{1}{3}\left(\frac{1}{3} - \frac{1}{4}\right) + \cdots .$$

At the moment we do not even know what an infinite sum means, let alone how to work with them, but the following likely looking manipulations are actually valid:

$$\frac{1}{2} + \frac{1}{2}\left(\frac{1}{2} - \frac{1}{3}\right) + \frac{1}{3}\left(\frac{1}{3} - \frac{1}{4}\right) + \cdots$$

$$= \left(\frac{1}{2^2} + \frac{1}{3^2} + \frac{1}{4^2} + \cdots\right) + \frac{1}{2} - \frac{1}{2 \cdot 3} - \frac{1}{3 \cdot 4} - \cdots$$

$$= \left(1 + \frac{1}{2^2} + \frac{1}{3^2} + \frac{1}{4^2} + \cdots\right) - \frac{1}{2} - \frac{1}{2 \cdot 3} - \frac{1}{3 \cdot 4} - \cdots$$

$$= \left(1 + \frac{1}{2^2} + \frac{1}{3^2} + \frac{1}{4^2} + \cdots\right) - \left(\frac{1}{1 \cdot 2} + \frac{1}{2 \cdot 3} + \frac{1}{3 \cdot 4} + \cdots\right).$$

From the fact that

$$\frac{1}{1 \cdot 2} + \cdots + \frac{1}{n(n+1)} = \frac{n}{n+1},$$

derived in Problem 2-6, we might guess that

$$\frac{1}{1 \cdot 2} + \frac{1}{2 \cdot 3} + \frac{1}{3 \cdot 4} + \cdots = 1.$$

The other infinite sum happens to equal $\pi^2/6$ (but we will not get to a proof of this fact anywhere in the text), so the integral of f is $\pi^2/6 - 1$.)

8. (i)

$$\int_{-2}^{2} \left(\frac{x^2}{2} + 2\right) - x^2 \, dx = \frac{16}{3}.$$

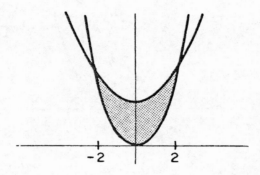

(ii)

$$\int_{-1}^{1} x^2 - (-x^2) \, dx = \frac{4}{3}.$$

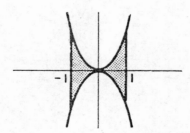

(iii)

$$\int_{-\sqrt{2}/2}^{\sqrt{2}/2} (1 - x^2) - x^2 \, dx = \frac{2\sqrt{2}}{3}.$$

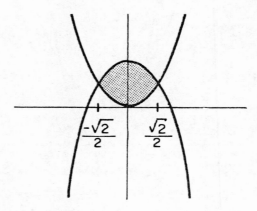

(iv)

$$\int_{-\sqrt{2}}^{-\sqrt{2}/2} 2 - x^2 \, dx + \int_{-\sqrt{2}/2}^{\sqrt{2}/2} 2 - (1 - x^2) \, dx + \int_{\sqrt{2}/2}^{\sqrt{2}} 2 - x^2 \, dx = 2\sqrt{2}.$$

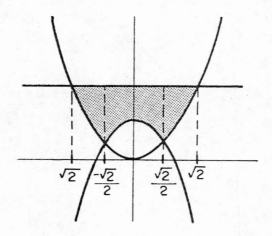

(v)

$$\int_0^2 (x^2 - 2x + 4) - x^2 \, dx = 4.$$

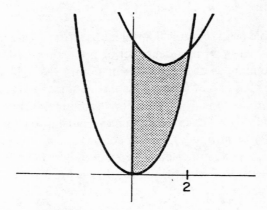

(vi) The area should be

$$2\sqrt{2} - \int_0^{\sqrt{2}} x^2 \, dx = \frac{4\sqrt{2}}{3}.$$

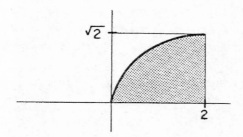

10. The first inequality is a special case of Problem 8-13, and the second inequality follows from the fact that $\{f(x_1) + g(x_2) : t_{i-1} \le x_1, x_2 \le t_i\}$ contains all numbers in $\{f(x) + g(x) : t_{i-1} \le x \le t_i\}$, and possibly some smaller ones.

11. (a) If $L(f, P) = U(f, P)$ for even one partition P, then each $m_i = M_i$, so f is constant on each $[t_{i-1}, t_i]$. Since these closed intervals overlap, f must be constant on all of $[a, b]$.

(b) If $L(f, P_1) = U(f, P_2)$ and P contains both P_1 and P_2, then $L(f, P_1) \le L(f, P) \le U(f, P) \le U(f, P_2) = L(f, P_1)$, so $L(f, P) = U(f, P)$. It follows from part (a) that f is constant on $[a, b]$.

(c) Only constant functions. For suppose f is not constant on $[a, b]$, and let m be the minimum value of f on $[a, b]$. Since $f(x) > m$ for some x, and since f is continuous, we can choose a partition $P = \{t_0, \ldots, t_n\}$ of $[a, b]$ so that $f > m$ on some interval $[t_{i-1}, t_i]$. Then $m_i > m$, so $L(f, P) > m(b-a)$. On the other hand, if Q is the partition $Q = \{a, b\}$, then $L(f, Q) = m(b-a)$.

(d) If f is integrable on $[a, b]$ and all lower sums are equal, then f takes on the value $m = \inf\{f(x) : a \leq x \leq b\}$ at a dense set of points in $[a, b]$. In fact, Problem 30 shows that f is continuous at a dense set of points. Now if f is continuous at x and $f(x) > m$, then, as in part (c), there is a partition P with $L(f, P) > m(b-a)$, while $L(f, Q) = m(b-a)$ if $Q = \{a, b\}$, contradicting the hypothesis. Conversely, it is easy to see that if f takes on its minimum value m on a dense set of points in $[a, b]$, then $L(f, P) = m(b-a)$, since each $m_i = m$. (The condition that f be integrable is essential in this problem. For example, if $f(x) = 1/q$ for $x = p/q$ in lowest terms, and $f(x) = 1$ for x irrational, then $L(f, P) = 0$ for all P, but f does not take on the value $0 = \inf\{f(x) : a \leq x \leq b\}$ anywhere.)

12. Theorem 4, applied to $a < b < d$, implies that f is integrable on $[b, d]$. Then Theorem 4, applied to $b < c < d$, implies that f is integrable on $[b, c]$.

14. Let $P = \{t_0, \ldots, t_n\}$ be a partition of $[a, b]$. If $g(x) = f(x - c)$, then

$$m_i = \inf\{f(x) : t_{i-1} \leq x \leq t_i\} = \inf\{g(x) : t_{i-1} + c \leq x \leq t_i + c\}$$

and similarly for M_i, so $L(f, P) = L(g, P')$ and $U(f, P) = U(g, P')$. If f is integrable, so that for every $\varepsilon > 0$ we have $U(f, P) - L(f, P) < \varepsilon$ for some P, then g is also integrable, since we have $U(g, P') - L(g, P') < \varepsilon$. Moreover,

$$\int_a^b f(x)\,dx = \sup\{L(f, P)\} = \sup\{L(g, P')\} = \int_{a+c}^{b+c} f(x - c)\,dx.$$

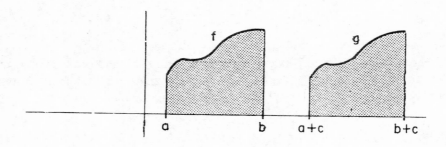

15. Notice that

$$b \cdot \inf\left\{\frac{1}{t} : t_{i-1} \leq t \leq t_i\right\} = \inf\left\{\frac{1}{t} : bt_{i-1} \leq x \leq bt_i\right\}.$$

Denoting the first inf by m_i and the second by m_i', we have

$$L(f, P') = \sum_{i=1}^{n} m_i'(bt_i - bt_{i-1})$$

$$= \sum_{i=1}^{n} bm_i'(t_i - t_{i-1})$$

$$= \sum_{i=1}^{n} m_i(t_i - t_{i-1})$$

$$= L(f, P).$$

So

$$\int_b^{ab} \frac{1}{t}\, dt = \sup\{L(f, P')\} = \sup\{L(f, P)\} = \int_1^a \frac{1}{t}\, dt.$$

16. If $P = \{t_0, \ldots, t_n\}$ is a partition of $[a, b]$, and $P' = \{ct_0, \ldots, ct_n\}$, then

$$m_i = \inf\{f(ct) : t_{i-1} \le t \le t_i\} = \inf\{f(t) : ct_{i-1} \le t \le ct_i\} = m_i'.$$

So if $g(t) = f(ct)$, then

$$cL(g, P') = c \sum_{i=1}^{n} m_i(t_i - t_{i-1})$$

$$= \sum_{i=1}^{n} m_i(ct_i - ct_{i-1})$$

$$= L(f, P').$$

So

$$\int_{ca}^{cb} f(t)\, dt = \sup\{L(f, P')\} = c \cdot \sup\{L(g, P)\} = c \cdot \int_a^b f(ct)\, dt.$$

(Actually, this proof is valid only for $c \ge 0$, but the case $c < 0$ can then be deduced easily.)

17. The upper half of the unit circle is the graph of

$$f(x) = \sqrt{1 - x^2}$$

while the upper half of the ellipse is the graph of

$$g(x) = b\sqrt{1 - \left(\frac{x}{a}\right)^2},$$

so the area enclosed by the ellipse is

$$2 \int_{-a}^{a} b \sqrt{1 - \left(\frac{t}{a}\right)^2}\, dt = 2b \int_{-a}^{a} \sqrt{1 - \left(\frac{t}{a}\right)^2}\, dt$$

$$= 2ab \int_{-1}^{1} \sqrt{1 - t^2}\, dt$$

$$= 2ab \cdot \pi/2 = \pi ab.$$

18. (a) We have

$$\int_{0}^{a} x^n\, dx = a \int_{0}^{1} (ax)^n\, dx = a^{n+1} \int_{0}^{1} x^n\, dx = c_n a^{n+1}.$$

(b) From part (a) we have

$$2^{n+1} c_n a^{n+1} = \int_{0}^{2a} x^n\, dx = \int_{-a}^{a} \sum_{k=0}^{n} \binom{n}{k} x^k a^{n-k}\, dx$$

$$= 2 \sum_{k \text{ even}} \binom{n}{k} a^{n-k} \int_{0}^{a} x^k\, dx \qquad \text{(compare Problem 5)}$$

$$= 2 \sum_{k \text{ even}} \binom{n}{k} a^{n-k} a^{k+1} c_k = 2a^{n+1} \sum_{k \text{ even}} \binom{n}{k} c_k.$$

(c) The proof is by complete induction. We know that $c_1 = 1/2$. Assume that $c_k = 1/(k+1)$ for $k < n$. Then

$$2^n c_n = 2 \sum_{k \text{ even}} \frac{\binom{n}{k}}{k+1}$$

$$= \frac{2}{n+1} \sum_{k \text{ even}} \frac{n+1}{k+1} \binom{n}{k}$$

$$= \frac{2}{n+1} \sum_{k \text{ even}} \binom{n+1}{k+1}$$

$$= \frac{2}{n+1} \sum_{k \text{ odd}} \binom{n+1}{k}$$

$$= \frac{2^n}{n+1} \qquad \text{by Problem 2-3(e)(iii)}$$

19. Choose $M \geq 1$ so that $|f(x)| \leq M$ for all x in $[a, b]$. Given $\varepsilon > 0$, let $\delta = \varepsilon/3M$. Since f is continuous on $[a, x_0 - \delta/2]$ and $[x_0 + \delta/2, b]$ there are partitions $P_1 = \{t_0, \ldots, t_n\}$ of $[a, x_0 - \delta/2]$ and $P_2 = \{s_0, \ldots, s_m\}$ of $[x_0 + \delta/2, b]$ such that $U(f, P_1) - L(f, P_1) < \varepsilon/3$ and $U(f, P_2) - L(f, P_2) < \varepsilon/3$. If $P = \{t_0, \ldots, t_n, s_0, \ldots, s_m\}$, then

$$U(f, P) - L(f, P) \leq [U(f, P_1) - L(f, P_1)] + \delta \cdot M + [U(f, P_2) - L(f, P_2)]$$
$$< \varepsilon/3 + \varepsilon/3 + \varepsilon/3 = \varepsilon.$$

20. (a)

$$L(f, P) = \sum_{i=1}^{n} f(t_{i-1})(t_i - t_{i-1}),$$

$$U(f, P) = \sum_{i=1}^{n} f(t_i)(t_i - t_{i-1}).$$

(b) If $t_i - t_{i-1} = \delta$ for each i, then

$$U(f, P) - L(f, P) = \sum_{i=1}^{n} [f(t_i) - f(t_{i-1})](t_i - t_{i-1})$$

$$= \delta \sum_{i=1}^{n} f(t_i) - f(t_{i-1})$$

$$= \delta [f(b) - f(a)].$$

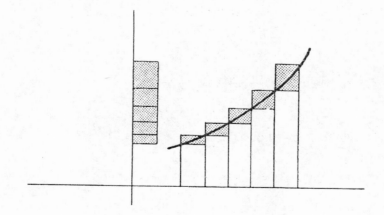

(c) For every $\varepsilon > 0$ we have $U(f, P) - L(f, P) < \varepsilon$ if $t_i - t_{i-1} = \delta < \varepsilon/[f(b) - f(a)]$.

(d) The function in Problem 7(vi) is an example (on the interval $[0, 1]$).

21. (a)

$$L(f^{-1}, P) + U(f, P')$$

$$= \sum_{i=1}^{n} f^{-1}(t_{i-1})(t_i - t_{i-1}) + \sum_{i=1}^{n} t_i (f^{-1}(t_i) - f^{-1}(t_{i-1}))$$

$$= \sum_{i=1}^{n} [t_i f^{-1}(t_i) - t_{i-1} f^{-1}(t_{i-1})]$$

$$= b f^{-1}(b) - a f^{-1}(a).$$

(b) It follows from (a) that

$$\int_a^b f^{-1} = \sup\{L(f^{-1}, P)\} = \sup\{b f^{-1}(b) - a f^{-1}(a) - U(f, P')\}$$

$$= b f^{-1}(b) - a f^{-1}(a) - \inf\{U(f, P')\}$$

$$= b f^{-1}(b) - a f^{-1}(a) - \int_{f^{-1}(a)}^{f^{-1}(b)} f.$$

(c) If $f(x) = x^n$ for $x \geq 0$, then for $0 \leq a < b$ we have

$$\int_a^b \sqrt[n]{x}\, dx = \int_a^b f^{-1} = b f^{-1}(b) - a f^{-1}(a) - \int_{f^{-1}(a)}^{f^{-1}(b)} x^n\, dx$$

$$= b \sqrt[n]{b} - a \sqrt[n]{a} - \left[\frac{(\sqrt[n]{b})^{n+1}}{n+1} - \frac{(\sqrt[n]{a})^{n+1}}{n+1} \right]$$

$$= \frac{n \sqrt[n]{b}}{n+1} - \frac{n \sqrt[n]{a}}{n+1}.$$

22. The Figure below shows the case $b < f(a)$.

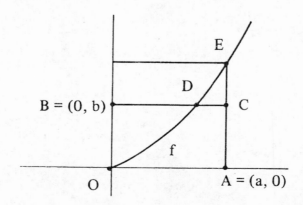

We have

$$ab = \text{area } OACB < \text{area } OAE + \text{area } OBD$$
$$= \int_0^a f(x)\,dx + \int_0^b f^{-1}(x)\,dx.$$

If $b = f(a)$ we clearly have equality. It is easy to see that we have the same inequality if $b > f(a)$ [or simply apply the first inequality to f^{-1}].

23. (b) To show that continuity is necessary, first choose any continuous one-one function f on $[a, b]$. Then $\int_a^b f(x)\,dx = (b-a)f(\xi)$ for a *unique* ξ. Now let $g(x) = f(x)$ for $x \neq \xi$, but $g(\xi) \neq f(\xi)$.

(c) From the inequality $mg(x) \leq f(x)g(x) \leq Mg(x)$, we obtain

$$m \int_a^b g(x)\,dx \leq \int_a^b f(x)g(x)\,dx \leq M \int_a^b g(x)\,dx.$$

Consequently

$$\int_a^b f(x)g(x)\,dx = \mu \int_a^b g(x)\,dx$$

for some μ with $m \leq \mu \leq M$. This $\mu = f(\xi)$ for some ξ in $[a, b]$.

(d) Replace g by $-g$.

(e) If $g(x) = x$ on $[-1, 1]$ and $f(x) = x$, then

$$\int_{-1}^1 f(x)g(x)\,dx = \int_{-1}^1 x^2\,dx = \frac{2}{3} \neq \mu \cdot \int_{-1}^1 x\,dx.$$

24. If $P = \{t_0, \ldots, t_n\}$ is a partition of $[\theta_0, \theta_1]$, then

$$L(f^2/2, P) = \sum_{i=1}^n m_i{}^2 \frac{t_i - t_{i-1}}{2} \qquad \text{and} \qquad U(f^2/2, P) = \sum_{i=1}^n M_i{}^2 \frac{t_i - t_{i-1}}{2}$$

represent the total area of sectors contained in A and containing A, respectively. So

$$L(f^2/2, P) \leq \text{area } A \leq U(f^2/2, P)$$

for all P. It follows that area A must be $\displaystyle\int_{\theta_0}^{\theta_1} f^2/2$.

25. (a) If $f(x) = \alpha x + \beta$, then for every P we have

$$\ell(f, P) = \sum_{i=1}^{n} \sqrt{(t_i - t_{i-1})^2 + \alpha^2(t_i - t_{i-1})^2}$$

$$= \sum_{i=1}^{n} (t_i - t_{i-1})\sqrt{1 + \alpha^2}$$

$$= (b - a)\sqrt{1 + \alpha^2},$$

and the distance from $(a, \alpha a + \beta)$ to $(b, \alpha b + \beta)$ is

$$\sqrt{[\alpha(a - b)]^2 + (a - b)^2} = (b - a)\sqrt{1 + \alpha^2}.$$

(b) If f is not linear, then there is some t in $[a, b]$ such that $(a, f(a))$, $(t, f(t))$ and $(b, f(b))$ do not lie on a straight line. Thus if $P = \{a, t, b\}$, then

$$\ell(f, P) = \sqrt{(t - a)^2 + [f(t) - f(a)]^2} + \sqrt{(b - t)^2 + [f(b) - f(t)]^2}$$

$$> \sqrt{(b - a)^2 + [f(b) - f(a)]^2}, \qquad \text{by Problem 4-9.}$$

(c) follows immediately from part (b).

(d) For each i there is some x_i in (t_{i-1}, t_i) with

$$f'(x_i)(t_i - t_{i-1}) = f(t_{i-1}) - f(t_i).$$

So

$$L\left(\sqrt{1 + (f')^2}, P\right) \leq \sum_{i=1}^{n} (t_i - t_{i-1})\sqrt{1 + [f'(x_i)]^2} \leq U\left(\sqrt{1 + (f')^2}, P\right)$$

and

$$\sum_{i=1}^{n} (t_i - t_{i-1})\sqrt{1 + [f'(x_i)]^2} = \sum_{i=1}^{n} \sqrt{(t_i - t_{i-1})^2 + [f'(x_i)(t_i - t_{i-1})]^2}$$

$$= \sum_{i=1}^{n} \sqrt{(t_i - t_{i-1})^2 + [f(t_i) - f(t_{i-1})]^2}$$

$$= \ell(f, P).$$

(e) Since $\sup\{\ell(f, P)\}$ is an upper bound for the set of all $\ell(f, P)$, it is also an upper bound for the set of all $L\left(\sqrt{1 + (f')^2}, P\right)$ by part (a).

(f) It suffices to show that

$$\sup\{\ell(f, P)\} \leq U\left(\sqrt{1 + (f')^2}, P''\right)$$

for any partition P'', and to prove this it suffices to show that

$$\ell(f, P') \leq U\left(\sqrt{1 + (f')^2}, P''\right)$$

for any partition P'. If P contains the points of P', then

$$\ell(f, P) \geq \ell(f, P');$$

the proof is similar for the proof for lower sums, putting in one point at a time and using Problem 4-9 to see that this increases ℓ. This if P contains the points of both P' and P'', then

$$\ell(f, P') \leq \ell(f, P) \leq U\left(\sqrt{1 + (f')^2}, P\right) \leq U\left(\sqrt{1 + (f')^2}, P''\right).$$

(g) We are considering

$$\lim_{x \to a} \frac{\int_a^x \sqrt{1 + (f')^2}}{\sqrt{(x-a)^2 + [f(x) - f(a)]^2}}.$$

By the Mean Value Theorem, $f(x) - f(a) = (x - a)f'(\xi)$ for some ξ in (a, b), and by the Mean Value Theorem for Integrals (Problem 23), the numerator is $(x - a)\sqrt{1 + f'(\eta)^2}$ for some η in $[a, b]$. So we are considering

$$\frac{(x - a)\sqrt{1 + f'(\eta)^2}}{\sqrt{(x-a)^2 + f'(\xi)^2(x-a)^2}} = \frac{\sqrt{1 + f'(\eta)^2}}{\sqrt{1 + f'(\xi)^2}},$$

which approaches 1 as $x \to a$ (we need to assume that f' is continuous at a).

26. (a) If $P = \{t_0, \dots, t_n\}$ is a partition of $[a, b]$ with $U(f, P) - L(f, P) < \varepsilon$, then $U(f, P) - \int_a^b f < \varepsilon$ and $\int_a^b f - L(f, P) < \varepsilon$. Let $s_1(x)$ be m_i for x in (t_{i-1}, t_i) and 0, say, for $x = t_0, \dots, t_n$; similarly let $s_2(x)$ be M_i for x in (t_{i-1}, t_i) and 0 for $x = t_0, \dots, t_n$.

(b) The existence of such step functions implies the existence of partitions P_1 and P_2 with $U(f, P_2) - L(f, P_1) < \varepsilon$.

(c) The function in Problem 34 is an example.

27. It obviously suffices to show that for any $\varepsilon > 0$ there are $g \leq f$ with $\int_a^b f - \int_a^b g < \varepsilon$ and $h \geq f$ with $\int_a^b h - \int_a^b f < \varepsilon$. Moreover, the second follows from the first by considering $-f$, so we just have to find the desired $g \leq f$.

   Choose a step function $s \leq f$ with $\int_a^b f - \int_a^b s < \varepsilon/2$, by Problem 26(a). Choose $M \geq 1$ so that $|f(x)| \leq M$ for all x in $[a, b]$, and if s is constant on (t_{i-1}, t_i) for $i = 1, \dots, n$, choose $\delta < \varepsilon/2nM$. Let $g = s$ on $[t_{i-1} + \delta/2, t_i - \delta/2]$ and let g be a linear function on $[t_i - \delta/2, t_i]$ and $[t_i, t_i + \delta/2]$ with $g(t_i) = -M$.

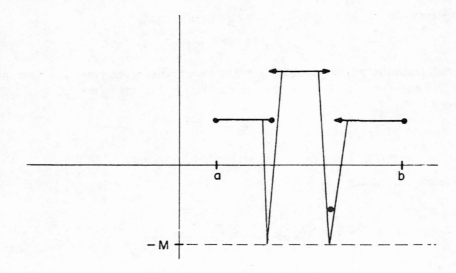

Then $g \leq s \leq f$ and $\int_a^b s - \int_a^b g \leq nM\delta < \varepsilon/2$, so $\int_a^b f - \int_a^b g < \varepsilon$.

28. (a) If s_1 (respectively s_2) is constant on each subinterval for a partition P_1 (respectively P_2), then $s_1 + s_2$ is constant on the intervals for the partition P which contains P_1 and P_2.

(b) Part (a) shows that there is a partition $P = \{t_0, \ldots, t_n\}$ such that s_1 and s_2 are constant on each (t_{i-1}, t_i), with values a_i and b_i, say. Then

$$\int_a^b (s_1 + s_2) = \sum_{i=1}^n (a_i + b_i)(t_i - t_{i-1})$$

$$= \sum_{i=1}^n a_i(t_i - t_{i-1}) + \sum_{i=1}^n b_i(t_i - t_{i-1})$$

$$= \int_a^b s_1 + \int_a^b s_2.$$

(c) Given $\varepsilon > 0$, choose step functions s_1, s_2 and t_1, t_2 with $s_1 \leq f \leq s_2$ and $t_1 \leq g \leq t_2$ and $\int_a^b s_2 - \int_a^b s_1 < \varepsilon/2$ and $\int_a^b t_2 - \int_a^b t_1 < \varepsilon/2$. Part (b) implies that

$$\int_a^b (s_1 + t_1) = \int_a^b s_1 + \int_a^b t_1 \leq \int_a^b f + \int_a^b g \leq \int_a^b s_2 + \int_a^b t_2 = \int_a^b (s_2 + t_2)$$

and that

$$\int_a^b (s_2 + t_2) - \int_a^b (s_1 + t_1) < \varepsilon.$$

This shows that f is integrable, and also that $\int_a^b (f + g) = \int_a^b f + \int_a^b g$, since there is only one number between all such $\int_a^b (s_1 + t_1)$ and $\int_a^b (s_2 + t_2)$.

29. Let $g(x) = \int_a^x f - \int_x^b f$. Then g is continuous and $g(a) = -\int_a^b f$ and $g(b) = \int_a^b f$; so $g(a)$ and $g(b)$ have different signs and consequently $g(x) = 0$ for some x in $[a, b]$, unless $g(a) = 0$, in which case we can choose $x = a$.

For the function f shown below, only $x = a$ or $x = b$ will work; f has been chosen so that $\int_a^c f = -\int_c^b f$.

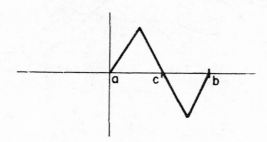

30. **(a)** Clearly if $M_i = m_i \geq 1$ for all i, then $U(f, P) - L(f, P) \geq b - a$.

(b) If $i = 1$, let $b_1 = t_1$ and choose any a_1 with $t_0 < a_1 < t_1$. Similarly if $i = n$.

(c) Choose a partition P of $[a_1, b_1]$ with $U(f, P) - L(f, P) < (b_1 - a_1)/2$. Then $M_i - m_i < 1/2$ for some i. Choose $[a_2, b_2] = [t_{i-1}, t_i]$ unless $i = 1$ or n, in which case use the modification of part (b).

(d) Let x be a point in each I_n. Notice that we cannot have $x = a_n$ or b_n, since x is also in $[a_{n+1}, b_{n+1}]$ and $a_n < a_{n+1} < b_{n+1} < b_n$. If $\varepsilon > 0$, there is some n such that
$$\sup\{f(x) : x \text{ in } I_n\} - \inf\{f(x) : x \text{ in } I_n\} < \varepsilon/2.$$
Then $|f(y) - f(x)| < \varepsilon$ for all y in I_n; since x is in (a_n, b_n), this means that $|f(y) - f(x)| < \varepsilon$ for all y satisfying $|y - x| < \delta$ where $\delta > 0$ is the minimum of $x - a_n$ and $b_n - x$. Thus f is continuous at x.

(e) f must be continuous at some point in every interval contained in $[a, b]$, since f is integrable on every such interval.

31. **(a)** Choose x_0 in $[a, b]$ and let $f(x) = 0$ for all $x \neq x_0$, and $f(x_0) = 1$. (The function in Problem 34 is another example.)

(b) There is a partition P of $[a, b]$ such that $f(x) > x_0/2$ for all x in some $[t_{i-1}, t_i]$. Then $L(f, P) \geq x_0(t_i - t_{i-1})/2$.

(c) This follows from part (b), since f is continuous at some x_0, by Problem 20.

32. **(a)** Choose $g = f$. Then $\int_a^b f^2 = 0$. Since f is continuous, this implies that $f = 0$.

(b) If $f(x_0) > 0$, then $f(x) > 0$ for all x in $(x_0 - d, x_0 + \delta)$ for some $\delta > 0$. Choose a continuous g with $g > 0$ on $(x_0 - \delta, x_0 + \delta)$ and $g = 0$ elsewhere. Then $\int_a^b fg > 0$, a contradiction.

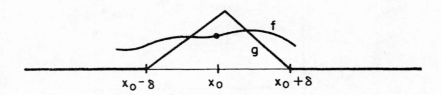

34. Let $\varepsilon > 0$. Choose n so that $1/n < \varepsilon/2$. Let $x_0 < x_1 < \cdots < x_m$ be those rational points p/q in $[0, 1]$ with $q < n$. Choose a partition $P = \{t_0, \dots, t_k\}$ such that the intervals $[t_{i-1}, t_i]$ which contain some x_j have total length $< \varepsilon/2$. On each of the other intervals we have $M_i \leq 1/n < \varepsilon/2$. Let I_1 denote all those i from $1, \dots, n$ for which $[t_{i-1}, t_i]$ contains some x_j, and let I_2 denote all other i from $1, \dots, n$. Since $f \leq 1$ everywhere, we have

$$U(f, P) = \sum_{i \text{ in } I_1} M_i(t_i - t_{i-1}) + \sum_{i \text{ in } I_2} M_i(t_i - t_{i-1})$$

$$\leq 1 \cdot \sum_{i \text{ in } I_1} (t_i - t_{i-1}) + \frac{\varepsilon}{2} \sum_{i \text{ in } I_2} (t_i - t_{i-1})$$

$$\leq 1 \cdot \frac{\varepsilon}{2} + \frac{\varepsilon}{2} \cdot 1 = \varepsilon.$$

35. Let f be the function in Problem 34, and let $g(x) = 0$ for $x = 0$, and $g(x) = 1$ for $x \neq 0$. Then $(g \circ f)(x) = 0$ if x is irrational, and 1 if x is rational.

36. (a) If $f \geq 0$ on $[t_{i-1}, t_i]$, then $M_i' = M_i$ and $m_i' = m_i$. If $f \leq 0$ on $[t_{i-1}, t_i]$, then $M_i' = -m_i$ and $m_i' = -M_i$, so again $M_i' - m_i' = M_i - m_i$. Now suppose that f has both positive and negative values on $[t_{i-1}, t_i]$, so that $m_i < 0 < M_i$. There are two cases to consider. If $-m_i \leq M_i$, then

$$M_i' = M_i,$$

so

$$M_i' - m_i' \leq M_i' = M_i \leq M_i - m_i, \qquad \text{since } m_i < 0.$$

A similar argument works if $-m_i \geq M_i$ (or consider $-f$).

(b) If P is a partition of $[a, b]$, then

$$U(|f|, P) - L(|f|, P) = \sum_{i-1}^{n}(M_i' - m_i')(t_i - t_{i-1})$$

$$\leq \sum_{i=1}^{n}(M_i - m_i)(t_i - t_{i-1})$$

$$= U(f, P) - L(f, P).$$

So integrability of f implies integrability of $|f|$, by Theorem 2.

(c) This follows from part (b) and the formulas

$$\max(f, g) = \frac{f + g + |f - g|}{2}, \qquad \min(f, g) = \frac{f + g - |f - g|}{2}.$$

(d) If f is integrable, then $\max(f, 0)$ and $\min(f, 0)$ are integrable, by part (d). Conversely, if $\max(f, 0)$ and $\min(f, 0)$ are integrable, then $f = \max(f, 0) + \min(f, 0)$ is integrable, by Theorem 5.

38. (a) Since

$$0 \leq m_i' \leq f(x) \leq M_i' \quad \text{and} \quad 0 \leq m_i'' \leq g(x) \leq M_i'' \qquad \text{for all } x \text{ in } [t_{i-1}, t_i],$$

we have

$$m_i'm_i'' \leq f(x)g(x) \leq M_i'M_i'' \qquad \text{for all } x \text{ in } [t_{i-1}, t_i],$$

which implies that $m_i'm_i'' \leq m_i$ and $M_i \leq M_i'M_i''$.

(b) This follows immediately from part (a).

(c) By part (b),

$$U(fg, P) - L(fg, P)$$

$$\leq \sum_{i=1}^{n}[M_i'M_i'' - m_i'm_i''](t_i - t_{i-1})$$

$$= \sum_{i=1}^{n}M_i''[M_i' - m_i'](t_i - t_{i-1}) + \sum_{i=1}^{n}m_i'[M_i'' - m_i''](t_i - t_{i-1})$$

$$\leq M\left\{\sum_{i=1}^{n}[M_i' - m_i'](t_i - t_{i-1}) + \sum_{i=1}^{n}[M_i'' - m_i''](t_i - t_{i-1})\right\}.$$

(d) Integrability of fg follows immediately from part (c) and Theorem 2.

(e) The same result clearly holds if $f \leq 0$ and/or $g \leq 0$ on $[a, b]$. Now write $f = \max(f, 0) + \min(f, 0)$ and $g = \max(g, 0) + \min(g, 0)$, so that fg is the sum of four products, each of which is integrable.

39. (a) Given $x_1, \ldots, x_n$ and $y_1, \ldots, y_n$, let f and g be defined on $[0, 1]$ by

$$f(x) = \begin{cases} x_i, & \dfrac{i-1}{n} \leq x < \dfrac{i}{n} \\ 0, & x = 1, \end{cases}$$

$$g(x) = \begin{cases} y_i, & \dfrac{i-1}{n} \leq x < \dfrac{i}{n} \\ 0, & x = 1. \end{cases}$$

Then

$$\int_0^1 fg = \frac{1}{n^2} \sum_{i=1}^n x_i y_i,$$

$$\int_0^1 f^2 = \frac{1}{n^2} \sum_{i=1}^n x_i^2,$$

$$\int_0^1 g^2 = \frac{1}{n^2} \sum_{i=1}^n y_i^2,$$

so

$$\left(\sum_{i=1}^n x_i y_i \right)^2 \leq \left(\sum_{i=1}^n x_i^2 \right) \left(\sum_{i=1}^n y_i^2 \right);$$

this is the Schwarz inequality.

(b) *First proof*: If $g = 0$, then equality holds. Otherwise for all λ we have

$$0 \leq \int_a^b (f - \lambda g)^2 = \int_a^b f^2 - 2\lambda \int_a^b fg + \lambda^2 \int_a^b g^2,$$

so

$$\int_a^b f^2 - \frac{4 \left(\displaystyle\int_a^b fg \right)^2}{4 \left(\displaystyle\int_a^b g^2 \right)} \geq 0.$$

Second proof: Using $2xy \leq x^2 + y^2$ with

$$x = \frac{f(x)}{\sqrt{\displaystyle\int_a^b f^2}}, \qquad y = \frac{g(x)}{\sqrt{\displaystyle\int_a^b g^2}}$$

we obtain

$$\frac{2f(x)g(x)}{\sqrt{\left(\displaystyle\int_a^b f^2 \right) \left(\displaystyle\int_a^b g^2 \right)}} \leq \frac{f(x)^2}{\displaystyle\int_a^b f^2} + \frac{g(x)^2}{\displaystyle\int_a^b g^2}.$$

So

$$\frac{2\int_a^b f(x)g(x)\,dx}{\sqrt{\left(\int_a^b f^2\right)\left(\int_a^b g^2\right)}} \leq \frac{\int_a^b f(x)^2\,dx}{\int_a^b f^2} + \frac{\int_a^b g(x)^2\,dx}{\int_a^b g^2} = 2.$$

Third proof: The analogue of the formula in the solution to Problem 2-21(c) is:

$$\left(\int_a^b f^2\right)\left(\int_a^b g^2\right) = \left(\int_a^b fg\right)^2 + \frac{1}{2}\int_a^b\left\{\int_a^b [f(x)g(y) - f(y)g(x)]^2\,dx\right\}dy.$$

To check this equality we simply compute that

$$\int_a^b\left\{\int_a^b [f(x)^2 g(y)^2 + f(y)^2 g(x)^2 - 2f(x)g(x)f(y)g(y)]\,dx\right\}dy$$

$$= \int_a^b\left\{g(y)^2\int_a^b f^2 + f(y)^2\int_a^b g^2 - 2f(y)g(y)\int_a^b fg\right\}dy$$

$$= \left(\int_a^b f^2\right)\left(\int_a^b g^2\right) + \left(\int_a^b f^2\right)\left(\int_a^b g^2\right)$$

$$- 2\left(\int_a^b fg\right)\left(\int_a^b fg\right).$$

(c) If $f = g$ except at one point, then equality holds, even though $f = \lambda g$ is false. But if f and g are continuous, then equality in the Cauchy-Schwarz inequality does imply that $f = \lambda g$ for $g \neq 0$. This follows from all of the above proofs: In the first proof, we will have

$$0 < \int_a^b (f - \lambda g)^2$$

since $(f - \lambda g)^2$ is a continuous non-negative function that is somewhere positive.

Similarly, in the second proof, we have equality only if we have

$$\frac{f(x)}{\sqrt{\int_a^b f^2}} = \frac{g(x)}{\sqrt{\int_a^b g^2}} \qquad \text{for all } x,$$

so we can choose

$$\lambda = \sqrt{\int_a^b g^2}\Bigg/\sqrt{\int_a^b f^2}.$$

In the third proof, equality implies that

$$\int_a^b\left\{\int_a^b [f(x)g(y) - f(y)g(x)]^2\,dx\right\}dy = 0.$$

This means that for all y,

$$\int_a^b [f(x)g(y) - f(y)g(x)]^2 \, dx = 0,$$

which means that for all x,

$$f(x)g(y) = f(y)g(x).$$

So if $g(y_0) \neq 0$, then

$$f(x) = \frac{f(y_0)}{g(y_0)} g(x) \qquad \text{for all } x.$$

(d) Apply the Cauchy-Schwarz inequality to f and $g(x) = 1$ on $[0, 1]$. The correct result for $[a, b]$ is

$$\left(\int_a^b f \right)^2 \leq (b - a) \left(\int_a^b f^2 \right).$$

40. (a) If $\varepsilon > 0$, pick $N \geq 0$ so that $|f(t) - a| < \varepsilon$ for $t \geq N$. Then for $N \geq 0$ we have

$$\left| \int_N^{N+M} f(t) \, dt - Ma \right| < \varepsilon M,$$

so

$$\left| \frac{N}{N + M} \int_N^{N+M} f(t) \, dt - \frac{Ma}{N + M} \right| < \frac{\varepsilon M}{N + M} < \varepsilon.$$

Choose M so that

$$\left| \frac{Ma}{N + M} - a \right| < \varepsilon \qquad \text{and} \qquad \left| \frac{1}{N + M} \int_1^N f(t) \, dt \right| < \varepsilon.$$

Then

$$\left| \frac{1}{N + M} \int_1^{N+M} f(t) \, dt - a \right| < 3\varepsilon.$$

1. Let $|f(x)| \leq M$ for x in $[a, b]$, and choose $\delta > 0$ so that $|g(x) - g(y)| < \varepsilon/M(b-a)$ for $|x - y| < \delta$. If all $t_i - t_{i-1} < \delta$, then

$$\left| \sum_{i=1}^{n} f(x_i)g(x_i)(t_i - t_{i-1}) - \sum_{i=1}^{n} f(x_i)g(u_i)(t_i - t_{i-1}) \right|$$

$$= \left| \sum_{i=1}^{n} f(x_i)[g(x_i) - g(u_i)](t_i - t_{i-1}) \right|$$

$$\leq M \cdot \frac{\varepsilon}{M(b-a)} \sum_{i=1}^{n} (t_i - t_{i-1})$$

$$= \varepsilon.$$

So by making $t_i - t_{i-1}$ small enough we can make $\sum_{i=1}^{n} f(x_i)g(u_i)(t_i - t_{i-1})$ as close

to $\sum_{i=1}^{n} f(x_i)g(x_i)(t_i - t_{i-1})$ as we like, and hence as close to $\int_a^b fg$ as we like.

2. Let $f(x) + g(x) \leq M$ on $[a, b]$, and choose $\delta > 0$ so that $\sqrt{x} - \sqrt{y} < \varepsilon/(b-a)$ for x, y in $[0, M]$ with $|x - y| < \delta$. Then choose $\delta' > 0$ so that $|g(x_i) - g(u_i)| < \delta$ for $|x_i - u_i| < \delta'$. If all $t_i - t_{i-1} < \delta'$, then $|x_i - u_i| < \delta'$, so

$$\left| [f(x_i) + g(u_i)] - [f(x_i) + g(x_i)] \right| = |g(u_i) - g(x_i)| < \delta,$$

hence

$$\left| \sqrt{f(x_i) + g(u_i)} - \sqrt{f(x_i) + g(x_i)} \right| < \frac{\varepsilon}{b - a}$$

and consequently

$$\left| \sum_{i=1}^{n} \sqrt{f(x_i) + g(u_i)}(t_i - t_{i-1}) - \sum_{i=1}^{n} \sqrt{f(x_i) + g(x_i)}(t_i - t_{i-1}) \right|$$

$$= \left| \sum_{i=1}^{n} \left[\sqrt{f(x_i) + g(u_i)} - \sqrt{f(x_i) + g(x_i)} \right] (t_i - t_{i-1}) \right|$$

$$< \frac{\varepsilon}{b - a} \sum_{i=1}^{n} (t_i - t_{i-1}) = \varepsilon.$$

3. By the Mean Value Theorem we have

$$\ell(c, P) = \sum_{i=1}^{n} \sqrt{[u'(x_i)]^2 + [v'(u_i)]^2} \, (t_i - t_{i-1})$$

for some x_i, u_i in $[t_{i-1}, t_i]$. By Problem 2, these can be made as close as we like to $I = \int_a^b \sqrt{u'^2 + v'^2}$ by choosing $t_i - t_{i-1}$ small enough. This means, first of all, that I must be an upper bound for all $\ell(c, P)$: for if some $\ell(c, P) > I$, refining the partition P would only increase ℓ, and hence never make it close to I. Since I is an upper bound and we can make $\ell(c, P)$ as close as we like to I, it follows that I must be the least upper bound.

4. The graph of f is given parametrically by

$$u(\theta) = f(\theta) \cos \theta, \qquad v(\theta) = f(\theta) \sin \theta.$$

So its length is

$$\int_{\theta_0}^{\theta_1} \sqrt{u'^2 + v'^2} = \int_{\theta_0}^{\theta_1} \sqrt{[f' \cos - f \sin]^2 + [f' \sin + f \cos]^2}$$

$$= \int_{\theta_0}^{\theta_1} \sqrt{f^2 + f'^2}.$$

5. Let $\{t_0, \ldots, t_n\}$ be a partition of $[a, b]$, and choose x_i in $[t_{i-1}, t_i]$. Then the Schwarz inequality shows that

$$\sum_{i=1}^n (fg)(x_i)(t_i - t_{i-1}) = \sum_{i=1}^n f(x_i)\sqrt{t_i - t_{i-1}} \, g(x_i)\sqrt{t_i - t_{i-1}}$$

$$\leq \left(\sum_{i=1}^n f(x_i)^2 (t_i - t_{i-1}) \right) \cdot \left(\sum_{i=1}^n g(x_i)^2 (t_i - t_{i-1}) \right).$$

But the left-hand side can be made as close as desired to $\int_a^b fg$ by making $t_i - t_{i-1}$ small enough, while the two factors on the right side can be made as close as desired to $\int_a^b f^2$ and $\int_a^b g^2$. Hence we must have

$$\int_a^b fg \leq \left(\int_a^b f^2 \right) \left(\int_a^b g^2 \right).$$

CHAPTER 14

1. (ii)

$$\frac{1}{1+\sin^6\left(\int_1^x \sin^3 t\,dt\right)+\left(\int_1^x \sin^3 t\,dt\right)^2}\cdot \sin^3 x.$$

(iv)

$$\frac{-1}{1+x^2+\sin^2 x}.$$

(vi)

$$\cos\left(\int_0^x \sin\left(\int_0^y \sin^3 t\,dt\right)dy\right)\cdot \sin\left(\int_0^x \sin^3 t\,dt\right).$$

(viii)

$$(F^{-1})'(x) = \frac{1}{F'(F^{-1}(x))} = \frac{1}{\dfrac{1}{\sqrt{1-[F^{-1}(x)]^2}}} = \sqrt{1-[F^{-1}(x)]^2}.$$

2. (ii) All $x \neq 1$.

(iv) All irrational x.

(vi), (viii) All x not of the form $1/n$ for some natural number n.

3. (a) Since f is differentiable at c it is continuous at c, so F is differentiable at c.

(b) If we assume that f is continuous in an interval around c, then F' will be continuous at c, since we will have $F'(x) = f(x)$ in this interval, and differentiability of f at c implies continuity of f at c. But without this assumption F' may not even exist at all points near c. For example, f could be the function shown below.

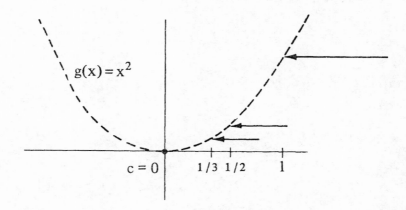

(c) Since f' is continuous at c, $f'(x)$ must exist for all x in an interval around c, so f is continuous in an interval around c. So, as in part (b), F' is continuous at c.

4. (i) If we let $F(x)$ be this expression, then

$$F'(x) = \frac{1}{1+x^2} + \frac{-1}{x^2} \cdot \frac{1}{1 + \left(\dfrac{1}{x}\right)^2} = 0.$$

(ii) In this case

$$F'(x) = \cos x \cdot \frac{1}{\sqrt{1 - \sin^2 x}}$$

$$- \sin x \frac{1}{\sqrt{1 - \cos^2 x}}$$

(a minus sign because the derivative of $-\cos$ is $\sin$, but it appears as a lower limit)

$$= 1 - 1 = 0.$$

(The meaning of these facts will become clear in the next chapter).

5. (ii)

$$(f^{-1})'(0) = \frac{1}{f'(f^{-1}(0))} = \frac{1}{\cos(\cos(f^{-1}(0)))}$$

$$= \frac{1}{\cos(\cos(1))}.$$

6. (i) Differentiating the equation $\int_0^x t g(t)\, dt = x + x^2$, we find that at points where g is continuous it must satisfy

$$x g(x) = 1 + 2x.$$

Now if we simply define

$$g(t) = \begin{cases} \dfrac{1}{t} + 2, & t \neq 0 \\ 0, & t = 0, \end{cases}$$

then $t g(t) = 1 + 2t$ for all $t \neq 0$, so $\int_0^x t g(t)\, dt = \int_0^x 1 + 2t\, dt = x + x^2$.

(ii) We must have

$$x^2 g(x^2) \cdot 2x = 1 + 2x.$$

Let

$$g(t) = \begin{cases} \frac{1}{2} t^{-\frac{3}{2}} + t^{-1}, & t > 0 \\ 0, & t \leq 0. \end{cases}$$

Then

$$t g(t) = \frac{1}{2} t^{-\frac{1}{2}} + 1 \qquad \text{for } t > 0,$$

so for all x,

$$\int_0^{x^2} tg(t)\,dt = \int_0^{x^2} \frac{1}{2}t^{-\frac{1}{2}} + 1\,dt$$

$$= (x^2)^{\frac{1}{2}} + x^2 = x + x^2.$$

7. Clearly f^2 is differentiable everywhere (its derivative at x is $f(x)$). So f is differentiable at x whenever $f(x) \neq 0$, and

$$f(x) = 2f(x)f'(x),$$

so $f'(x) = 0$ at such points. Thus, f is constant on any interval where it is non-zero. Since f is continuous, it must be constant (proof left to the reader). So if $f(x) = K$ for all x, then

$$\int_0^x K = K^2 + C$$

so for all x we have

$$Kx = K^2 + C.$$

This is possible only if $K = 0$, which is possible only if $C = 0$.

8. Since the two sides of the desired inequality are equal for $x = 0$, we just need to prove the same inequality for their derivatives, i.e.,

$$f(x)^3 \leq 2f(x)\int_0^x f.$$

We have $f(x) > 0$ for $x > 0$, since $f(0) = 0$ and $0 < f'$, so this inequality is equivalent to

$$f(x)^2 \leq 2\int_0^x f.$$

But both sides of *this* inequality are true for $x = 0$, so we just need to prove the inequality for *their* derivatives:

$$2f(x)f'(x) \leq 2f(x).$$

This is true since $f(x) > 0$ and $0 < f'(x) \leq 1$.

9. If

$$g(x) = \begin{cases} x^2 \sin\dfrac{1}{x}, & x \neq 0 \\ 0, & x = 0, \end{cases}$$

then

$$g'(x) = \begin{cases} 2x\sin\dfrac{1}{x} = \cos\dfrac{1}{x}, & x \neq 0 \\ 0, & x = 0. \end{cases}$$

So if we define

$$h = \begin{cases} 2x \sin \dfrac{1}{x}, & x \neq 0 \\[2mm] 0, & x = 0 \end{cases}$$

we have

$$f(x) = h(x) - g'(x) \qquad \text{for all } x.$$

Hence

$$F(x) = \int_0^x (h - g')$$

$$= \left(\int_0^x h \right) - g,$$

using the *Second* Fundamental Theorem of Calculus (and not merely the Corollary of the First Fundamental Theorem). Since h is continuous we can then apply the First Fundamental Theorem to conclude that

$$F'(0) = h(0) - g'(0)$$

$$= 0.$$

10. (i) In Problem 13-23(c), choose

$$f(x) = \frac{1}{\sqrt{1+x^2}}, \qquad g(x) = x^6.$$

Then

$$\int_0^1 \frac{x^6}{\sqrt{1+x^2}}\, dx = \frac{1}{\sqrt{1+\xi^2}} \int_0^1 x^6\, dx = \frac{1}{7\sqrt{1+\xi^2}},$$

where $0 \leq \xi \leq 1$, and hence

$$\frac{1}{7\sqrt{2}} \leq \frac{1}{\sqrt{1+\xi^2}} \leq \frac{1}{7}.$$

(ii) Write this integral as

$$\int_0^{1/2} \frac{\sqrt{1-x}}{\sqrt{1-x}} \sqrt{\frac{1-x}{1+x}} = \int_0^{1/2} \frac{1-x}{\sqrt{1-x^2}}\, dx,$$

and choose

$$f(x) = \frac{1}{\sqrt{1-x^2}}, \qquad g(x) = 1 - x.$$

Then

$$\int_0^{1/2} \frac{1-x}{\sqrt{1-x^2}}\, dx = \frac{1}{\sqrt{1-\xi^2}} \int_0^{1/2} 1 - x\, dx = \frac{1}{\sqrt{1-\xi^2}} \cdot \frac{3}{8},$$

where $0 \le \xi \le 1/2$, and hence

$$1 \le \frac{1}{\sqrt{1 - \xi^2}} \le \frac{2}{\sqrt{3}}.$$

12. If

$$F(x) = \int_0^x f(u)(x - u)\, du = \int_0^x xf(u)\, du - \int_0^x uf(u)\, du,$$

then

$$F'(x) = \left[xf(x) + \int_0^x f(u)\, du \right] - xf(x) \qquad \text{by Problem 11}$$

$$= \int_0^x f(u)\, du.$$

Consequently, there is some number c such that

$$\int_0^x f(u)(x - u)\, du = \int_0^x \left(\int_0^u f(t)\, dt \right) du + c \qquad \text{for all } x.$$

Clearly $c = 0$, since each of the other two terms is 0 for $x = 0$.

13. Applying Problem 12 to $g(u) = f(u)(x - u)$, we obtain

$$\int_0^x f(u)(x - u)^2\, du = \int_0^x [f(u)(x - u)](x - u)\, du$$

$$= \int_0^x \left(\int_0^u f(t)(x - t)\, dt \right) du.$$

Therefore we must show that

$$\int_0^x \left(\int_0^u f(t)(x - t)\, dt \right) du = 2 \int_0^x \left(\int_0^{u_2} \left(\int_0^{u_1} f(t)\, dt \right) du_1 \right) du_2.$$

Now $x - t = (u - t) + (x - u)$, so

(1) $$\int_0^u f(t)(x - t)\, dt = \int_0^u f(t)(u - t)\, dt + \int_0^u f(t)(x - u)\, dt.$$

For the first integral on the right we have

(2) $$\int_0^u f(t)(u - t)\, dt = \int_0^u \left(\int_0^{u_1} f(t)\, dt \right) du_1$$

by Problem 12. The second can be written

(3) $$\int_0^u f(t)(x - u)\, dt = (x - u) \int_0^u f(t)\, dt.$$

From (1), (2), and (3) we have

$$\int_0^x \left(\int_0^u f(t)(x-t)\,dt \right) du = \int_0^x \left(\int_0^u \left(\int_0^{u_1} f(t)\,dt \right) du_1 \right) du$$

$$+ \int_0^x \left[(x-u) \int_0^u f(t)\,dt \right] du.$$

On the other hand, applying Problem 12 to $g(u) = \int_0^u f(t)\,dt$ we obtain

$$\int_0^x \left[(x-u) \int_0^u f(t)\,dt \right] du = \int_0^x \left(\int_0^u \left(\int_0^{u_1} f(t)\,dt \right) du_1 \right) du.$$

15. (a) This follows from Problem 13-14, since $f(x-a) = f(x)$ for all x.

(b) Let g be periodic and continuous with $g \geq 0$ (for example, $g(x) = \sin^2 x$). If $f(x) = \int_0^x g$, then $f' = g$ is periodic, but f is increasing, so it is not periodic.

(c) Let $g(x) = f(x+a)$. Then $g'(x) = f'(x+a) = f'(x)$. If $f(a) = f(0)$, then we also have $g(0) = f(a) = f(0)$. Consequently $g = f$, i.e., $f(x+a) = f(x)$ for all x.

Conversely, suppose that f is periodic (with some period not necessarily $= a$). Let $g(x) = f(x+a) - f(x)$. Then $g'(x) = f'(x+a) - f'(x) = 0$, so g has the constant value $g(0) = f(a) - f(0)$. I.e.,

$$f(x+a) = f(x) + f(a) - f(0).$$

It follows that

$$f(na) = nf(a) - (n-1)f(0)$$
$$= \frac{n}{n+1}[(n-1)f(a) - f(0)].$$

Now if $f(a) \neq f(0)$, then this would be unbounded. But f is bounded since it is periodic.

17. Let $F = \int_a^x f$. Then Problem 13-21 states that

$$\int_a^x f^{-1} = xf^{-1}(x) - af^{-1}(a) - \int_{f^{-1}(a)}^{f^{-1}(x)} f$$
$$= xf^{-1}(x) - af^{-1}(a) - F(f^{-1}(x)) + F(f^{-1}(a)).$$

So if $G(x) = xf^{-1}(x) - F(f^{-1}(x))$, then $G'(x) = f^{-1}(x)$.

18. (a) For each point $(x, 2x^2) = (x, f(x))$ on C, we have

$$\text{area } A = \int_0^x 2t^2 - t^2\,dt = \frac{x^3}{3}.$$

It is simplest to consider C_2 as the graph of g^{-1}, for then

$$\text{area } B = \int_0^{2x^2} f^{-1} - g^{-1}.$$

Clearly (compare Problem 13-21)

$$\int_0^{2x^2} f^{-1} = x \cdot 2x^2 - \int_0^x f(t)\, dt$$

$$= 2x^3 - \int_0^x 2t^2\, dt$$

$$= \frac{4}{3}x^3.$$

So

$$\text{area } B = \frac{4}{3}x^3 - \int_0^{2x^2} g^{-1}.$$

Thus we require that for all $x \geq 0$,

$$\frac{x^3}{3} = \frac{4}{3}x^3 - \int_0^{2x^2} g^{-1},$$

and hence that

$$3x^2 = g^{-1}(2x^2) \cdot 4x$$

$$g^{-1}(2x^2) = \frac{3x}{4},$$

and thus

$$g^{-1}(y) = \frac{3\sqrt{2y}}{8}.$$

Finally,

$$g(x) = \frac{32}{9}x^2.$$

(b) Now for $f(x) = cx^m$ we have

$$\text{area } A = \int_0^x cx^m - t^m\, dt = \frac{c-1}{m+1}x^{m+1},$$

and

$$\text{area } B = \int_0^{cx^m} f^{-1} - g^{-1}$$

$$= x \cdot cx^m - \int_0^x ct^m\, dt - \int_0^{cx^m} g^{-1}$$

$$= c\frac{m}{m+1}x^{m+1} - \int_0^{cx^m} g^{-1}.$$

So we require that

$$\frac{c-1}{m+1}x^{m+1} = c\frac{m}{m+1}x^{m+1} - \int_0^{cx^m} g^{-1}$$

and thus

$$(c-1)x^m = cmx^m - g^{-1}(cx^m) \cdot cmx^{m-1}$$

$$g^{-1}(xc^m) = \frac{(cm-c+1)x^m}{cmx^{m-1}} = \frac{cm-c+1}{cm}x.$$

so

$$g^{-1}(y) = \frac{cm-c+1}{cm}\left(\frac{y}{c}\right)^{1/m}.$$

Finally,

$$g(x) = c\left(\frac{cm-c+1}{cm}\right)^{-m}x^m.$$

19. (a) $F'(x) = 1/x$; $G'(x) = (1/bx) \cdot b = 1/x$.

(b) It follows from part (a) that there is some c such that $F(x) = G(x) + c$ for all $x > 0$. Since $F(1) = 0 = G(1)$, we have $F(x) = G(x)$ for all $x > 0$.

20. Suppose f is continuous on $[a, b]$ and $f(a) < 0 < f(b)$. The Fundamental Theorem of Calculus shows that $f = F'$ for some F (namely $F(x) = \int_0^x f$). Darboux's Theorem then implies that $f(x) = 0$ for some x in $[a, b]$.

21. We have

$$F(x) = \int_{f(x)}^a h(t)\,dt + \int_a^{g(x)} h(t)\,dt,$$

so

$$F'(x) = -h(f(x)) \cdot f'(x) + h(g(x)) \cdot g'(x).$$

22. Applying the Cauchy-Schwarz inequality to f' and 1 on $[0, 1]$ we have

$$\left(\int_0^1 f' \cdot 1\right)^2 \le \int_0^1 (f')^2 \cdot \int_0^1 1;$$

since $f(0) = 0$ this gives

$$f(1)^2 \le \int_0^1 (f')^2.$$

To show that the hypotheses $f(0) = 0$ is needed just take $f(x) = 1$ for all x.

23. (a) Equation $(*)$ just says that $(G \circ y)' = F'$ in the interval, so there is a c such that $G \circ y = F + c$ in this interval, i.e., $G(y(x)) = F(x) + c$ for all x in the interval.

(b) Conversely, if y satisfies (∗∗), then differentiation yields (∗).

(c) If

$$y'(x) = \frac{1+x^2}{1+y(x)},$$

so that

$$[1 + y(x)]y'(x) = 1 + x^2,$$

then there is some c such that

$$y(x) + \frac{y^2(x)}{2} = x + \frac{x^3}{3} + c$$

for all x in the interval on which y is defined. So

$$y^2(x) + 2y(x) - 2x - \frac{2}{3}x^3 - c = 0, \qquad \text{(calling } 2c \text{ simply } c\text{)}$$

so

$$y(x) = \frac{-2 + \sqrt{4 + 4(x + \frac{2}{3}x^2 + c)}}{2}$$

or

$$y(x) = -1 - \sqrt{1 + x + \frac{2}{3}x^2 + c}.$$

These solutions are never defined on all of **R**, since $1 + x + 2x^3/3 + c < 0$ for $x < 0$ with $|x|$ sufficiently large.

(d) If

$$(1 + 5[y(x)]^4)y'(x) = -1,$$

then there is some constant c such that

$$[y(x)]^5 + y(x) + x = c.$$

(e) If $y(x)y'(x) = -x$, then there is some c such that

$$\frac{[y(x)]^2}{2} = \frac{-x^2 + c}{2},$$

so

$$y(x) = \sqrt{c - x^2}$$

or

$$y(x) = -\sqrt{c - x^2}.$$

If $y(0) = -1$, then clearly $y(x) = -\sqrt{1 - x^2}$ (for $|x| < 1$).

24. (a) If the Schwarzian derivative is 0 then

$$2f'^2 f''' - 3f' f''^2 = 0.$$

But then

$$\left(\frac{f''^2}{f'^3}\right)' = \frac{f'^3 \cdot 2f''f''' - f''^2 \cdot 3f'^2 f''}{f'^6}$$

$$= \frac{f'f''[2f'^2 f''' - 3f'f''^2]}{f'^6}$$

$$= 0,$$

so f''^2/f'^3 is constant.

(b) Hence $u = f'$ satisfies

$$u^{-3/2} \cdot u' = C \qquad \text{for some } C.$$

By Problem 23,

$$2u^{-1/2} = Cx + d \qquad \text{for some } D$$

so

$$f'(x) = u(x) = \frac{4}{(Cx + D)^2}.$$

This implies that

$$f(x) = \frac{-4}{Cx + D} + E \qquad \text{for some } E,$$

which is the desired form.

25. (a)

$$\int_1^\infty x^r \, dx = \lim_{N \to \infty} \frac{N^{r+1}}{r+1} - \frac{1}{r+1} = \frac{-1}{r+1}$$

(because $r + 1 < 0$ so $\lim_{N \to \infty} N^{r+1} = 0$).

(b) Problem 13-15 implies that

$$\int_1^{2^n} \frac{1}{x} \, dx = \underbrace{\int_1^2 \frac{1}{x} \, dx + \cdots + \int_1^2 \frac{1}{x} \, dx}_{n \text{ times}}$$

$$= n \int_1^2 \frac{1}{x} \, dx.$$

Since $\int_1^2 1/x \, dx > 0$, we have $\lim_{n \to \infty} \int_1^{2^n} 1/x \, dx = \infty$.

(c) The function $I(N) = \int_0^N g$ is clearly increasing, and it is bounded above by $\int_0^\infty f$. Consequently, $\lim_{N \to \infty} I(N)$ exists (it is the least upper bound of $\{I(N) : N \geq 0\}$).

(d) Clearly $\int_0^\infty 1/(1+x^2)\,dx$ exists if $\int_1^\infty 1/(1+x^2)\,dx$ exists; the latter integral exists by part (c), because $\int_1^\infty 1/x^2\,dx$ exists, by part (a), and we have $1/(1+x^2) \le 1/x^2$.

26. (i) Since

$$\frac{1}{\sqrt{1+x^3}} \le \frac{1}{x^{3/2}}$$

and

$$\int_1^\infty \frac{dx\cdot}{x^{3/2}} \quad \text{exists,}$$

the integral $\int_0^\infty 1/\sqrt{1+x^3}\,dx$ also exists.

(ii) For $x \ge 1$ we have

$$1 + x^{3/2} \le 2x^{3/2}$$

so

$$\frac{x}{1+x^{3/2}} \ge \frac{1}{2}\cdot\frac{x}{x^{3/2}} = \frac{1}{2}x^{1/2}.$$

Since $\int_1^\infty x^{1/2}\,dx$ does not exist, neither does $\int_0^\infty x/(1+x^{3/2})\,dx$.

(iii) For large x the integrand looks like $1/x\sqrt{x} = x^{-3/2}$, which causes no problem, but for $x \le 1$ we have $\sqrt{1+x} \le 2$, so

$$\frac{1}{x\sqrt{1+x}} \ge \frac{1}{2x}$$

and $\int_0^\infty dx/x$ does not exist (this is really an integral of the sort considered in Problem 28).

27. (a) Clearly $\int_{-\infty}^0 1/(1+x^2)\,dx$ exists; in fact, it equals $\int_0^\infty 1/(1+x^2)\,dx$.

(b) $\int_0^\infty x\,dx$ does not exist.

(c) If $\lim\limits_{N\to\infty} h(N) = \infty$ and $\lim\limits_{N\to-\infty} g(N) = -\infty$ and $\int_{-\infty}^\infty f$ exists, then

$$\lim_{N\to\infty}\int_{g(N)}^{h(N)} f = \int_{-\infty}^\infty f.$$

Proof. Given $\varepsilon > 0$ choose M_0 so that

$$\left|\int_0^\infty f - \int_0^M f\right| < \frac{\varepsilon}{2} \quad \text{and} \quad \left|\int_{-\infty}^0 f - \int_{-M}^0 f\right| < \frac{\varepsilon}{2} \qquad \text{for all } M > M_0.$$

Now choose N so that $h(N) > M$ and $g(N) < -M$ for all $N > N_0$. Then for $N > N_0$ we have

$$\left|\int_{-\infty}^\infty f - \int_{g(N)}^{h(N)} f\right| \le \left|\int_0^\infty f - \int_0^{h(N)} f\right| + \left|\int_{-\infty}^0 f - \int_{g(N)}^0 f\right| < \frac{\varepsilon}{2} + \frac{\varepsilon}{2} = \varepsilon.$$

28. (a)

$$\lim_{\varepsilon \to 0^+} \int_\varepsilon^a \frac{1}{\sqrt{x}}\, dx = \lim_{\varepsilon \to 0^+} 2\sqrt{a} - 2\sqrt{\varepsilon} = 2\sqrt{a}.$$

(b)

$$\int_0^a x^r\, dx = \lim_{\varepsilon \to 0^+} \int_\varepsilon^a x^r\, dx = \lim_{\varepsilon \to 0^+} \frac{a^{r+1}}{r+1} - \frac{\varepsilon^{r+1}}{r+1} = \frac{a^{r+1}}{r+1}$$

(because $r + 1 > 0$, so $\lim\limits_{\varepsilon \to 0^+} \varepsilon^{r+1} = 0$.)

(c) Problem 13-15 implies that

$$\underbrace{\int_{1/2}^1 \frac{1}{x}\, dx + \cdots + \int_{1/2}^1 \frac{1}{x}\, dx}_{n \text{ times}} = \int_{1/2^n}^1 \frac{1}{x}\, dx,$$

so

$$\int_{1/2^n}^1 \frac{1}{x}\, dx = n \int_{1/2}^1 \frac{1}{x}\, dx,$$

so $\lim\limits_{\varepsilon \to 0^+} \int_\varepsilon^1 1/x\, dx$ does not exist. Of course, this implies that $\lim\limits_{\varepsilon \to 0^+} \int_\varepsilon^a 1/x\, dx$ does not exist for any $a > 0$.

(d)

$$\int_a^0 |x|^r\, dx = \lim_{\varepsilon \to 0^-} \int_a^\varepsilon |x|^4\, dx$$

$$= -\lim_{\varepsilon \to 0^+} \int_\varepsilon^a x^r\, dx$$

$$= -\frac{a^{r+1}}{r+1}.$$

(e) Since $\lim\limits_{x \to 1} 1/\sqrt{1 - x^2} = \lim\limits_{x \to -1} 1/\sqrt{1 - x^2} = \infty$, we define

$$\int_{-1}^1 \frac{1}{\sqrt{1 - x^2}}\, dx = \int_{-1}^0 \frac{1}{\sqrt{1 - x^2}}\, dx + \int_0^1 \frac{1}{\sqrt{1 - x^2}}\, dx$$

$$= \lim_{\varepsilon \to -1^+} \int_\varepsilon^0 \frac{1}{\sqrt{1 - x^2}}\, dx + \lim_{\varepsilon \to 1^-} \int_0^\varepsilon \frac{1}{\sqrt{1 - x^2}}\, dx$$

$$= 2 \lim_{\varepsilon \to -1^+} \int_\varepsilon^0 \frac{1}{\sqrt{1 - x^2}}\, dx.$$

Now the limit

$$\lim_{\varepsilon \to -1^+} \int_\varepsilon^0 \frac{1}{\sqrt{1 + x}}\, dx = \lim_{\varepsilon \to 0^+} \int_\varepsilon^1 \frac{1}{\sqrt{x}}\, dx$$

exists by part (a). For $-1 < x < 0$ we have

$$x(1+x) < 0,$$
$$x < -x^2,$$
$$1+x < 1-x^2,$$
$$\sqrt{1+x} < \sqrt{1-x^2},$$
$$\frac{1}{\sqrt{1+x}} > \frac{1}{\sqrt{1-x^2}}.$$

It follows that

$$\lim_{\varepsilon \to -1^+} \int_\varepsilon^0 \frac{1}{\sqrt{1-x^2}}\,dx$$

also exists.

29. (a) By the version of l'Hôpital's Rule given in Problem 11-52 we have

$$\lim_{x \to 0^+} x \int_x^1 \frac{dt}{d} = \lim_{x \to 0^+} \frac{\dfrac{-1}{x}}{-\dfrac{1}{x^2}} = 0.$$

(Note that we have the necessary hypothesis $\lim\limits_{x \to 0^+} \int_x^1 dt/t = \infty$ by Problem 28(c). Actually, in the solution for Problem 28(c) we showed that

$$\int_{1/2^n}^1 \frac{1}{x}\,dx = n \int_{1/2}^1 \frac{1}{x}\,dx,$$

which implies that

$$\frac{1}{2^n} \int_{1/2^n}^1 \frac{1}{x}\,dx = \frac{n}{2^n} \int_{1/2}^1 \frac{1}{x}\,dx,$$

from which we could immediately deduce the result.)

Now if $|f| \leq M$ on $[0, 1]$, then

$$\left| \int_x^1 \frac{f(t)}{t}\,dt \right| \leq M \int_x^1 \frac{dt}{t},$$

so we still have

$$\lim_{x \to 0^+} x \int_x^1 \frac{f(t)}{t}\,dt = 0.$$

(b) For $f = 1$ we have

$$\lim_{x \to 0^+} x \int_x^1 \frac{dt}{t^2} = \lim_{x \to 0^+} x \cdot \left(\frac{1}{x} - 1 \right) = 1.$$

In general, let $l = \lim\limits_{x \to 0^+} f(x)$. Given $\varepsilon > 0$ choose $\delta > 0$ so that $|l - f(t)| < \varepsilon$ for $0 < t < \delta$. Then

$$\left| x \int_x^1 \frac{f(t) - l}{t^2} \, dt \right| \le x \int_x^\delta \frac{|f(t) - l|}{t^2} \, dt + x \left| \int_\delta^1 \frac{f(t) - l}{t^2} \, dt \right|$$

$$\le x\varepsilon \int_x^\delta \frac{dt}{t^2} + x \left| \int_\delta^1 \frac{f(t) - l}{t^2} \, dt \right|,$$

or

$$\left| x \int_x^1 \frac{f(t)}{t^2} \, dt + xl - l \right| \le \varepsilon - \frac{\varepsilon x}{\delta} + x \left| \int_\delta^1 \frac{f(t) - l}{t^2} \, dt \right|,$$

or, finally,

$$\left| x \int_x^1 \frac{f(t)}{t^2} \, dt - l \right| \le \varepsilon - \frac{\varepsilon x}{\delta} + x \left| \int_\delta^1 \frac{f(t) - l}{t^2} \, dt \right| + xl.$$

This shows that by making x small enough we can make

$$\left| x \int_x^1 \frac{f(t)}{t^2} \, dt - l \right|$$

as close to ε as we like. Since this is true for every $\varepsilon > 0$,

$$\lim_{x \to 0^+} x \int_x^1 \frac{f(t)}{t^2} = l.$$

30. (a)

$$\int_0^\infty f(x) \, dx = \lim_{\varepsilon \to 0^+} \int_\varepsilon^1 \frac{1}{\sqrt{x}} \, dx + \lim_{N \to \infty} \int_1^N \frac{1}{x^2} \, dx$$

$$= 2 + 1 = 3,$$

by Problem 28(a) and 25(a).

(b) If $-1 < r < 0$, then $\int_1^\infty x^r \, dx$ does not exist, since $x^r \ge x^{-1}$ for $x \ge 1$ and $\int_1^\infty x^{-1} \, dx$ does not exist. If $x < -1$, then $\int_0^1 x^r \, dx$ does not exist, since $x^r \ge x^{-1}$ for $0 < x \le 1$ and $\int_0^1 x^{-1} \, dx$ does not exist. (Of course, if $r > 0$, then $\int_1^\infty x^r \, dx$ does not exist.)

CHAPTER 15

1. (ii)

$$\frac{1}{\sqrt{1 - [\arctan(\arccos x)]^2}} \cdot \frac{1}{1 + (\arccos x)^2} \cdot \frac{-1}{\sqrt{1 - x^2}}.$$

(iv)

$$\frac{1}{\sqrt{1 - \left(\dfrac{1}{\sqrt{1 + x^2}}\right)^2}} \cdot \frac{-x}{(1 + x^2)^{3/2}} = \frac{1}{\sqrt{\dfrac{x^2}{1 + x^2}}} \cdot \frac{-x}{(1 + x^2)^{3/2}}$$

$$= \frac{-1}{1 + x^2}.$$

(This result is not surprising, since $f(x) = \arctan 1/x = \pi/2 - \arctan x$.)

2. (ii)

$$\lim_{x \to 0} \frac{\sin x - x + x^3/6}{x^4} = \lim_{x \to 0} \frac{\cos x - 1 + x^2/2}{4x^3} = \lim_{x \to 0} \frac{-\sin x + x}{12x^2}$$

$$= \lim_{x \to 0} \frac{-\cos x + 1}{24x} = \lim_{x \to 0} \frac{\sin x}{24} = 0.$$

(iv)

$$\lim_{x \to 0} \frac{\cos x - 1 + x^2/2}{x^4} = \lim_{x \to 0} \frac{-\sin x + x}{4x^3} = \lim_{x \to 0} \frac{-\cos x + 1}{12x^2}$$

$$= \lim_{x \to 0} \frac{\sin x}{24x} = \lim_{x \to 0} \frac{\cos x}{24} = \frac{1}{24}.$$

(vi)

$$\lim_{x \to 0} \left(\frac{1}{x} - \frac{1}{\sin x}\right) = \lim_{x \to 0} \frac{\sin x - x}{x \sin x} = \lim_{x \to 0} \frac{\cos x - 1}{\sin x + x \cos x}$$

$$= \lim_{x \to 0} \frac{-\sin x}{2\cos x - x \sin x} = 0.$$

3. (a)

$$f'(0) = \lim_{h \to 0} \frac{\dfrac{\sin h}{h} - 1}{h} = \lim_{h \to 0} \frac{\sin h - h}{h^2}$$

$$= \lim_{h \to 0} \frac{\cos h - 1}{h} = 0.$$

(b) Since

$$f'(x) = \frac{x \cos x - \sin x}{x^2} \qquad \text{for } x \neq 0,$$

we have

$$\begin{aligned}
f''(0) &= \lim_{h \to 0} \frac{h \cos h - \sin h}{h^3} \\
&= \lim_{h \to 0} \frac{\cos h - h \sin h - \cos h}{3h^2} \\
&= -\frac{1}{3}.
\end{aligned}$$

4. (a)

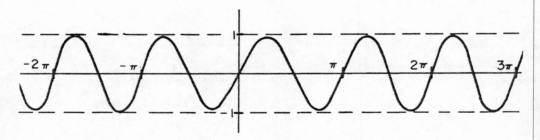

(b) Clearly $f(x) = 0$ for $x = \sqrt{k\pi}$. The numbers $\sqrt{k\pi}$ become arbitrarily large, of course (since $\sqrt{k^2\pi} > k$), but the also cluster closer and closer together, because, for example,

$$\sqrt{k+1} - \sqrt{k} = \frac{1}{2\sqrt{x}} \qquad \begin{array}{l}\text{for some } x \text{ in } (k, k+1), \text{ by the Mean}\\ \text{Value Theorem}\end{array}$$

$$< \frac{1}{2\sqrt{k}}.$$

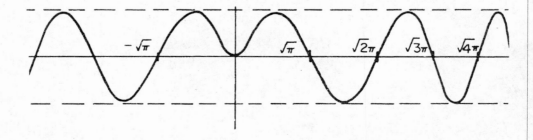

(c) If

$$0 = f'(x) = \cos x + 2\cos 2x = \cos x + 2(\cos^2 x - [1 - \cos^2 x])$$
$$= \cos x + 2(2\cos^2 x - 1)$$
$$= 4\cos^2 x + \cos x - 2,$$

then

$$\cos x = \frac{-1 \pm \sqrt{1 + 32}}{8} = \frac{-1 \pm \sqrt{33}}{8}.$$

Since $0 < \left[-1 + \sqrt{33}\right]/8 < 1$ and $-1 < \left[-1 - \sqrt{33}\right]/8 < 0$, there will be four such x in $[0, 2\pi]$:

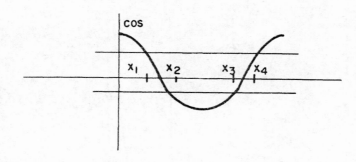

The critical points x_1 and x_4, with $\cos x_1 = \cos x_4 = \left[-1 + \sqrt{33}\right]/8$, satisfy $0 < x_1 < \pi/2$ and $3\pi/4 < x_4 < 2\pi$; so $f(x_1) > 0$ and $f(x_2) < 0$, since $\sin x$ and $\sin 2x$ are both positive on $(0, \pi/2)$ and both negative on $(3\pi/4, 1)$. To determine the sign of $f(x_3)$ and $f(x_4)$ notice that

$$f(x) = \sin x + \sin 2x$$
$$= \sin x + 2\sin x \cos x$$
$$= \sin x (1 + 2\cos x).$$

Now $\sin(x_2) > 0$, since $0 < x_2 < \pi$, but $1 + 2\cos(x_2) < 0$, so $f(x_1) < 0$. Similarly, $f(x_3) > 0$.

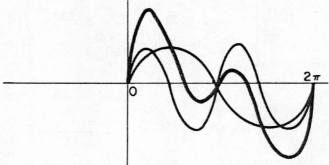

(d) $f'(x) = \sec^2 x - 1 = \tan^2 x \geq 0$ for all x, so f is always increasing. On $(-\pi/2, \pi/2)$ clearly f increases from $-\infty$ to ∞. On $(k\pi - \pi/2, k\pi + \pi/2)$ the derivative f' is the same as on $(-\pi/2, \pi2)$, so f differs by a constant from f on $(-\pi/2, \pi/2)$ The constant is clearly $-\pi$.

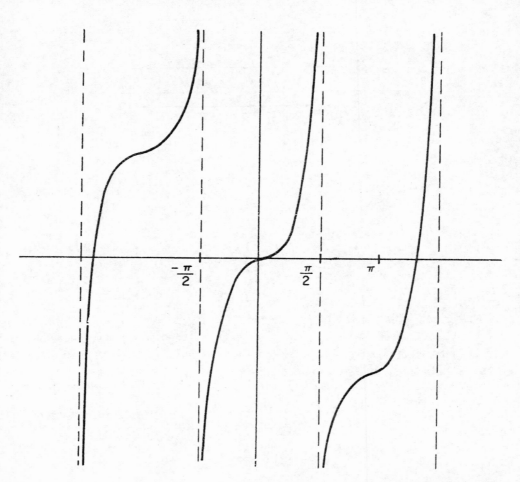

(e) $f'(x) = \cos x - 1 \le 0$ for all x, so f is decreasing. Moreover f' is periodic, so f is the same on $[2\pi, 4\pi]$ as on $[0, 2\pi]$ except moved down by 2π. Since $f''(x) = -\sin x$, it follows from the Appendix to Chapter 11 that f is concave on $[0, \pi]$ and convex on $[\pi, 2\pi]$.

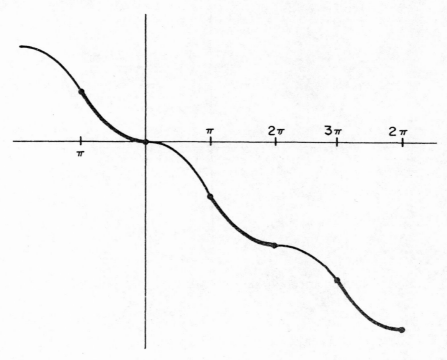

(f) If $0 = f'(x) = (x \cos x - \sin x)/x^2$, then $x = \tan x$. The graph in part (d) shows that on the right side of the vertical axis this happens for $0 < x_1 < x_2 < \cdots$, where x_n is slightly smaller than $n\pi + \pi/2$.

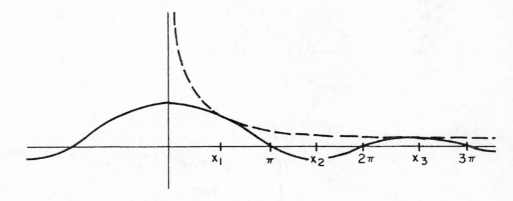

(g) $f'(x) = \sin x + x \cos x$, so $0 = f'(x)$ when $\tan x = -x$. Comparing the graphs of $\tan$ and $-I$ we see that this happens for $x = 0$ and for x sightly larger than $n\pi + \pi/2$ $(n > 0)$ or sightly smaller than $n\pi + \pi/2$ $(n < 0)$. The graph is even, and $f(x) = 0$ at multiples of π.

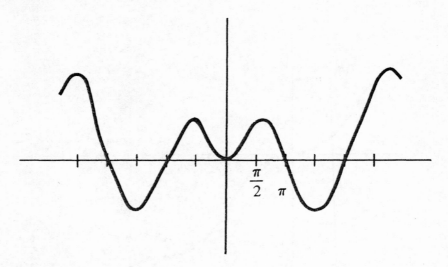

5. The point with polar coordinates $(\theta, a/\theta)$ has cartesian coordinates

$$x = \frac{a}{\theta} \cdot \cos\theta, \qquad y = \frac{a}{\theta} \cdot \sin\theta.$$

For θ close to 0, x is large, but y is close to a.

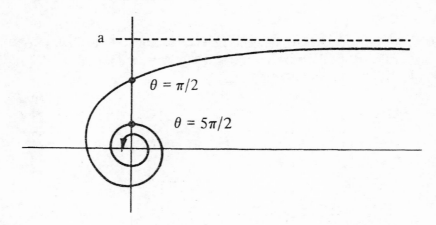

6. For any particular number y, define $f(x) = \cos(x + y)$. Then

$$f'(x) = -\sin(x + y),$$
$$f''(x) = -\cos(x + y),$$

so

$$f'' + f = 0,$$
$$f(0) = \cos y$$
$$f'(0) = -\sin y.$$

So

$$f = (-\sin y) \cdot \sin + (\cos y) \cdot \cos,$$

so

$$\cos(x + y) = \cos y \cos x - \sin y \sin x.$$

8. (a) Clearly $f(x) = A \sin(x + B)$ satisfies $f + f'' = 0$. (Moreover, $a = f'(0) = A \cos B$ and $b = f(0) = A \sin B$.)

(b) It clearly suffices to choose A and B so that $a = A \cos B$ and $b = A \sin B$. Since we want

$$a^2 + b^2 = (A \cos B)^2 + (A \sin B)^2$$

we must clearly choose

$$A = \sqrt{a^2 + b^2}.$$

If $a \neq 0$, we can choose

$$B = \arctan \frac{b}{a}.$$

If $a = 0$, we can choose $B = \pi/2$.

(c) $\sqrt{3} \sin x + \cos x = A \sin(x + B)$, where

$$A = \sqrt{\left(\sqrt{3}\right)^2 + 1} = 2$$
$$B = \arctan \frac{1}{\sqrt{3}} = \pi/6,$$

so $\sqrt{3} \sin x + \cos x = 2 \sin(x + \pi/6)$.

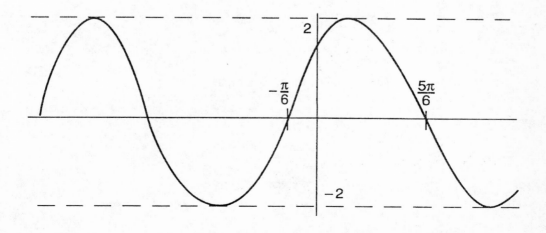

10. From the addition formula for sin we obtain, for $|\alpha| \leq 1$ and $|\beta| \leq 1$,

$$\sin(\arcsin \alpha + \arcsin \beta) = \sin(\arcsin \alpha) \cos(\arcsin \beta)$$
$$+ \cos(\arcsin \alpha) \sin(\arcsin \beta)$$
$$= \alpha\sqrt{1 - \beta^2} + \beta\sqrt{1 - \alpha^2}.$$

Consequently

$$\arcsin \alpha + \arcsin \beta = \arcsin(\alpha\sqrt{1 - \beta^2} + \beta\sqrt{1 - \alpha^2}),$$

provided that $-\pi/2 \leq \arcsin \alpha + \arcsin \beta \leq \pi/2$. $\left[\text{If } \pi/2 < \arcsin \alpha + \arcsin \beta \leq \pi,\right.$ the right side must be replaced by $\pi - \arcsin(\alpha\sqrt{1 - \beta^2} + \beta\sqrt{1 - \alpha^2})$, and if $-\pi \leq$ $\arcsin \alpha + \arcsin \beta < \pi/2$, replaced by $-\pi - \arcsin(\alpha\sqrt{1 - \beta^2} + \beta\sqrt{1 - \alpha^2}).\big]$

13. **(a)** If

$$H(a) = \int_{-\pi}^{\pi} (f(x) - a \cos nx)^2 \, dx$$
$$= a^2 \int_{-\pi}^{\pi} \cos^2 nx \, dx - 2a \int_{-\pi}^{\pi} f(x) \cos nx \, dx + \int_{-\pi}^{\pi} f(x)^2 \, dx,$$

then the minimum occurs for

$$0 = H'(a) = 2a \int_{-\pi}^{\pi} \cos^2 nx \, dx - 2 \int_{-\pi}^{\pi} f(x) \cos nx \, dx,$$

so

$$a = \frac{\displaystyle\int_{-\pi}^{\pi} f(x) \cos nx \, dx}{\displaystyle\int_{-\pi}^{\pi} \cos^2 nx \, dx} = \frac{1}{\pi} \int_{-\pi}^{\pi} f(x) \cos nx \, dx,$$

by Problem 12. The proof for $\sin nx$ is similar.

(b)

$$\int_{-\pi}^{\pi} \left(f(x) - \left[\frac{c_0}{2} + \sum_{n=1}^{N} c_n \cos nx + d_n \sin nx \right] \right)^2 \, dx =$$

$$= \int_{-\pi}^{\pi} [f(x)]^2 \, dx - 2 \int_{-\pi}^{\pi} f(x) \left[\frac{c_0}{2} + \sum_{n=1}^{N} c_n \cos nx + d_n \sin nx \right] dx$$

$$+ \int_{-\pi}^{\pi} \left[\frac{c_0^2}{4} + \sum_{n=1}^{N} c_n^2 \cos^2 nx + d_n^2 \sin^2 nx \right] dx$$

$$+ \int_{-\pi}^{\pi} \sum_{n,m=1}^{N} c_n d_m \cos nx \sin mx \, dx$$

$$+ \int_{-\pi}^{\pi} \frac{c_0}{2} \sum_{n=1}^{N} c_n \cos nx + d_n \sin nx \, dx$$

$$= \int_{-\pi}^{\pi} [f(x)]^2 \, dx$$

$$- 2\pi \left(\frac{a_0 c_0}{2} + \sum_{n=1}^{N} a_n c_n + b_n d_n \right) + \pi \left(\frac{c_0^2}{2} + \sum_{n=1}^{N} c_n^2 + d_n^2 \right),$$

using Problem 12, the definition of a_n and b_n, and the fact that the last integral vanishes because $\int_{-\pi}^{\pi} \cos nx \, dx = \int_{-\pi}^{\pi} \sin nx \, dx = 0$. The second equality follows by algebra.

14. (a) Substituting $a = (x + y)/2$, $b = (x - y)/2$ in

$$\sin(a + b) + \sin(a - b) = \sin a \cos b + \cos a \sin b$$
$$+ \sin a \cos(-b) + \cos a \sin(-b)$$
$$= 2 \sin a \cos b$$

yields

$$\sin x + \sin y = 2 \sin \left(\frac{x + y}{2} \right) \cos \left(\frac{x - y}{2} \right).$$

(b) Using the same substitution in the equation

$$\cos(a + b) + \cos(a - b) = \cos a \cos b - \sin a \sin b$$
$$+ \cos a \cos(-b) - \sin a \sin(-b)$$
$$= 2 \cos a \cos b$$

we obtain

$$\cos x + \cos y = 2 \cos \left(\frac{x + y}{2} \right) \cos \left(\frac{x - y}{2} \right).$$

Similarly, from the equation

$$\cos(a + b) - \cos(a - b) = \cos a \cos b - \sin a \sin b$$
$$- \cos a \cos(-b) + \sin a \sin(-b)$$
$$= -2 \sin a \sin b$$

we obtain

$$\cos x - \cos y = -2 \sin \left(\frac{x+y}{2} \right) \sin \left(\frac{x-y}{2} \right).$$

15. (d)

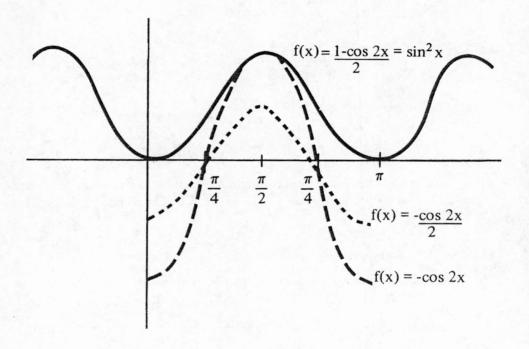

$$f(x) = \frac{1 - \cos 2x}{2} = \sin^2 x$$

$$f(x) = \frac{-\cos 2x}{2}$$

$$f(x) = -\cos 2x$$

16. If $y = \arctan x$ then

$$x = \tan y = \frac{\sin y}{\cos y} = \frac{\sin y}{\sqrt{1 - \sin^2 y}},$$

so

$$x\sqrt{1 - \sin^2 y} = \sin y,$$
$$x^2(1 - \sin^2 y) = \sin^2 y$$
$$\sin^2 y = \frac{x^2}{1 + x^2},$$

so

$$\sin(\arctan x) = \sin y = \frac{x}{\sqrt{1 + x^2}},$$
$$\cos(\arctan x) = \cos y = \sqrt{1 - \sin^2 y} = \frac{1}{\sqrt{1 + x^2}}.$$

17. If $x = \tan u/2$, then $u = 2 \arctan x$, so by Problem 16

$$\sin u = \sin(2 \arctan x)$$
$$= 2 \sin(\arctan x) \cos(\arctan x)$$
$$= \frac{2x}{1 + x^2},$$
$$\cos u = \sqrt{1 - \sin^2 u} = \frac{1 - x^2}{1 + x^2}.$$

18. (a) By the addition formula,

$$\sin(x + \pi/2) = \sin x \cos \pi/2 + \cos x \sin \pi/2 = \cos x.$$

(b) Part (a) implies that $x + \pi/2 = \arcsin(\cos x)$ for $-\pi/2 \leq x + \pi/2 \leq \pi/2$, or equivalently $-\pi \leq x \leq 0$. If $x = 2k\pi + x'$ for $-\pi \leq x \leq 0$, then $\cos x = \cos x'$, and if $x = 2k\pi + x'$ for $0 \leq x' \leq \pi$, then $\cos x = \cos x' = \cos(-x')$. So

$$\arcsin(\cos x) = \begin{cases} x - 2k\pi + \pi/2, & (2k-1)\pi \leq x \leq 2k\pi \\ 2k\pi + \pi/2 - x, & 2k\pi \leq x \leq (2k+1)\pi. \end{cases}$$

Similarly, from

$$\cos(x - \pi/2) = \sin x,$$

we conclude that

$$\arccos(\sin x) = \begin{cases} x - 2k\pi - \pi/2, & 2k\pi + \pi/2 \leq x \leq (2k+1)\pi + \pi/2 \\ (2k+1)\pi - \pi/2 - x, & 2k\pi - \pi/2 \leq x \leq 2k\pi + \pi/2. \end{cases}$$

22. If (x, y) is on the unit circle, then $x^2 + y^2 = 1$. In particular, $|x^2| \leq 1$, so $-1 \leq x \leq 1$. On the intervals $[0, \pi]$ and $[-\pi, 0]$ the function cos takes on all values between -1 and 1 so there is some θ in $[0, \pi]$ with $x = \cos \theta$, and also some θ in $[-\pi, 0]$ with $x = \cos \theta$. If $y \geq 0$, then $y = \sin \theta$ when θ is in $[0, \pi]$, and if $y \leq 0$, then $y = \sin \theta$ when θ is in $[-\pi, 0]$.

23. (a) If $a < 2k\pi + \pi/2 < b$, then sin is not one-one on $[a, b]$, because sin has a maximum at $2k\pi + \pi/2$, so sin takes on all values slightly less than 1 on both sides of $2k\pi + \pi/2$. Similarly, we cannot have $a < 2k\pi - \pi/2 < b$. Since the numbers of the form $2k\pi \pm \pi/2$ are within π of each other, π is the maximum length of an interval $[a, b]$ on which sin is one-one, and in this case $[a, b]$ must be of the form $[2k\pi - \pi/2, 2k\pi + \pi/2]$ or $[2k\pi + \pi/2, 2(k+1)\pi - \pi/2]$.

(b) $(g^{-1})'(x) = 1/\sqrt{1 - x^2}$, since $g^{-1}(x) = \arcsin x + 2k\pi$.

24. The domain of f^{-1} is $(-\infty, 1] \cup [1, \infty)$.

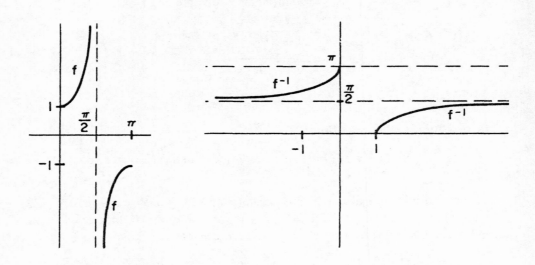

25. By the Mean Value Theorem,

$$|\sin x - \sin y| = |x - y| \cdot |\cos \theta| \qquad \text{form some } \theta \text{ between } x \text{ and } y$$
$$\leq |x - y|.$$

Strict inequality holds unless $\theta = 2k\pi$. But in any case, if $x < y$, say, then we can choose $x < z < y$ so that (x, z) does not contain any number of the form $2k\pi$. Then

$$\sin y - \sin x = (\sin y - \sin z) + (\sin z - \sin x)$$
$$= (y - z)\cos\theta_1 + (z - x)\cos\theta_2$$

for some θ_1 in (y, z) and θ_2 in (x, z). Since $|\cos\theta_1| \leq 1$ and $|\cos\theta_2| < 1$, it follows that

$$|\sin y - \sin x| < |y - x|.$$

26. (a)

$$\lim_{\lambda \to \infty} \int_c^d \sin \lambda x \, dx = \lim_{\lambda \to \infty} \frac{\cos \lambda c}{\lambda} - \frac{\cos \lambda d}{\lambda} = 0.$$

(b) If s has the values s_i on (t_{i-1}, t_i), then

$$\lim_{\lambda \to \infty} \int_a^b s(x) \sin \lambda x \, dx = \lim_{\lambda \to \infty} \sum_{i=1}^n s_i \int_{t_{i-1}}^{t_i} \sin \lambda x \, dx$$
$$= 0, \qquad \text{by part (a).}$$

(c) For any $\varepsilon > 0$ there is, by Problem 13-16, a step function $s \leq f$ with

$$\int_a^b [f(x) - s(x)] \, dx < \varepsilon.$$

Now

$$\left| \int_a^b f(x) \sin \lambda x \, dx - \int_a^b s(x) \sin \lambda x \, dx \right| = \left| \int_a^b [f(x) - s(x)] \sin \lambda x \, dx \right|$$

$$\leq \int_a^b [f(x) - s(x)] \cdot |\sin \lambda x| \, dx$$

$$\leq \int_a^b [f(x) - s(x)] \, dx < \varepsilon.$$

Part (b) then shows that

$$\lim_{\lambda \to \infty} \left| \int_a^b f(x) \sin \lambda x \, dx \right| < \varepsilon.$$

Since this is true for every $\varepsilon > 0$, the limit must be 0.

27. (a) We have

$$\text{area } OAB < \frac{x}{2} < \text{area } OCB,$$

so

$$\frac{\sin x}{2} < \frac{x}{2} < \frac{\sin x}{2 \cos x}.$$

(b) From

$$\frac{\sin x}{2} < \frac{x}{2}$$

we obtain

$$\frac{\sin x}{2} < 1;$$

from

$$\frac{x}{2} < \frac{\sin x}{2 \cos x}$$

we obtain

$$\cos x < \frac{\sin x}{x}.$$

Since $\lim_{x \to 0} \cos x = 1$, it follows that $\lim_{x \to 0} (\sin x)/x = 1$.

(c)

$$\lim_{x \to 0} \frac{1 - \cos x}{x} = \lim_{x \to 0} \frac{1 - \cos^2 x}{x(1 + \cos x)}$$

$$= \lim_{x \to 0} \frac{\sin x}{x} \cdot \frac{\sin x}{1 + \cos x} = 1 \cdot 0 = 0.$$

(d)

$$\sin'(x) = \lim_{h \to 0} \frac{\sin(x+h) - \sin x}{h}$$

$$= \lim_{h \to 0} \frac{\sin x \cos h + \cos x \sin h - \sin x}{h}$$

$$= \lim_{h \to 0} \frac{\sin h}{h} \cos x + \lim_{h \to 0} \frac{\cos h - 1}{h} \sin x$$

$$= \cos x.$$

28. (a) Problem 13-25 shows that

$$\mathcal{L}(x) = \int_x^1 \sqrt{1 + [f'(t)]^2}\, dt = \int_x^1 \sqrt{1 + \left(\frac{t}{\sqrt{1 - t^2}}\right)^2}\, dt$$

$$= \int_x^1 \frac{1}{\sqrt{1 - t^2}}\, dt.$$

[Actually, a more detailed argument is necessary, because $\int_x^1 1/\sqrt{1 - t^2}\, dt$ is not an ordinary integral, but an improper integral. It does follow immediately from Problem 13-25 that

$$\text{length of } f \text{ on } [x, 1 - \varepsilon] = \int_x^{1-\varepsilon} \frac{1}{\sqrt{1 - t^2}}\, dt.$$

To obtain the desired expression for $\mathcal{L}(x)$ we must then use the fact that

$$\lim_{\varepsilon \to 0} \left(\text{length of } f \text{ on } [x, 1 - \varepsilon]\right) = \text{length of } f \text{ on } [x, 1].$$

This is proved as follows. First of all, the following figure shows that the "length of f on $[x, 1]$" does make sense; in fact, the length of f on $[0, 1]$ is ≤ 2.

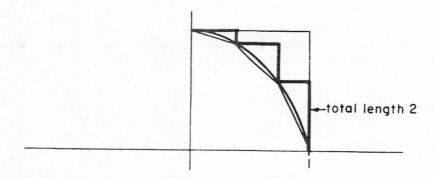

—total length 2

The same sort of figure also shows that the length of f on $[1 - \varepsilon, 1]$ is $\leq 2\varepsilon$. The desired limit then follows from this inequality and the fact that

$$\text{length of } f \text{ on } [x, 1] = \text{length of } f \text{ on } [x, 1 - \varepsilon] + \text{length of } f \text{ on } [1 - \varepsilon, 1].$$

The proof of this latter fact is very similar to the corresponding assertion for integrals.]

(b) This follows from part (a) and the Fundamental Theorem of Calculus.

(c) By the definition given, $\cos = \mathcal{L}^{-1}$, so

$$\cos'(x) = (\mathcal{L}^{-1})'(x) = \frac{1}{\mathcal{L}'(\mathcal{L}^{-1}(x))}$$

$$= \frac{1}{\dfrac{-1}{\sqrt{1-\cos^2 x}}} = -\sin x.$$

The proof for $\sin'(x)$ is the same as the one in the text.

29. (a) Clearly α is odd and increasing. The limit $\lim_{x \to \infty} \alpha(x)$, i.e., the improper integral $\int_0^\infty (1+t^2)^{-1} \, dt$, exists by Problem 14-25.

(b)

$$(\alpha^{-1})'(t) = \frac{1}{\alpha'(\alpha^{-1}(x))}$$

$$= \frac{1}{\dfrac{1}{1+[\alpha^{-1}(x)]^2}} = 1 + [\alpha^{-1}(x)]^2.$$

(c) If $-\pi/2 < x < \pi/2$, then

$$\cos x = \frac{1}{\sqrt{1+[\alpha^{-1}(x)]^2}} = (1 + [\alpha^{-1}(x)]^2)^{-1/2},$$

so

$$\cos'(x) = -\alpha^{-1}(x)(\alpha^{-1})'(x)(1 + [\alpha^{-1}(x)]^2)^{-3/2}$$

$$= -\alpha^{-1}(x)(1 + [\alpha^{-1}(x)]^2)^{-1/2}$$

$$= -\tan x \cos x.$$

Naturally the same result hold if x is not of the form $k\pi + \pi/2$ or $k\pi - \pi/2$. (For x which are of this form we have, by Theorem 11-7,

$$\cos'(x) = \lim_{y \to x} \cos'(y)$$

$$= \lim_{y \to x} -\tan y \cos y$$

$$= \lim_{y \to x} \frac{-\tan y}{\sqrt{1 + \tan^2 y}}$$

$$= -1 \qquad \text{since } \lim_{y \to x} \tan x = \infty.)$$

Now for x not of the form $k\pi + \pi/2$ or $k\pi - \pi/2$ we have

$$\cos''(x) = -\tan x \cos'(x) - \tan'(x)\cos x$$
$$= -\tan^2 x \cos x - [1 + \tan^2 x]\cos x \qquad \text{by part (b)}$$
$$= -\cos x.$$

For x which are of this form we have

$$\cos''(x) = \lim_{y \to x} \cos''(y) = \lim_{y \to x} -\cos y,$$

$$= \lim_{y \to x} \frac{-1}{\sqrt{1 + \tan^2 y}} = 0, \qquad \text{since } \lim_{y \to x} \tan y = \infty.$$

30. (a) $(y_0{}^2 + (y_0')^2)' = 2y_0 y_0' + 2y_0' y_0'' = 2y_0'(y_0' + y_0'') = 0$, so $y_0{}^2 + (y_0')^2$ is a constant. The constant is non-zero, since y_0 is not always 0, so $y_0(0)^2 + y_0'(0)^2 \neq 0$, so either $y_0(0) \neq 0$ or $y_0'(0) \neq 0$.

(b) Any function $s = ay_0 + by_0'$ satisfies $s'' + s = 0$, so we just have to choose a and b such that

$$ay_0(0) + by_0'(0) = 0$$
$$ay_0'(0) - by_0(0) = 1.$$

This is always possible, since

$$-y_0(0)^2 - y_0'(0)^2 \neq 0.$$

(c) Suppose that $\cos x > 0$ for all $x > 0$. Then $\sin$ would be increasing, since $\sin' = \cos$. Since $\sin 0 = 0$, this would mean that $\sin x > 0$ for all $x > 0$. Thus we would have $\cos'(x) = -\sin x < 0$ for all $x > 0$, so $\cos$ would be decreasing. Thus $\cos$ would satisfy all the hypotheses for f in Problem 7 of the Appendix to Chapter 11. But then the problem implies that $\cos''(x) = -\cos x = 0$ for some $x > 0$, a contradiction.

(d) Suppose $\cos x > 0$ for $0 < x < x_0 = \pi/2$, the function $\sin$ is increasing on $[0, \pi/2]$. Since $\sin 0 = 0$, it follows that $\sin \pi/2 > 0$, so $\sin \pi/2 = 1$.

(e)

$$\cos \pi = \cos(\pi/2 + \pi/2) = \cos^2 \pi/2 - \sin^2 \pi/2 = 0 - 1 = -1.$$
$$\sin \pi = \sin(\pi/2 + \pi/2) = 2 \sin \pi/2 \cos \pi/2 = 0.$$
$$\cos 2\pi = \cos(\pi + \pi) = \cos^2 \pi - \sin^2 \pi = 1.$$
$$\sin 2\pi = \sin(\pi + \pi) = 2 \sin \pi \cos \pi = 0.$$

(f)

$$\sin(x + 2\pi) = \sin x \cos 2\pi + \cos x \sin 2\pi = \sin x.$$
$$\cos(x + 2\pi) = \cos x \cos 2\pi - \sin x \sin 2\pi = \cos x.$$

31. (a) A rational function cannot be 0 at infinitely many points unless it is 0 everywhere.

(b) The assumed equation implies that $f_0(x) = 0$ for $x = 2k\pi$, so $f_0 = 0$. So

$$(\sin x)[(\sin x)^{n-1} + f_{n-1}(x)(\sin x)^{n-2} + \cdots + f_1(x)] = 0.$$

The term in brackets in continuous and 0 except perhaps at multiples of 2π, so it is 0 everywhere. We have just shown that if sin does not satisfy such an equation for $n - 1$, then it does not satisfy it for n. Since it clearly does not satisfy such an equation for $n = 1$, it does not satisfy it for any n.

32. (a) Multiplying the equation for g_1 by ϕ_2 and the equation for g_2 by ϕ_1 we obtain

$$\phi_1''\phi_2 + g_1\phi_1\phi_2 = 0$$
$$\phi_2''\phi_1 + g_2\phi_1\phi_2 = 0.$$

Subtraction yields the desired equation.

(b)

$$\int_a^b [\phi_1''\phi_2 - \phi_2''\phi_1] = \int_a^b (g_2 - g_2)\phi_1\phi_2 > 0,$$

since $g_2 > g_1$ and $\phi_1\phi_2 > 0$ by assumption. Since

$$(\phi_1'\phi_2 - \phi_1\phi_2')' = \phi_1''\phi_2 + \phi_1'\phi_2' - \phi_1'\phi_2' - \phi_1\phi_2''$$
$$= \phi_1''\phi_2 - \phi_1\phi_2'',$$

we have

$$0 < \int_a^b [\phi_1''\phi_2 - \phi_2''\phi_1]$$

$$= [\phi_1'(b)\phi_2(b) - \phi_1(b)\phi_2'(b)] - [\phi_1'(a)\phi_2(a) - \phi_1(a)\phi_2'(a)]$$
$$= \phi_1'(b)\phi_2(b) - \phi_1'(a)\phi_2(a) + [\phi_1(b)\phi_2'(b) - \phi_1(a)\phi_2'(a)].$$

(c) If $\phi_1(a) = \phi_1(b) = 0$, then it follows from part (b) that

$$\phi_1'(b)\phi_2(b) - \phi_1'(a)\phi_2(a) > 0.$$

But clearly

$$\phi_2(a) \geq 0, \qquad \phi_2(b) \geq 0$$
$$\phi_1'(a) \geq 0, \qquad \phi_1'(b) \leq 0.$$

This implies that

$$\phi_1'(b)\phi_2(b) - \phi_1'(a)\phi_2(a) \leq 0,$$

a contradiction.

(d) This follows from part (c) by replacing ϕ_1 by $-\phi_1$ and/or ϕ_2 by $-\phi_2$.

33. (a) Substitute $(k + \frac{1}{2})x$ for x and $(k - \frac{1}{2})x$ for y in the formula

$$\sin x - \sin y = \sin x + \sin(-y) = 2 \sin \left(\frac{x-y}{2} \right) \cos \left(\frac{x+y}{2} \right).$$

(b) We have

$$\frac{1}{2} + \cos x + \cos 2x + \ldots + \cos nx$$

$$= \frac{1}{2} + \frac{1}{2 \sin \dfrac{x}{2}} \left[\sum_{k=1}^{n} \sin(k + \tfrac{1}{2})x - \sin(k - \tfrac{1}{2})x \right]$$

$$= \frac{1}{2} + \frac{1}{2 \sin \dfrac{x}{2}} \left[\sin(n + \tfrac{1}{2})x - \sin(\tfrac{1}{2})x \right]$$

$$= \frac{\sin(n + \tfrac{1}{2})x}{2 \sin \dfrac{x}{2}}.$$

(c) Substituting $(k + \frac{1}{2})x$ for x and $(k - \frac{1}{2})x$ for y in the formula

$$(*) \qquad \cos x - \cos y = -2 \sin \left(\frac{x+y}{2} \right) \sin \left(\frac{x-y}{2} \right)$$

from Problem 14 we obtain

$$\cos(k + \tfrac{1}{2})x - \cos(k - \tfrac{1}{2})x = -2 \sin kx \sin \frac{x}{2}.$$

So

$$\sin x + \ldots + \sin nx$$

$$= -\frac{1}{2 \sin \dfrac{x}{2}} \left[\sum_{k=1}^{n} \cos(k + \tfrac{1}{2})x - \cos(k - \tfrac{1}{2})x \right]$$

$$= -\frac{1}{2 \sin \dfrac{x}{2}} \left[\cos(n + \tfrac{1}{2})x - \cos(\tfrac{1}{2}x) \right]$$

$$= -\frac{1}{2 \sin \dfrac{x}{2}} \left[-2 \sin \left(\frac{n}{2}x \right) \sin \left(\frac{n+1}{2}x \right) \right] \qquad \text{by } (*) \text{ again.}$$

(d) It obviously suffices to compute the integral for $b \leq \pi/2$, which makes things sightly easier, since sin is increasing and cos is decreasing on this interval. Let

$P = \{t_0, \ldots, t_n\}$ be the partition of $[0, b]$ with $t_i = ib/n$. Then

$$L(\cos, P) = \frac{b}{n} \sum_{i=1}^{n} \cos\left(\frac{ib}{n}\right)$$

$$= \frac{\sin\left([n + \frac{1}{2}]\frac{b}{n}\right)}{\frac{2n}{b} \sin \frac{b}{2n}} - \frac{b}{2n}.$$

For n large, $\sin([n + \frac{1}{2}]\frac{b}{n}) = \sin([1 + \frac{1}{2n}]b)$ is close to $\sin b$, and $\frac{2n}{b} \sin \frac{b}{2n} = (\sin b/2n)/(b/2n)$ is close to 1. So $L(\cos, P)$ can be made as close as desired to $\sin b$, which means that $\int_0^b \cos = \sin b$. For $\int_0^b \sin$ it is best to use the next-to-last equation in the derivation of part (c):

$$U(\sin, P) = \frac{b}{n} \sum_{i=1}^{n} \sin\left(\frac{ib}{n}\right)$$

$$= -\frac{1}{\frac{2n}{b} \sin \frac{b}{2n}} \left[\cos\left([n + \frac{1}{2}]\frac{b}{n}\right) - \cos\left(\frac{b}{2n}\right) \right].$$

For n large this is close to $-\cos b + 1$, so $\int_0^b \sin = 1 - \cos b$.

CHAPTER 16

1. (a) Let $\Delta = $ area OAB. Since

$$xy = 2\Delta,$$
$$x^2 + y^2 = 1,$$

we have

$$\left(\frac{2\Delta}{y}\right)^2 + y^2 = 1,$$
$$4\Delta^2 + y^4 = y^2,$$
$$y^4 - y^2 + 4\Delta^2 = 0,$$
$$y^2 = \frac{1 \pm \sqrt{1 - 16\Delta^2}}{2}.$$

We have

$$y^2 = \frac{1 - \sqrt{1 - 16\Delta^2}}{2},$$

provided that $y^2 < 1/2$, or $y < \sqrt{2}/2$. So

$$\text{area } OAC = \frac{y}{2} = \frac{1}{2}\sqrt{\frac{1 - \sqrt{1 - 16(\text{area } OAB)^2}}{2}}.$$

(b) Let P_m be the union of m triangles congruent to the triangle OAA' in the figure below.

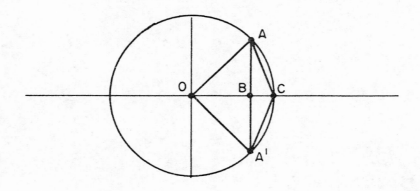

238

Each such triangle has area A_m/m so triangle OAB had area $A_m/2m$. Now P_{2m} is the union of $2m$ triangles congruent to OAC. So by part (a),

$$A_{2m} = 2m \text{ area } OAC = m\sqrt{\frac{1 - \sqrt{1 - 16(\text{area } OAB)^2}}{2}}$$

$$= m\sqrt{\frac{1 - \sqrt{1 - 16A_m{}^2/4m^2}}{2}}$$

$$= \frac{m}{2}\sqrt{2 - 2\sqrt{1 - (2A_m/m)^2}}.$$

2. (a)

$$\frac{A_m}{A_{2m}} = \frac{2m \text{ area}(OAB)}{2m \text{ area}(OAC)} = OB = \alpha_m.$$

(b)

$$\frac{2}{A_{2^k}} = \frac{2}{A_8} \cdot \frac{A_8}{A_{16}} \cdots \frac{A_{2^{k-1}}}{A_{2^k}}$$

$$= \frac{A_4}{A_8} \cdots \frac{A_{2^{k-1}}}{A_{2^k}} = \alpha_4 \cdots \alpha_{2^{k-1}}.$$

(c)

$$\alpha_4 = \cos\frac{\pi}{4} = \frac{\sqrt{2}}{2} = \sqrt{\frac{1}{2}},$$

$$\alpha_8 = \cos\left(\frac{\pi/4}{2}\right) = \sqrt{\frac{1}{2} + \frac{\cos\pi/4}{2}}$$

$$= \sqrt{\frac{1}{2} + \frac{1}{2}\sqrt{\frac{1}{2}}},$$

etc.

CHAPTER 18

1. (ii)

$$\frac{1}{1 + \log(1 + \log(1 + e^{1+e^{1+x}}))} \cdot \frac{1}{1 + \log(1 + e^{1+e^{1+x}})}$$

$$\cdot \frac{1}{1 + e^{1+e^{1+x}}} \cdot e^{1+e^{1+x}} \cdot e^{1+x}.$$

(iv)

$$\left(\int_0^x e^{-t^2} \, dt \right) \cdot e^{-x^2}.$$

(vi) We have

$$f(x) = \frac{\log(\sin x)}{\log e^x} = \frac{\log(\sin x)}{x},$$

so

$$f'(x) = \frac{x \cdot \dfrac{\cos x}{\sin x} - \log(\sin x)}{x^2}.$$

(viii)

$$4 \log(3 + e^4) e^{4x} + (\log 3)(\arcsin x)^{(\log 3) - 1} \frac{1}{\sqrt{1 - x^2}}.$$

(x) $f(x) = e^{x \log x}$, so

$$f'(x) = e^{x \log x} \left(x \cdot \frac{1}{x} + 1 \cdot \log x \right) = x^x (1 + \log x).$$

2. (a) $(\log \circ f)' = (\log' \circ f) \cdot f' = (1/f) \cdot f'$.

(b) (i) $\log(f(x)) = \log(1 + x) + \log(1 + e^{x^2})$ so

$$(\log \circ f)'(x) = \frac{1}{1 + x} + \frac{2x e^{x^2}}{1 + e^{x^2}}$$

so

$$f'(x) = (1 + x)(1 + e^{x^2}) \left[\frac{1}{1 + x} + \frac{2x e^{x^2}}{1 + e^{x^2}} \right].$$

(ii)

$$(\log \circ f)'(x) = -\frac{1}{3(3 - x)} + \frac{2}{x} + \frac{1}{1 - x} - \frac{2}{3(3 + x)},$$

$$f'(x) = \frac{(3 - x)^{1/3} x^2}{(1 - x)(3 + x)^{2/3}} \left[-\frac{1}{3(3 - x)} + \frac{2}{x} + \frac{1}{1 - x} - \frac{2}{3(3 + x)} \right].$$

(iii)

$$f'(x) = (\sin x)^{\cos x} \left[\cos x \cdot \frac{\cos x}{\sin x} - \sin x \log \cos x \right]$$

$$+ (\cos x)^{\sin x} \left[\sin x \cdot \frac{-\sin x}{\cos x} + \cos x \log \sin x \right].$$

(iv)

$$f(x) = \frac{1}{e^x(1+x^3)} - \frac{1}{e^{3x}(1+x^3)},$$

$$f'(x) = \frac{1}{e^x(1+x^3)} \left[-1 - \frac{3x^2}{1+x^3} \right] - \frac{1}{e^{3x}(1+x^3)} \left[-3 - \frac{3x^2}{1+x^3} \right]$$

3.

$$\int_a^b \frac{f'(t)}{f(t)}\, dt = \int_a^b (\log \circ f)'(t)\, dt = \log(f(b)) - \log(f(a)).$$

4. (a)

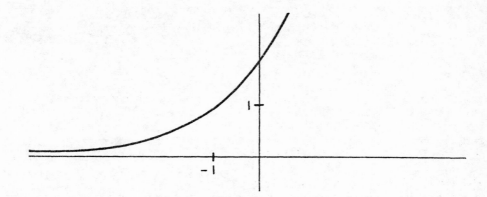

(b)

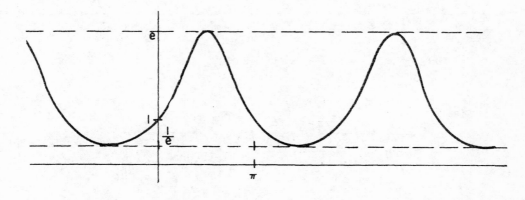

(c), (d)

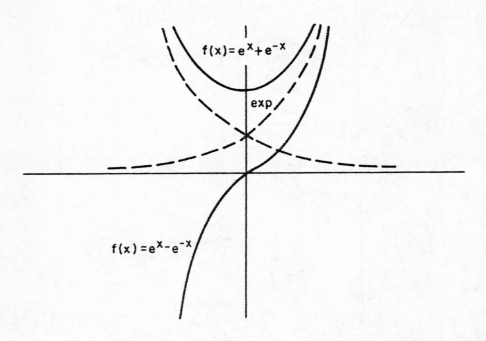

(e)

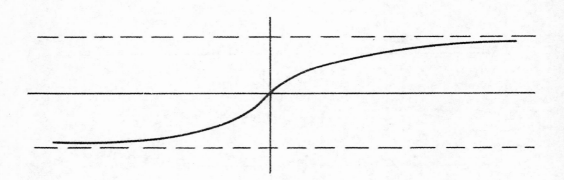

5. (ii)

$$\lim_{x \to 0} \frac{e^x - 1 - x - x^2/2 - x^3/6}{x^3} = \lim_{x \to 0} \frac{e^x - 1 - x - x^2/2}{3x^2}$$

$$= \lim_{x \to 0} \frac{e^x - 1 - x}{6x} = \lim_{x \to 0} \frac{e^x - 1}{6} = 0.$$

(iv)

$$\lim_{x \to 0} \frac{\log(1+x) - x - x^2/2}{x^2} = \lim_{x \to 0} \frac{\dfrac{1}{1+x} - 1 + x}{2x}$$

$$= \lim_{x \to 0} \frac{-\dfrac{1}{(1+x)^2} + 1}{2} = 0.$$

(vi)

$$\lim_{x \to 0} \frac{\log(1+x) - x + x^2/2 - x^3/3}{x^3}$$

$$= \lim_{x \to 0} \frac{\dfrac{1}{1+x} - 1 + x - x^2}{3x^2}$$

$$= \lim_{x \to 0} \frac{-\dfrac{1}{(1+x)^2} + 1 - 2x}{6x} = \lim_{x \to 0} \frac{\dfrac{2}{(1+x)^3} - 2}{6} = 0.$$

6.

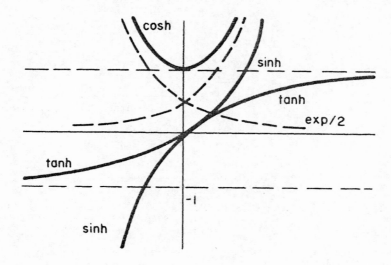

7. (b) Since $\cosh^2 - \sinh^2 = 1$ by part (a), we have

$$1 - \frac{\sinh^2}{\cosh^2} = \frac{1}{\cosh^2}.$$

(d)

$$\cosh x \cosh y + \sinh x \sinh y$$

$$= \frac{e^x + e^{-x}}{2} \cdot \frac{e^y + e^{-y}}{2} + \frac{e^x - e^{-x}}{2} \cdot \frac{e^y - e^{-y}}{2}$$

$$= \frac{e^{x+y}}{4} + \frac{e^{-x-y}}{4} + \frac{e^{x-y}}{4} + \frac{e^{y-x}}{4} + \frac{e^{x+y}}{4} + \frac{e^{-x-y}}{4} - \frac{e^{y-x}}{4} - \frac{e^{x-y}}{4}$$

$$= \frac{e^{x+y} + e^{-(x+y)}}{2} = \cosh(x+y).$$

(f) Since

$$\cosh x = \frac{e^x + e^{-x}}{2},$$

we have

$$\cosh'(x) = \frac{e^x - e^{-x}}{2} = \sinh x.$$

8. (b) It follows from Problem 7(a) that

$$\cosh^2(\sinh^{-1} x) = 1 + \sinh^2(\sinh^{-1} x) = 1 + x^2,$$

so

$$\cosh(\sinh^{-1} x) = \sqrt{1 + x^2},$$

since $\cosh y \geq 0$ for all y.

(d)

$$(\cosh^{-1})'(x) = \frac{1}{\cosh'(\cosh^{-1} x)}$$

$$= \frac{1}{\sinh(\cosh^{-1} x)}$$

$$= \frac{1}{\sqrt{x^2 - 1}}.$$

10. Since $0 < \log t < t$ for $t > 1$ we have

$$\int_2^x \frac{1}{\log t} \, dt > \int_2^x \frac{1}{t} \, dt = \log x,$$

and $\log$ is not bounded on $[2, \infty)$.

11. If $|f| \leq M$ on $[1, \infty)$, then

$$|F(x)| \leq \int_1^x \frac{|f(t)|}{t} \, dt \leq M \int_1^x \frac{1}{t} \, dt = M \log x,$$

so $|F(x)|/\log x \le M$ for all $x \ge 1$. To prove the converse, first suppose $f \ge 0$ on $[1, \infty)$. Then since f is nondecreasing,

$$F(x) = \int_1^x \frac{f(t)}{t}\, dt \le f(x) \int_1^x \frac{1}{t}\, dt = f(x) \log x,$$

so $|F/\log|$ bounded implies $|f|$ bounded. For the general case note that since f is nondecreasing it is certainly bounded on any interval to the right of 1 on which it is negative. If $f(b) = 0$ for some $b > 1$, then for $x \ge b$

$$F(x) = \int_1^b \frac{f(t)}{t}\, dt + \int_b^x \frac{f(t)}{t}\, dt$$

$$\le \int_1^b \frac{f(t)}{t}\, dt + f(x) \int_b^x \frac{1}{t}\, dt$$

$$= \int_1^b \frac{f(t)}{t}\, dt + f(x)[\log x - \log b],$$

so

$$\frac{F(x)}{\log x} = \frac{1}{\log x} \cdot \int_1^b \frac{f(t)}{t}\, dt + \frac{[\log x - \log b]}{\log x} f(x)$$

$$= A(x) + B(x)f(x), \quad \text{say}.$$

The for $x \ge b$ we have

$$|f(x)| \le \frac{1}{|B(x)|} \left[\left| \frac{F(x)}{\log x} \right| + |A(x)| \right].$$

Now $|A(x)|$ is bounded [it $\to 0$ as $x \to \infty$] and $1/|B(x)|$ is bounded [$B(x) \to 1$ as $x \to \infty$], so if $|F/\log|$ is bounded, then so is $|f|$.

12. (b)

$$\lim_{x \to \infty} \frac{x}{(\log x)^n} = \lim_{y \to \infty} \frac{e^y}{y^n} = \infty.$$

(d)

$$\lim_{x \to 0^+} x(\log x)^n = \lim_{x \to 0^+} \frac{(-1)^n \left(\log \dfrac{1}{x} \right)^n}{\dfrac{1}{x}}$$

$$= \lim_{y \to \infty} \frac{(-1)^n (\log y)^n}{y} = 0.$$

13. f is convex, since

$$f'(x) = x^x(1 + \log x),$$

$$f''(x) = x^x(1 + \log x)^2 + \frac{x^x}{x} \geq 0.$$

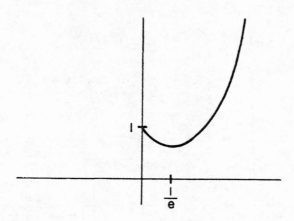

14. (a) If $x > 0$ and

$$0 = f'(x) = \frac{x^n e^x - n x^{n-1} e^x}{x^{2n}} = \frac{e^x(x - n)}{x^{n+1}},$$

then $x = n$, so the minimum is at n, since $\lim_{x \to 0^+} f(x) = \infty = \lim_{x \to \infty} f(x)$. So for $x > n$ we have $f(x) > f(n) = e^n/n^n$.

(b) If $x > n + 1$, then

$$f'(x) > \frac{e^x}{x^{n+1}} > \frac{e^{n+1}}{(n+1)^{n+1}}$$

by part (a) applied in the case $n + 1$. It follows immediately that $\lim_{x \to \infty} f(x) = \infty$ (merely using the fact that $f'(x) > \varepsilon > 0$ for some ε and all sufficiently large x).

15. f is convex, since

$$f'(x) = e^x x^{-n} - n e^x x^{-n-1},$$

$$f''(x) = e^x x^{-n} - n e^x x^{-n-1} - n e^x x^{-n-1} + n(n+1) e^x x^{-n-2}$$

$$= \frac{e^x}{x^{n+2}}[x^2 - 2nx + n^2 + n] \geq 0 \qquad \text{for all } x.$$

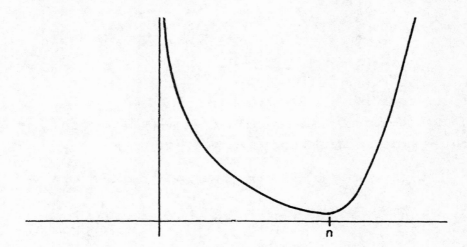

16. (e) If $f(x) = e^{bx}$, then $f'(0) = b$, so

$$\lim_{y \to 0} \frac{e^{by} - 1}{y} = b.$$

Thus

$$\lim_{x \to \infty} x(e^{b/x} - 1) = b.$$

So

$$\log b = \lim_{x \to \infty} x(e^{(\log b)/x} - 1)$$
$$= \lim_{x \to \infty} x(b^{1/x} - 1).$$

17. We have $\lim_{x \to \infty} f(x) = e$ by Problem 16(c) and

$$\lim_{x \to 0^+} f(x) = \lim_{x \to 0^+} \left(1 + \frac{1}{x}\right)^x = \exp\left(\lim_{x \to 0^+} x \log\left(1 + \frac{1}{x}\right)\right)$$
$$= \exp\left(\lim_{x \to 0^+} x \log\left(\frac{x+1}{x}\right)\right)$$
$$= \exp\left(\lim_{x \to 0^+} [x \log(x + 1) - x \log x]\right)$$
$$= \exp 0 = 1, \qquad \text{using Problem 12(d).}$$

Moreover,

$$f'(x) = \left(1 + \frac{1}{x}\right)^x \left[\log\left(1 + \frac{1}{x}\right) - \frac{1}{x+1}\right] = \left(1 + \frac{1}{x}\right)^x \cdot g(x), \quad \text{say.}$$

To analyze f', we notice that

$$g'(x) = \frac{1}{1 + \dfrac{1}{x}} \cdot \frac{-1}{x^2} + \frac{1}{(x+1)^2}$$

$$= \frac{-1}{x(x+1)^2} < 0.$$

Thus g is decreasing. Since $\lim\limits_{x \to \infty} g(x) = 0$, we must have $g(x) > 0$ for all $x > 0$. So f is increasing. We also have

$$\lim\limits_{x \to 0^+} f'(x) = 1 \cdot \lim\limits_{x \to 0^+} \left[\log\left(1 + \frac{1}{x}\right) - \frac{1}{x+1} \right]$$

$$= \lim\limits_{x \to 0^+} \left[\log(x+1) - \log x - \frac{1}{x+1} \right]$$

$$= \infty.$$

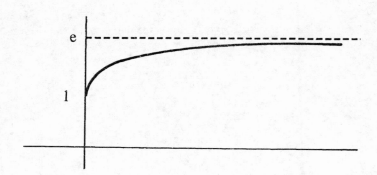

20. (a) We have

$$(\log|f|)' = \frac{f'}{f} = c$$

so

$$\log|f(x)| = cx + d$$

for some number d, so

$$|f(x)| = e^d e^{cx}, \qquad e^d > 0.$$

(b) On an interval where f is non-zero it has the form $f(x) = ke^{cx}$. But this can't approach 0 at the endpoint of the interval; so f couldn't be 0 at the endpoints. This proves that if f is non-zero at one point x_0, then it is nowhere 0 (consider $\sup\{x > x_0 : f(x) \neq 0\}$ and $\inf\{x < x_0 : f(x) \neq 0\}$).

(d) Let $h(x) = f(x)/e^{g(x)}$. Then

$$h'(x) = \frac{e^{g(x)} f'(x) - f(x) g(x) e^{g(x)}}{e^{2g(x)}}$$

$$= \frac{e^{g(x)} [f'(x) - f(x) g(x)]}{e^{2g(x)}} = 0,$$

so $f(x)/e^{g(x)} = k$ for some constant k.

23. Notice that f is continuous, by Theorem 13-8. We therefore have $f'(x) = f(x)$, so there is a number c such that $f(x) = ce^x$. But $f(0) = 0$, so $c = 0$.

24. (i) Differentiating $\int_0^x f = e^x$, we see that f must satisfy $f(x) = e^x$, but this f doesn't work, since

$$\int_0^x e^t \, dt = e^x - e^0 = e^x - 1.$$

So there is no such f (easier proof: set $x = 0$, to get $0 = e^0$!).

(ii) Differentiating, we obtain

$$2xf(x^2) = -4xe^{2x^2}$$

so

$$f(y) = -2e^{2y} \qquad y \geq 0.$$

This f does work.

25. We need

(1) $$f(f(x)) \cdot g(f(x)) \cdot f'(x) = g'(f(x)) \cdot f'(x)$$

or

(2) $$f(y)g(y) = g'(y).$$

According to Problem 20(d), if F is a function with $F' = f$, then we must have

$$g(y) = ke^{F(y)}.$$

For this g to work we need

$$\int_0^{f(x)} kf(t)e^{F(t)} \, dt = ke^{F(f(x))} - 1.$$

Since fe^F is the derivative of e^F, this means

$$k\left[e^{F(f(x))} - e^{F(0)}\right] = ke^{F(f(x))} - 1,$$

or $ke^{F(0)} = 1$. We can choose $F(0)$ arbitrarily; we might as well make $F(0) = 0$. Then we need $k = 1$.

This derivation is not complete, because we assumed $f'(x) \neq 0$ in order to go from (1) to (2). In fact, if $f(x) = c$ for all x, then our equation simply says

$$\int_0^c cg = g(c) - 1,$$

and any g satisfying this will work. Even if we assume that $f'(x) \neq 0$ for all x, there is the further problem that equation (2) only holds for y in the range of f! In order not to get too involved, let's simply assume that f is defined on $[0, \infty)$, with $f' > 0$ everywhere on $[0, \infty)$, with $f(0) = a > 0$ and $[a, \infty)$ being the range of f. Then

$$g(y) = ke^{F(y)} \qquad y \geq a.$$

For this g to work we need

$$\int_0^a fg + \int_a^{f(x)} kf(t)e^{F(t)}\, dt = ke^{F(f(x))} - 1$$

or

$$\left(\int_0^a fg \right) + k\left[e^{F(f(x))} - e^{F(a)} \right] = ke^{F(f(x))} - 1$$

or

$$\left(\int_0^a fg \right) - ke^{F(a)} = -1.$$

Choosing $F(a) = 0$, this says

(∗) $$k = \int_0^a fg + 1.$$

On $[0, a)$ we can choose g arbitrarily, and then let $g(y) = ke^{F(y)}$ for $y \geq a$, where F is the function with $F(a) = 0$, $F' = f$, and k is determined by (∗).

26. We have $f''(t) = f'(t)$, so $f'(t) = ce^t$ for some c, so $f(t) = a + ce^t$ for some a. So

$$ce^t = (a + ce^t) + \int_0^1 (a + ce^t)\, dt$$

$$= a + ce^t + a + ce - c,$$

so $a = c(1 - e)/2$.

27. The given equation implies that f^2 is differentiable, so f is also differentiable at any x with $f(x) \neq 0$, and for such x we have

$$2f(x)f'(x) = f(x)\frac{x}{1 + x^2},$$

so

$$f'(x) = g'(x), \qquad \text{where } g(x) = \frac{\log(1 + x^2)}{4}.$$

So on any interval where $f \neq 0$ we have

$$f(x) = \frac{\log(1 + x^2)}{4} + C$$

for some C.

If $x = 0$ is in the interval we immediately have $0 = f(0)^2 = C$. But it is possible to have a pieced-together solution like

$$f(x) = \begin{cases} 0 & x \leq 0 \\ \dfrac{\log(1 + x^2)}{4} & x \geq 0. \end{cases}$$

28. (a) Let

$$h(x) = C + \int_a^x fg \geq C > 0 \qquad \text{on } [a, b].$$

Then $h'(x) = f(x)g(x) \geq 0$ on $[a, b]$ and

$$h'(x) = f(x)g(x)$$
$$\leq g(x)\left[C + \int_a^x fg\right]$$
$$= g(x)h(x),$$

so

$$(\log \circ h)'(x) = \frac{h'(x)}{h(x)} \leq g(x).$$

If we set

$$G(x) = \int_a^x g,$$

then since $\log(h(a)) = \log C$, we can write

$$[\log \circ h - \log c]' \leq G'$$

where both G and $\log \circ h - \log C$ are 0 at a. It follows that

$$\log(h(x)) - \log C \leq G(x) = \int_a^x g \qquad \text{for } x \geq a,$$

or

$$h(x) \leq Ce^{\int_a^x g}.$$

(b) If

$$f(x) \leq \int_a^x fg$$

then for every $\varepsilon > 0$ we have

$$f(x) \leq \varepsilon + \int_a^x fg$$

so by part (a)

$$0 \le f(x) \le \varepsilon^{\int_a^x g};$$

since this is true for every $\varepsilon > 0$, we must have $f(x) = 0$.

(c) We have

$$f(x) = \int_0^x f' = \int_0^x fg;$$

by part (b), this proves that $f(x) = 0$ for all x (but we need to assume $f, g \ge 0$).

29. (a) For $n = 0$ the inequality reads $1 \le e^x$ for $x \ge 0$, which is certainly true, since $e^0 = 1$ and exp is increasing. Suppose the inequality is true for n. Let

$$f(x) = 1 + x + \frac{x^2}{2!} + \cdots + \frac{x^{n+1}}{(n+1)!}.$$

Then

$$f'(x) = 1 + x + \frac{x^2}{2!} + \cdots + \frac{x^n}{n!} \le e^x,$$

while $f(0) = e^0$. It follows that $f(x) \le e^x$ for $x \ge 0$.

(b)

$$\lim_{n \to \infty} \frac{e^x}{x^n} \ge \lim_{n \to \infty} \frac{1 + x + x^2/2! + \cdots + x^{n+1}/(n+1)!}{x^n}$$

$$= \lim_{n \to \infty} \frac{1}{x^n} + \frac{1}{x^{n-1}} + \frac{1}{2! \, x^{n-2}} + \cdots + \frac{x}{(n+1)!}$$

$$= \infty.$$

30. Using the form of l'Hôpital's Rule which was proved in the answer to Problem 11-53, we have

$$\lim_{x \to \infty} \frac{e^x}{x^n} = \lim_{x \to \infty} \frac{e^x}{nx^{n-1}} = \cdots = \lim_{x \to \infty} \frac{e^x}{n!} = \infty.$$

31. (a) A good guess is that the limit is 0. Reason: On $[0, x]$ the maximum value of e^{t^2} is e^{x^2}; on most of the interval the value is much smaller, so that integral should be much smaller than e^{x^2}. We can easily evaluate the limit by the form of l'Hôpital's Rule that appears in Problem 11-52:

$$\lim_{x \to \infty} \frac{\displaystyle\int_0^x e^{t^2} \, dt}{e^{x^2}} = \lim_{x \to \infty} \frac{e^{x^2}}{2xe^{x^2}} = 0.$$

(b) (i) Using l'Hôpital's Rule, we obtain

$$\lim_{x \to \infty} \frac{\displaystyle\int_x^{x+\frac{1}{x}} e^{t^2}\, dt}{e^{x^2}} = \lim_{x \to \infty} \frac{e^{[x+\frac{1}{x}]^2} - e^{x^2}}{2x e^{x^2}}$$

$$= \lim_{x \to \infty} \frac{e^2 e^{\frac{1}{x^2}} - 1}{2x}$$

$$= 0.$$

(ii)

$$\lim_{x \to \infty} \frac{\displaystyle\int_x^{x+\frac{\log x}{x}} e^{t^2}\, dt}{e^{x^2}} = \lim_{x \to \infty} \frac{e^{\left(x+\frac{\log x}{x}\right)^2} - e^{x^2}}{2x e^{x^2}}$$

$$= \lim_{x \to \infty} \frac{e^{2\log x}\, e^{\left(\frac{\log x}{x}\right)^2} - 1}{2x}$$

$$= \lim_{x \to \infty} \frac{x^2 e^{\left(\frac{\log x}{x}\right)^2} - 1}{2x}$$

$$= \infty.$$

(iii)

$$\lim_{x \to \infty} \frac{\displaystyle\int_x^{x+\frac{\log x}{2x}} e^{t^2}\, dt}{e^{x^2}} = \lim_{x \to \infty} \frac{e^{\left(x+\frac{\log x}{2x}\right)^2} - e^{x^2}}{2x e^{x^2}}$$

$$= \lim_{x \to \infty} \frac{e^{\log x}\, e^{\left(\frac{\log x}{2x}\right)^2} - 1}{2x}$$

$$= \frac{1}{2}.$$

32. (a) We have

$$\log_a'(x) = \lim_{h \to 0} \frac{\log_a(x+h) - \log_a x}{h}$$

$$= \lim_{h \to 0} \frac{\log_a\left(\dfrac{x+h}{x}\right)}{h} = \lim_{h \to 0} \log\left(1 + \frac{h}{x}\right)^{\frac{1}{h}}$$

$$= \lim_{\frac{h}{x} \to 0} \log\left(1 + \frac{h}{x}\right)^{\frac{x}{h} \cdot \frac{1}{x}} = \lim_{k \to 0} \log(1+k)^{\frac{1}{k} \cdot \frac{1}{x}}$$

$$= \lim_{k \to 0} \frac{1}{x} \log(1+k)^{\frac{1}{k}} = \frac{1}{x} \lim_{k \to 0} \log(1+k)^{\frac{1}{k}}$$

$$= \frac{1}{x} \log \left(\lim_{k \to 0} (1+k)^{\frac{1}{k}} \right), \qquad \text{by continuity of log.}$$

(b) By the binomial theorem,

$$a_n = \left(1 + \frac{1}{n}\right)^n = 1 + n \cdot 1 \cdot \frac{1}{n} + \sum_{k=2}^{n} \binom{n}{k} \left(\frac{1}{n}\right)^k$$

$$= 2 + \sum_{k=2}^{n} \frac{1}{k!} n(n-1) \cdots (n-k+1) \cdot \frac{1}{n^k}$$

$$= 2 + \sum_{k=2}^{n} \frac{1}{k!} \left(\frac{n-1}{n}\right) \cdots \left(\frac{n-k+1}{n}\right)$$

$$= 2 + \sum_{k=2}^{n} \frac{1}{k!} \left(1 - \frac{1}{n}\right) \left(1 - \frac{2}{n}\right) \cdots \left(1 - \frac{k-1}{n}\right).$$

Similarly,

$$a_{n+1} = 2 + \sum_{k=2}^{n+1} \frac{1}{k!} \left(1 - \frac{1}{n+1}\right) \left(1 - \frac{1}{n+1}\right) \cdots \left(1 - \frac{k-1}{n+1}\right).$$

All terms in these sums are positive, and for each $k \leq n$ we have

$$\frac{1}{k!} \left(1 - \frac{1}{n+1}\right) \left(1 - \frac{2}{n+1}\right) \cdots \left(1 - \frac{k-1}{n+1}\right)$$

$$> \frac{1}{k!} \left(1 - \frac{1}{n}\right) \left(1 - \frac{2}{n}\right) \cdots \left(1 - \frac{k-1}{n}\right),$$

since each $\left(1 - \frac{i}{n+1}\right) > \left(1 - \frac{i}{n}\right)$. So $a_{n+1} \geq a_n$.

(c) Since each $\left(1 - \frac{i}{n}\right) < 1$ and $1/k! \leq 1/2^{k-1}$ for $k \geq 2$, we have

$$a_n < 2 + \sum_{k=2}^{n} \frac{1}{2^{k-1}} = 2 + \left(\frac{1}{2} + \frac{1}{4} + \cdots + \frac{1}{2^{n-1}}\right)$$

$$= 2 + \left(1 - \frac{1}{2^{n-1}}\right)$$

$$= 3 - \frac{1}{2^{n-1}},$$

so $a_n < 3$.

For any $\varepsilon > 0$, there is some n with $e - a_n < \varepsilon$, since e is the least upper bound. Since $a_n < a_{n+1} < \cdots$ we have $e - a_k < \varepsilon$ for all $k \geq n$.

(d) If $n \leq x \leq n + 1$, then

$$\frac{1}{n+1} \leq \frac{1}{x} \leq \frac{1}{n}$$

so

$$\left(1 + \frac{1}{n+1}\right)^n \leq \left(1 + \frac{1}{x}\right)^n \leq \left(1 + \frac{1}{x}\right)^x \leq \left(1 + \frac{1}{x}\right)^{n+1} \leq \left(1 + \frac{1}{n}\right)^{n+1}.$$

Now

$$\left(1 + \frac{1}{n+1}\right)^n = \left[\left(1 + \frac{1}{n+1}\right)^{n+1}\right]^{\frac{n+1}{n}}$$

For large enough n the terms in brackets $[\quad]$ is close to e, and $(n+1)/n$ is close to 1, so the whole expression is close to e. Similarly, $(1 + 1/n)^{n+1}$ is close to e for large n. So $(1 + 1/x)^x$ is close to e for large x, i.e., $\lim_{x \to \infty} (1 + 1/x)^x = e$. This implies that $\lim_{h \to 0^+} (1 + h)^{1/h} = e$ (Problem 5-34).

We also have

$$\lim_{x \to \infty} \frac{\left(1 + \dfrac{1}{x}\right)^x}{\left(1 - \dfrac{1}{x}\right)^x} = \lim_{x \to \infty} \left(\frac{x+1}{x-1}\right)^x = \lim_{x \to \infty} \left(1 + \frac{2}{x-1}\right)^x$$

$$= \lim_{x \to \infty} \left[\left(1 + \frac{2}{x-1}\right)^{\frac{x-1}{2}}\right]^{\frac{2x}{x-1}}$$

$$= \left[\lim_{y \to \infty} \left(1 + \frac{1}{y}\right)^y\right]^2,$$

$$= e^2,$$

so

$$\lim_{x \to \infty} \left(1 - \frac{1}{x}\right)^x = \lim_{x \to \infty} \left(1 + \frac{1}{x}\right)^x \cdot e^{-2} = e^{-1}.$$

Consequently,

$$\lim_{x \to -\infty} \left(1 + \frac{1}{x}\right)^x = \lim_{x \to \infty} \left(1 - \frac{1}{x}\right)^{-x}$$

$$= \frac{1}{e^{-1}} = e.$$

It follows that $\lim_{h \to 0^-} (1 + h)^{1/h} = e$, and thus that $\lim_{h \to 0} (1 + h)^{1/h} = e$.

33. If $A(t) = P(t) = 10^7$, then

$$A'(t) = P'(t) = 10^7 - P(t) = -A(t).$$

So (by Problem 20) there is some number k such that

$$A(t) = ke^{-t}.$$

Since $A(0) = P(0) - 10^7 = -10^7$, we obtain $k = -10^7$, so

$$P(t) - 10^7 = -10^7 e^{-t},$$

so

$$10^7 t = \text{Nap } \log[10^7 - P(t)]$$
$$= \text{Nap } \log 10^7 e^{-t};$$

letting $x = 10^7 e^{-t}$, so that $t = \log(10^7/x)$, we obtain

$$\text{Nap } \log x = 10^7 \log \frac{10^7}{x}.$$

34. (a) We have $\lim\limits_{x \to 0^+} f(x) = -\infty$ and $\lim\limits_{x \to \infty} f(x) = 0$ by Problem 12.

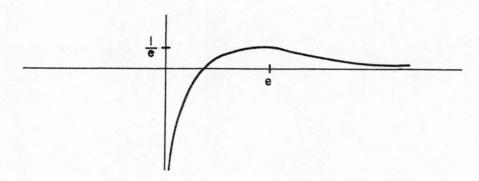

(b) Since f has its maximum at e, we have

$$\frac{\log e}{e} > \frac{\log \pi}{\pi},$$

so

$$e \log \pi > \pi \log e,$$

so

$$\pi^e > e^\pi.$$

(c) The equation $x^y = y^x$ is equivalent to $f(x) = f(y)$. The assertions in part (c) amount to the fact that the values $f(x)$ for $0 < x < 1$ or $x = e$ are taken on only once, while the values $f(x)$ for $1 < x < e$ are taken on for some $x' > e$ and vice versa.

(d) Part (c) shows that the only possible natural numbers $x < y$ with $x^y = y^x$ must involve $1 < x < e$, so $x = 2$.

(e), (f) If f_1 and f_2 are defined as in part (f), then $g = f_2^{-1} \circ f_1$. The curve in part (e) is the graph of g on $(1, e)$; the straight line is the graph of the identity function. They "intersect" at (e, e) [more precisely $\lim_{x \to e} g(x) = e$].

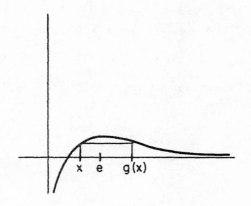
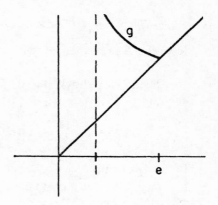

Moreover, g is differentiable, since f_1 and f_2 are differentiable and $f_2'(x) \neq 0$ for all x in the domain of f_2. In fact, we have

$$g'(a) = (f_2^{-1} \circ f_1)'(x) = (f_2^{-1})'(f_1(x)) \cdot f_1'(x)$$

$$= \frac{1}{f_2'\left(f_2^{-1}(f_1(x))\right)} \cdot f_1'(x)$$

$$= \frac{[g(x)]^2}{1 - \log g(x)} \cdot \frac{1 - \log x}{x^2}.$$

35. (a) exp is convex, since $\exp''(x) = \exp(x) > 0$ for all x. Similarly, log is concave, since $\log''(x) = -1/x^2 < 0$ for all $x > 0$.

(b) Naturally we are assuming that $z_i > 0$. Problem 9 of the Appendix to Chapter 11, applied to the convex function exp, shows that

$$\exp\left(\sum_{i=1}^{n} p_i \log z_i\right) > \sum_{i=1}^{n} p_i \exp(\log z_i)$$

or

$$z_1{}^{p_1} \cdots z_n{}^{p_n} \leq p_1 z_1 + \cdots + p_n z_n.$$

(c) Choose $p_i = 1/n$.

36. (a) If m_i is the inf of f on $[t_{i-1}, t_i]$, then

$$\frac{1}{b-a} L(\log f, P_n) = \frac{1}{b-a} \sum_{i=1}^{n} \log m_i \frac{b-a}{n} = \frac{1}{n} \sum_{i=1}^{n} \log m_i$$

$$= \log\left[(m_1 \cdots m_n)^{1/n}\right],$$

while

$$\log\left(\frac{1}{b-a}L(f, P_n)\right) = \log\left[\frac{1}{n} \cdot \sum_{i=1}^{n} m_i\right].$$

Since $(m_1 \cdots m_n)^{1/n} \leq (1/n) \cdot \sum_{i=1}^{n} m_i$ by Problem 2-22, and log is increasing, we have the desired inequality.

(b) Theorem 1 shows that if f is integrable then for every $\varepsilon > 0$ there is $\delta > 0$ such that

$$\left|\sum_{i=1}^{n} f(x_i)(t_i - t_{i-1}) - \int_a^b f(x)\,dx\right| < \varepsilon/2$$

for any partition $P = \{t_0, \ldots, t_n\}$ of $[a, b]$, and choices x_i in $[t_{i-1}, t_i]$, for which all $t_i - t_{i-1} < \delta$. It is easy to conclude that we then have

$$\left|L(f, P) - \int_a^b f(x)\,dx\right| < \varepsilon$$

for such partitions (we need to increase $\varepsilon/2$ to ε since m_i may not actually be $f(x_i)$ for any x_i in $[t_{i-1}, t_i]$). In particular,

$$\left|L(\log f, P_n) - \int_a^b \log f\right| < \varepsilon$$

$$\left|L(f, P_n) - \int_a^b f\right| < \varepsilon$$

for n sufficiently large. The desired result then follows easily from part (a).

(c) Let $P = \{t_0, \ldots, t_n\}$ be any partition of $[a, b]$, and let m_i be the inf of f on $[t_{i-1}, t_i]$. Letting $p_i = (t_i - t_{i-1})/(b - a)$, we have

$$\frac{1}{b-a}L(\log f, P) = \sum_{i=1}^{n} p_i \log m_i \leq \log\left(\sum_{i=1}^{n} p_i m_i\right)$$

$$= \log\left(\frac{1}{b-a}L(f, P)\right).$$

Since this is true for all partitions P, we have

$$\frac{1}{b-a}\int_a^b \log f \leq \log\left(\frac{1}{b-a}\int_a^b f\right).$$

(d) More generally, if g is concave and increasing, then

$$\frac{1}{b-a}\int_a^b g \circ f \leq g\left(\frac{1}{b-a}\int_a^b f\right).$$

37. From $f' = f$ we conclude that $f(x) = ce^x$ for some c. From

$$f(x + 0) = f(x)f(0)$$

we conclude that either $f(x) = 0$ for all x, or else that $f(0) = 1$, in which case $c = 1$.

38. Suppose that $f \neq 0$. From $f(x + 0) = f(x)f(0)$ it follows that $f(0) = 1$. Then from

$$1 = f(0) = f(x + (-x)) = f(x) \cdot f(-x)$$

it follows that $f(x) \neq 0$ for all x. Moreover $f(x) > 0$ for all x, since

$$f(x) = f(x/2 + x/2) = f(x/2)^2.$$

Now if n is a natural number, then

$$f(n) = f(\underbrace{1 + \cdots + 1}_{n \text{ times}}) = f(1)^n;$$

moreover,

$$1 = f(0) = f(n + (-n)) = f(n) \cdot f(-n),$$

so

$$f(-n) = \frac{1}{f(n)} = \frac{1}{f(1)^n} = f(1)^{-n}.$$

Similarly,

$$f(1) = f\left(\underbrace{\frac{1}{n} + \cdots + \frac{1}{n}}_{n \text{ times}}\right) = f\left(\frac{1}{n}\right)^n,$$

so

$$f\left(\frac{1}{n}\right) = \sqrt[n]{f(1)} = f(1)^{1/n}.$$

Finally

$$f\left(\frac{m}{n}\right) = f\left(\underbrace{\frac{1}{n} + \cdots + \frac{1}{n}}_{m \text{ times}}\right) = f\left(\frac{1}{n}\right)^m = f(1)^{m/n}.$$

Since f agrees with $g(x) = [f(1)]^x$ for rational x, it follows from Problem 8-6 that $f = g$.

39. If $g(x) = f(e^x)$, then

$$g(x + y) = f(e^{x+y}) = f(e^x \cdot e^y) = f(e^x) + f(e^y) = g(x) + g(y).$$

It follows from Problem 8-7 that $g(x) = cx$ for some c. If $c = 0$, then $f = 0$. If $c \neq 0$, then

$$f(e) = f(e^1) = g(1) = c,$$

so

$$f(e^x) = f(e)x$$

or

$$f(x) = f(e) \log x \qquad \text{for } x > 0.$$

40. The formulas for $f'(x)$ and $f''(x)$ (for $x \neq 0$) given in the text suggest the following conjecture, which is easy to prove by induction on k:

$$f^{(k)}(x) = e^{-1/x^2} \sum_{i=1}^{3k} \frac{a_i}{x^i} \qquad \text{for some numbers } a_1, \ldots, a_{3k}.$$

It is then clear that $f^{(k)}(0) = 0$ for all k, using the same argument as in the text.

41. The following conjecture is easy to verify:

$$f^{(k)}(x) = e^{-1/x^2} \left[\sum_{i=1}^{3k} \frac{a_i}{x^i} \sin\frac{1}{x} + \sum_{i=1}^{3k} \frac{b_i}{x^i} \cos\frac{1}{x} \right]$$

$$\text{for some numbers } a_1, \ldots, a_{3k}, b_1, \ldots, b_{3k}.$$

It is then clear that $f^{(k)}(0) = 0$ for all k, as in the previous example (note that $|\sin 1/x| \leq 1$ and $|\cos 1/x| \leq 1$ for all $x \neq 0$).

42. (a) If $y(x) = e^{\alpha x}$, then

$$a_n y^{(n)}(x) + a_{n-1} y^{(n-1)}(x) + \cdots + a_1 y'(x) + a_0 y(x)$$
$$= a_n \alpha^n e^{\alpha x} + a_{n-1} \alpha^{n-1} e^{\alpha x} + \cdots + a_1 \alpha e^{\alpha x} + a_0 e^{\alpha x}$$
$$= e^{\alpha x} \left(a_n \alpha^n + a_{n-1} \alpha^{n-1} + \cdots + a_1 \alpha + a_0 \right) = 0.$$

(b) If $y(x) = xe^{\alpha x}$, then

$$y^{(l)}(x) = \alpha^l x e^{\alpha x} + l\alpha^{l-1} e^{\alpha x}.$$

(This formula can be verified by induction, or deduced from Problem 10-18.) So

$$a_n y^{(n)}(x) + a_{n-1} y^{(n-1)}(x) + \cdots + a_1 y'(x) + a_0 y(x)$$
$$= xe^{\alpha x} \left[a_n \alpha^n + a_{n-1} \alpha^{n-1} + \cdots + a_1 \alpha + a_0 \right]$$
$$\quad + e^{\alpha x} \left[na_n \alpha^{n-1} + \cdots + a_1 \right]$$
$$= 0$$

(the second term in brackets is 0 because α is a double root of (*)).

(c) If $y(x) = x^k e^{\alpha x}$, then by Problem 10-18,

$$y^{(l)}(x) = \left[\sum_{s=0}^{k} \binom{l}{s} \frac{k!}{(k-s)!} x^{k-s} \alpha^{l-s} \right] e^{\alpha x}.$$

So

$$\sum_{l=0}^{n} a_l y^{(l)}(x) = \sum_{s=0}^{k} \left[\sum_{l=0}^{n} \binom{l}{s} a_l \alpha^{l-s} \right] \frac{k!}{(k-s)!} x^{k-s} e^{\alpha x} = 0$$

(the terms in brackets are 0 because α is a root of (*) of order $s+a$, for each $s \leq k$).

(d) If $y_1, \ldots, y_n$ satisfy (**), then

$$\sum_{l=0}^{n} a_l (c_1 y_1 + \cdots + c_n y_n)^{(l)} = \sum_{j=1}^{n} \left(c_j \sum_{l=0}^{n} a_l y_j^{(l)} \right) = 0.$$

43. (a) From

$$0 = f'(f'' - f) = f'f'' - ff' = \frac{1}{2} \left[(f')^2 - f^2 \right]'$$

it follows that $(f')^2 - f^2$ is constant. The constant must be 0, since $f(0) = f'(0) = 0$.

(b) Since $f(x) \neq 0$ for x in (a, b), it follows from part (a) that either $f'(x) = f(x)$ for all x in (a, b) or else $f'(x) = -f(x)$ for all x in (a, b). Thus either $f(x) = ce^s$ or else $f(x) = ce^{-x}$ for all x in (a, b).

(c) Let a be the largest number in $[0, x_0]$ with $f(a) = 0$. Then $f(x) \neq 0$ for x in (a, x_0). But then $f(x) = ce^x$ or $f(x) = ce^{-x}$ for all x in (a, x_0), where $c \neq 0$. This contradicts $f(a) = 0$, since f is continuous, because $f(a) = 0 \neq \lim_{x \to a} ce^x$ or $\lim_{x \to a} ce^{-x}$.

44. (a) Let

$$a = \frac{f(0) + f'(0)}{2}$$

$$b = \frac{f(0) - f'(0)}{2}.$$

If $g(x) = ae^x + be^{-x} - f(x)$, then $g'' - g = 0$, so $f(x) = ae^x + be^{-x}$.

(b) Note that

$$ae^x + be^{-x} = (a - b) \frac{e^x - e^{-x}}{2} + (a + b) \frac{e^x + e^{-x}}{2}$$
$$= (a - b) \sinh x + (a + b) \cosh x.$$

(Comparing with part (a) we see that

$$f(x) = f'(0) \sinh x + f(0) \cosh x,$$

in exact analogy with the trigonometric functions.)

45. (a) We have $f^{(n-1)}(x) = ce^x$, so

$$f(x) = a_0 + a_1 x + \cdots + a_{n-2} x^{n-2} + ce^x.$$

(b) We have

$$f^{(n-2)}(x) = ce^x + de^{-x}$$

by Problem 44, so

$$f(x) = a_0 + a_1 x + \cdots + a_{n-3} x^{n-3} + ce^x + de^{-x}.$$

46. (a) Since

$$\begin{aligned}
g'(x) &= f'(x_0 + x) f(x_0 - x) - f(x_0 + x) f'(x_0 - x) \\
&= f(x_0 + x) f(x_0 - x) - f(x_0 + x) f(x_0 - x) = 0,
\end{aligned}$$

the function g is constant. Moreover, $g(0) = f(x_0)^2 \neq 0$. So

$$f(x_0 + x) f(x_0 - x) \neq 0 \qquad \text{for all } x,$$

which implies that $f(x) \neq 0$ for all x.

(b) Let $f = f_1 / f_1(0)$, where $f_1 \neq 0$ and $f_1' = f_1$.

(c) Since

$$\begin{aligned}
g'(x) &= \frac{f(x) f'(x+y) - f(x+y) f'(x)}{f(x)^2} \\
&= \frac{f(x) f(x+y) - f(x+y) f(x)}{f(x)^2} = 0,
\end{aligned}$$

the function g is constant, and clearly $g(0) = f(y)$, so $f(x+y)/f(x) = f(y)$ for all x.

(d) f is increasing, since $f'(x) = f(x) = f(x/2 + x/2) = [f(x/2)]^2 > 0$. Moreover,

$$\begin{aligned}
(f^{-1})'(x) &= \frac{1}{f'(f^{-1}(x))} \\
&= \frac{1}{f(f^{-1}(x))} = \frac{1}{x}.
\end{aligned}$$

47. (a) No. For example, let $f(x) = x$ and let $g(x) = x(2 + \sin x)$.

(b) We have

$$\lim_{x\to\infty} \frac{f(x)+g(x)}{f(x)} = 1 + \lim_{x\to\infty}\frac{g(x)}{f(x)} = 1 + 0, \qquad \text{since } \lim_{x\to\infty}\frac{f(x)}{g(x)} = \infty.$$

(c) For sufficiently large x we have

$$\log f(x) \geq c \log g(x)$$

so

$$f(x) \geq g(x)^c \qquad c > 1$$

and therefore

$$\frac{f(x)}{g(x)} \geq g(x)^{c-1} \qquad c - 1 > 0.$$

Since we are assuming that $\lim_{x\to\infty} g(x) = \infty$, this implies that $\lim_{x\to\infty} f(x)/g(x) = \infty$.

(d) Yes. *Proof*: Given $N > 0$, choose x_0 such that $f(y) \geq 2Ng(y)$ for all $y \geq x_0$. Then

$$F(x_0 + x) = \int_0^{x_0+x} f = \int_0^{x_0} f + \int_{x_0}^{x_0+x} f \geq \int_0^{x_0} f + 2N \int_{x_0}^{x_0+x} g$$

$$= \int_0^{x_0} f + 2N \int_0^{x_0+x} g - 2N \int_0^{x_0} g$$

$$= \int_0^{x_0} f - 2N \int_0^{x_0} g + 2NG(x_0 + x)$$

$$= A + 2NG(x_0 + x), \qquad \text{say.}$$

So

$$\frac{F(x_0 + x)}{G(x_0 + x)} = 2N + \frac{A}{G(x_0 + x)}.$$

Since $G(x_0 + x) \to \infty$ as $x \to \infty$, it follows that $F(x_0 + x)/G(x_0 + x) > N$ for large enough x.

(e) (i) $\log 4x \ll x + e^{-5x} \ll x^3 + \log(x^3) \sim x^3$ [by part (b)] $\ll x^3 \log x \ll e^x \ll (\log x)^x \ll x^x$.

(ii) $\log(x^x) [= x \log x] \ll x \log^2 x [\ll x^3] \ll x^{\log x} \ll e^{5x} \ll (\log x)^x \ll x^x \ll e^{x^2}$ [the last four $\ll$ follow easily from part (c)].

(iii) $x^e \ll e^{x/2} \ll 2^x \ll e^x \ll (\log x)^{2x} \ll x^x \ll e^{x^2}$ [the second $\ll$ depends on the fact that $\log 2 > 1/2$ which is true since $2 > \sqrt{e}$].

48. Let M_n be the maximum of $|g_1| + \cdots + |g_n|$ on $[0, n]$ and choose f so that $f(x) \geq nM_n$ on $[0, n]$.

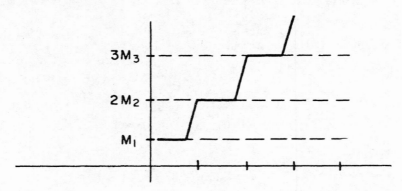

49. If there were natural numbers a and b with $\log_{10} 2 = a/b$, then $2 = 10^{a/b}$, so

$$2^b = 10^a.$$

This contradicts the fact, mentioned in Problem 2-17, that an integer can be factored uniquely into primes (since the product 2^b does not involve the prime 5, while the product 10^a does).

2. (ii) $-e^{-x^2}/2$. (Let $u = -x^2$.)

(iv) $-1/(e^x + 1)$. (Let $u = e^x$.)

(vi) $(\arcsin x^2)/2$. (Let $u = x^2$.)

(viii) $-(1 - x^2)^{3/2}/3$. (Let $u = 1 - x^2$.)

(x) $[\log(\log x)]^2/2$. (Let $u = \log(\log x)$.)

3. (ii)

$$\int x^3 e^{x^2}\, dx = \int x^2 \left(x e^{x^2}\right) dx = \frac{x^2 e^{x^2}}{2} - \int x e^{x^2}\, dx$$
$$= \frac{x^2 e^{x^2}}{2} - \frac{e^{x^2}}{2}.$$

(iv)

$$\int x^2 \sin x\, dx = x^2(-\cos x) + 2\int x \cos x\, dx$$
$$= -x^2 \cos x + 2\left[x \sin x - \int \sin x\, dx\right]$$
$$= -x^2 \cos x + 2x \sin x + 2\cos x.$$

(vi)

$$\int \log(\log x) \cdot \frac{1}{x}\, dx = (\log x) \cdot \log(\log x) - \int \log x \cdot \frac{1}{\log x} \cdot \frac{1}{x}\, dx$$
$$= (\log x) \cdot \log(\log x) - \log x.$$

(viii)

$$\int 1 \cdot \cos(\log x)\, dx = x \cos(\log x) + \int x \sin(\log x) \cdot \frac{1}{x}\, dx$$
$$= x \cos(\log x) + \int 1 \cdot \sin(\log x)\, dx$$
$$= x \cos(\log x) + x \sin(\log x) - \int x \cos(\log x) \cdot \frac{1}{x}\, dx,$$

so

$$\int \cos(\log x)\, dx = \frac{x \cos(\log x) + x \sin(\log x)}{2}.$$

(x)

$$\int \log x)^2 = \frac{x^2(\log x)^2}{2} - \int \frac{x^2}{2} \cdot 2\log x \cdot \frac{1}{x}\,dx$$

$$= \frac{x^2(\log x)^2}{2} - \int x\log x\,dx$$

$$= \frac{x^2(\log x)^2}{2} - \left(\frac{x^2\log x}{2} - \int \frac{x^2}{2} \cdot \frac{1}{x}\,dx\right)$$

$$= \frac{x^2(\log x)^2}{2} - \frac{x^2\log x}{2} + \frac{x^2}{4}.$$

4. (ii) Let $x = \tan u$, $dx = \sec^2 u\,du$. The integral becomes

$$\int \frac{\sec^2 u\,du}{\sqrt{1 + \tan^2 u}} = \int \sec u\,du = \log(\sec u + \tan u)$$

$$= \log\left(x + \sqrt{1 + x^2}\right).$$

(iv) Let $x = \sec u$, $dx = \sec u \tan u\,du$. The integral becomes

$$\int \frac{\sec u \tan u\,du}{\sec u\sqrt{\sec^2 u - 1}} = \int 1\,du = u = \operatorname{arcsec} x.$$

[This can be written in terms of more familiar functions as $\arctan\left(\sqrt{x^2 - 1}\right)$.]

(vi) Let $x = \tan u$, $dx = \sec^2 u\,du$. The integral becomes

$$\int \frac{\sec^2 u\,du}{\tan u\sqrt{1 + \tan^2 u}} = \int \frac{\sec u\,du}{\tan u} = \int \csc u\,du$$

$$= -\log(\csc u + \cot u) = -\log\left(\frac{1}{x} + \frac{\sqrt{1 + x^2}}{x}\right)$$

$$= \log\left(\frac{x}{1 + \sqrt{1 + x^2}}\right).$$

(viii) Let $x = \sin u$, $dx = \cos u\,du$. The integral becomes

$$\int \sqrt{1 - \sin^2 u}\,\cos u\,du = \int \cos^2 u\,du = \int \frac{1 + \cos 2u}{2}\,du$$

$$= \frac{u}{2} + \frac{\sin 2u}{4}$$

$$= \frac{u}{2} + \frac{\sin u \cos u}{2}$$

$$= \frac{\arcsin x}{2} + \frac{x\sqrt{1 - x^2}}{2}.$$

(x) Let $x = \sec u$, $dx = \sec u \tan u \, du$. The integral becomes

$$\int \sqrt{\sec^2 u - 1} \, \sec u \tan u \, du$$

$$= \int \sec u \tan^2 u \, du$$

$$= \int (\sec u)(\sec^2 u - 1) \, du = \int \sec^3 u \, du - \int \sec u \, du$$

$$= \frac{1}{2}[\tan u \sec u + \log(\sec u + \tan u)] \qquad \text{[Problem 3(vi)]}$$

$$- \log(\sec u + \tan u)$$

$$= \frac{1}{2} x \sqrt{x^2 - 1} - \frac{1}{2} \log\left(x + \sqrt{x^2 - 1}\right).$$

5. (ii) Let $u = e^x$, $x = \log u$, $dx = 1/u \, du$. The integral becomes

$$\int \frac{du}{u(1+u)} = \int \frac{1}{u} - \frac{1}{1+u} \, du$$

$$= \log u - \log(1 + u)$$

$$= x - \log(1 + e^x).$$

(iv) Let $u = \sqrt{1 + e^x}$, $x = \log(u^2 - 1)$, $dx = 2u/(u^2 - 1) \, du$. The integral becomes

$$\int \frac{2u \, du}{u(u^2 - 1)} = \int -\frac{1}{u+1} + \frac{1}{u-1} \, du$$

$$= -\log(u + 1) + \log(u - 1)$$

$$= -\log\left(1 + \sqrt{1 + e^x}\right) + \log\left(\sqrt{1 + e^x} - 1\right).$$

(vi) Let $u = \sqrt{\sqrt{x} + 1}$, $x = (u^2 - 1)^2$, $dx = 4u(u^2 - 1) \, du$. The integral becomes

$$\int \frac{4u(u^2 - 1) \, du}{u} = \frac{4}{3}u^3 - 4u$$

$$= \frac{4}{3}\left(\sqrt{x} + 1\right)^{3/2} - 4\left(\sqrt{x} + 1\right)^{1/2}.$$

(viii) Let $u = \sqrt{x}$, $x = u^2$, $dx = 2u \, du$. The integral becomes

$$\int 2u e^u \, du = 2u e^u - 2 \int e^u \, du$$

$$= 2\sqrt{x} e^{\sqrt{x}} - 2e^{\sqrt{x}}.$$

(x) Let $u = 1/x$, $x = 1/u$, $dx = -1/u^2 \, du$. The integral becomes

$$\int \sqrt{\frac{1/u - 1}{1/u + 1}} \cdot u^2 \cdot -\frac{1}{u^2} \, du = -\int \frac{\sqrt{1 - u}}{\sqrt{1 + u}} \, du = -\int \frac{1 - u}{\sqrt{1 - u^2}} \, du.$$

Now let $u = \sin t$, $du = \cos t\, dt$. The integral becomes

$$-\int \frac{1-\sin t}{\cos t} \cos t\, dt = -\int 1 - \sin t\, dt$$

$$= -t - \cos t$$

$$= -\arcsin u - \sqrt{1-u^2}$$

$$= -\arcsin \frac{1}{x} - \frac{\sqrt{x^2-1}}{x}.$$

6. In this answer set, I will denote the original integral.

(ii)

$$I = \int \frac{2}{(x-1)^2}\, dx + \frac{3}{(x-1)^3}\, dx = -\frac{2}{(x-1)} - \frac{3}{2(x-1)^2}.$$

(iv)

$$I = \int \frac{1}{x+3} + \frac{1}{x-1} + \frac{1}{(x-1)^2}\, dx = \log(x+3) + \log(x-1) - \frac{1}{x-1}.$$

(vi)

$$I = \int \frac{x}{x^2+1} + \frac{2}{(x^2+1)^2}\, dx$$

$$= \frac{\log(x^2+1)}{2} + 2\left[\frac{x}{2(x^2+1)} + \frac{1}{2}\int \frac{1}{x^2+1}\, dx \right]$$

$$= \frac{\log(x^2+1)}{2} + \frac{x}{x^2+1} + \arctan x.$$

(viii)

$$I = \int \frac{dx}{x^4 + 2x^2 + 1 - 2x^2} = \int \frac{dx}{(x^2+1)^2 - 2x^2}$$

$$= \int \frac{dx}{(x^2 + \sqrt{2}x + 1)(x^2 - \sqrt{2}x + 1)}$$

$$= \int \frac{\dfrac{\sqrt{2}}{4}x + \dfrac{1}{2}}{(x^2 + \sqrt{2}x + 1)} + \frac{-\dfrac{\sqrt{2}}{4}x + \dfrac{1}{2}}{(x^2 - \sqrt{2}x + 1)}\, dx$$

$$= \frac{\sqrt{2}}{8}\int \frac{(2x + \sqrt{2})\, dx}{(x^2 + \sqrt{2}x + 1)} + \frac{1}{4}\int \frac{dx}{(x^2 + \sqrt{2}x + 1)}$$

$$\qquad - \frac{\sqrt{2}}{8}\int \frac{(2x - \sqrt{2})\, dx}{(x^2 - \sqrt{2}x + 1)} + \frac{1}{4}\int \frac{dx}{(x^2 - \sqrt{2}x + 1)}$$

$$= \frac{\sqrt{2}}{8} \log(x^2 + \sqrt{2}x + 1) - \frac{\sqrt{2}}{8} \log(x^2 - \sqrt{2}x + 1)$$

$$+ \frac{1}{2} \int \frac{dx}{(\sqrt{2}x + 1)^2 + 1} + \frac{1}{2} \int \frac{dx}{(-\sqrt{2}x + 1)^2 + 1}$$

$$= \frac{\sqrt{2}}{8} \log(x^2 + \sqrt{2}x + 1) - \frac{\sqrt{2}}{8} \log(x^2 - \sqrt{2}x + 1)$$

$$+ \frac{\sqrt{2}}{4} \arctan(\sqrt{2}x + 1) - \frac{\sqrt{2}}{4} \arctan(-\sqrt{2}x + 1).$$

(x)

$$I = \frac{3}{2} \int \frac{2x + 1}{(x^2 + x + 1)^3} dx - \frac{3}{2} \int \frac{1}{(x^2 + x + 1)^3} dx$$

$$= \frac{-3}{4(x^2 + x + 1)^2} - \frac{3}{2} \int \frac{dx}{\left((x + \frac{1}{2})^2 + \frac{3}{4}\right)^3}$$

$$= \frac{-3}{4(x^2 + x + 1)^2} - \frac{3}{2} \left(\frac{4}{3}\right)^3 \int \frac{dx}{\left(\left(\dfrac{x + \frac{1}{2}}{\sqrt{3/4}}\right)^2 + 1\right)^3}$$

$$\left(\text{let } u = \frac{x + \frac{1}{2}}{\sqrt{3/4}}, \quad dx = \sqrt{3/4}\, du\right)$$

$$= \frac{-3}{4(x^2 + x + 1)^2} - 32\sqrt{3/4} \int \frac{du}{(u^2 + 1)^3}$$

$$= \frac{-3}{4(x^2 + x + 1)^2} - 32\sqrt{3/4} \left[\frac{1}{4} \frac{u}{(u^2 + 1)^2} + \frac{3}{4} \int \frac{du}{(u^2 + 1)^2}\right]$$

$$= \frac{-3}{4(x^2 + x + 1)^2} - \frac{8\sqrt{3/4}\, u}{(u^2 + 1)^2} - 24\sqrt{3/4} \left[\frac{1}{2} \frac{u}{(u^2 + 1)} + \frac{1}{2} \int \frac{1}{u^2 + 1} du\right]$$

$$= \frac{-3}{4(x^2 + x + 1)^2} - \frac{8x + 4}{(x^2 + x + 1)} \cdot \frac{4}{3} - \frac{12x + 6}{x^2 + x + 1} \sqrt{\frac{4}{3}} + \frac{12}{x^2 + x + 1}.$$

7. (i) $(\arctan)^2 / 2.$

(ii)

$$\int \frac{x}{(1 + x^2)^3} \arctan x\, dx = \frac{-1}{4(1 + x^2)^2} \arctan x - \int \frac{x\, dx}{(1 + x^2)^4}$$

$$= \frac{-\arctan x}{4(1 + x^2)^2} + \frac{1}{6(1 + x^2)^3}.$$

(iii)

$$\int \log \sqrt{1+x^2}\, dx = x \log \sqrt{1+x^2} - \int x \cdot \frac{1}{\sqrt{1+x^2}} \cdot \frac{2x}{\sqrt{1+x^2}}\, dx$$

$$= x \log \sqrt{1+x^2} - \int \frac{2x^2}{1+x^2}\, dx$$

$$= x \log \sqrt{1+x^2} - \int 2 + \frac{-2}{1+x^2}\, dx$$

$$= x \log \sqrt{1+x^2} - 2x + 2 \arctan x.$$

(iv)

$$\int x \log \sqrt{1+x^2}\, dx = \frac{x^2}{2} \log \sqrt{1+x^2} - \int \frac{x^2}{2} \cdot \frac{2x}{1+x^2}\, dx$$

$$= \frac{x^2}{2} \log \sqrt{1+x^2} - \int x - \frac{x}{1+x^2}\, dx$$

$$= \frac{x^2}{2} \log \sqrt{1+x^2} - \frac{x^2}{2} + \frac{\log(1+x^2)}{2}.$$

(v) Let

$$y = \frac{x^2 - 1}{x^2 + 1},$$

so that

$$yx^2 + y = x^2 - 1,$$

$$y + 1 = x^2(1 - y)$$

$$x = \sqrt{\frac{1+y}{1-y}},$$

$$dx = \sqrt{\frac{1-y}{1+y}}\, \frac{1}{(1-y)^2}\, dy.$$

The integral becomes

$$\int y \frac{1}{\sqrt{1 + \left(\frac{1+y}{1-y}\right)^2}} \frac{\sqrt{1-y}}{\sqrt{1+y}} \frac{1}{(1-y)^2}\, dy = \int \frac{y}{(1-y)^2} \frac{1}{\sqrt{\frac{2+2y^2}{(1-y)^2}}} \frac{\sqrt{1-y}}{\sqrt{1+y}}\, dy$$

$$= \int \frac{y}{(1-y)} \frac{1}{\sqrt{2+2y^2}} \frac{\sqrt{1-y}}{\sqrt{1+y}}\, dy$$

$$= \frac{1}{\sqrt{2}} \int \frac{y\, dy}{\sqrt{1-y^2}\sqrt{1+y^2}}.$$

Now let $u = y^2$, $du = 2y\,dy$. The integral becomes

$$\frac{1}{2\sqrt{2}} \int \frac{du}{\sqrt{1-u}\sqrt{1+u}} = \frac{1}{2\sqrt{2}} \int \frac{du}{\sqrt{1-u^2}}$$

$$= \frac{1}{2\sqrt{2}} \arcsin u$$

$$= \frac{1}{2\sqrt{2}} \arcsin \left(\frac{x^2-1}{x^2+1}\right)^2.$$

(vi) Let $u = \sqrt{x}$, $x = u^2$, $dx = 2u\,du$. The integral becomes

$$\int 2u \arcsin u\,du = u^2 \arcsin u - \int \frac{u^2}{\sqrt{1-u^2}}\,du.$$

Now let $u = \sin t$, $du = \cos t\,dt$. The integral becomes

$$\int \frac{\sin^2 t \cos t}{\sqrt{1-\sin^2 t}} = \int \sin^2 t\,dt = \int \frac{1-\cos 2t}{2}\,dt$$

$$= \frac{t}{2} - \frac{\sin 2t}{4}$$

$$= \frac{t}{2} - \frac{\sin t \cos t}{2}.$$

So the original integral is

$$u^2 \arcsin u - \frac{\arcsin u}{2} - \frac{u\sqrt{1-u^2}}{2}$$

$$= x \arcsin \sqrt{x} - \frac{\arcsin \sqrt{x}}{2} - \frac{\sqrt{x}\sqrt{1-x}}{2}.$$

(vii) Since

$$\int \frac{1}{1+\sin x}\,dx = \tan x - \sec x \qquad \text{(Problem 1(viii)))}$$

we have

$$\int x \cdot \frac{1}{1+\sin x}\,dx = x(\tan x - \sec x) - \int \tan x - \sec x\,dx$$

$$= x(\tan x - \sec x) + \log(\cos x) + \log(\sec x + \tan x).$$

(viii)

$$\int x \cos x e^{\sin x}\, dx - \int \sec x \tan x e^{\sin x}\, dx$$

$$= \left(x e^{\sin x} - \int e^{\sin x}\, dx \right)$$

$$- \left(\sec x e^{\sin x} - \int \sec x \cos x e^{\sin x}\, dx \right)$$

$$= x e^{\sin x} - \sec x e^{\sin x}.$$

(ix) Let

$$u = \sqrt{\tan x},$$

so that

$$u^2 = \tan x$$
$$x = \arctan u^2$$
$$dx = \frac{2u}{1 + u^4}\, du.$$

The integral becomes (compare with Problem 6(viii)))

$$\int \frac{2u^2\, du}{1 + u^4}$$

$$= \int \frac{-\dfrac{\sqrt{2}}{2} u}{u^2 + \sqrt{2}u + 1} + \frac{\dfrac{\sqrt{2}}{2} u}{u^2 - \sqrt{2}u + 1}\, du$$

$$= \frac{-\sqrt{2}}{4} \log(u^2 + \sqrt{2}u + 1) + \frac{\sqrt{2}}{4} \log(u^2 - \sqrt{2}u + 1)$$

$$+ \frac{\sqrt{2}}{2} \arctan(\sqrt{2}u + 1) - \frac{\sqrt{2}}{2} \arctan(-\sqrt{2}u + 1)$$

$$= \frac{-\sqrt{2}}{4} \log(\tan x + \sqrt{2 \tan x} + 1) + \frac{\sqrt{2}}{4} \log(\tan x - \sqrt{2 \tan x} + 1)$$

$$+ \frac{\sqrt{2}}{2} \arctan(\sqrt{2 \tan x} + 1) - \frac{\sqrt{2}}{2} \arctan(-\sqrt{2 \tan x} + 1).$$

(x)

$$\int \frac{dx}{x^6 + 1} = \int \frac{dx}{(x^2 + 1)(x^4 - x^2 + 1)} = \int \frac{dx}{(x^2 + 1)[(x^4 + 2x^2 + 1) - 3x^2]}$$

$$= \int \frac{dx}{(x^2 + 1)[(x^2 + 1)^2 - 3x^2]}$$

$$= \int \frac{dx}{(x^2 + 1)(x^2 + \sqrt{3}x + 1)(x^2 - \sqrt{3}x + 1)}$$

$$= \int \frac{\frac{1}{3}}{x^2+1} + \frac{\frac{\sqrt{3}}{6}+\frac{1}{3}}{x^2+\sqrt{3}x+1} + \frac{-\frac{\sqrt{3}}{6}x+\frac{1}{3}}{x^2-\sqrt{3}x+1}\,dx$$

$$= \frac{\arctan x}{3} + \frac{\sqrt{3}}{12}\int \frac{2x+\sqrt{3}}{x^2+\sqrt{3}+1}\,dx + \frac{1}{12}\int \frac{dx}{x^2+\sqrt{3}x+1}$$

$$- \frac{\sqrt{3}}{12}\int \frac{2x-\sqrt{3}}{x^2-\sqrt{3}x+1}\,dx + \frac{1}{12}\int \frac{dx}{x^2-\sqrt{3}x+1}$$

$$= \frac{\arctan x}{3} + \frac{\sqrt{3}}{12}\log\left(x^2+\sqrt{3}x+1\right) - \frac{\sqrt{3}}{12}\log\left(x^2-\sqrt{3}x+1\right)$$

$$+ \frac{1}{6}\arctan\left(2x+\sqrt{3}\right) + \frac{1}{6}\arctan\left(2x-\sqrt{3}\right).$$

8. (i)

$$\int \log(a^2+x^2)\,dx = x\log(a^2+x^2) - \int \frac{2x^2}{a^2+x^2}\,dx$$

$$= x\log(a^2+x^2) - \int 2 + \frac{-2a^2}{a^2+x^2}\,dx$$

$$= x\log(a^2+x^2) - \int 2 + \frac{-2}{1+\left(\frac{x}{a}\right)^2}\,dx$$

$$= x\log(a^2+x^2) - 2x + 2a\arctan(x/a).$$

(ii)

$$\int \frac{1+\cos x}{\sin^2 x}\,dx = \int \csc^2 dx + \int \frac{\cos x}{\sin^2 x}\,dx$$

$$= -\cot x - \frac{1}{\sin x}.$$

(iii)

$$\int \frac{x+1}{\sqrt{4-x^2}} = \int \frac{x\,dx}{\sqrt{4-x^2}} + \int \frac{dx}{2\sqrt{1-\left(\frac{x}{2}\right)^2}}$$

$$= -\sqrt{4-x^2} + \arcsin(x/2).$$

(iv)

$$\int x \arctan x \, dx = \frac{x^2 \arctan x}{2} - \frac{1}{2} \int \frac{x^2}{(1+x^2)} \, dx$$

$$= \frac{x^2 \arctan x}{2} - \frac{1}{2} \int 1 + \frac{-1}{1+x^2} \, dx$$

$$= \frac{x^2 \arctan x}{2} - \frac{x}{2} + \frac{1}{2} \arctan x.$$

(v)

$$\int \sin^3 x \, dx = \int \sin x (1 - \cos^2 x) \, dx$$

$$= -\cos x + \frac{\cos^3 x}{3}.$$

(vi)

$$\int \frac{\sin^3 x}{\cos^2 x} \, dx = \int \frac{\sin x (1 - \cos^2 x)}{\cos^2 x}$$

$$= \int \frac{\sin x}{\cos^2 x} - \sin x \, dx$$

$$= \frac{1}{\cos x} + \cos x.$$

(vii)

$$\int x^2 \arctan x \, dx = \frac{x^3 \arctan x}{3} - \frac{1}{3} \int \frac{x^3}{1+x^2} \, dx$$

$$= \frac{x^3 \arctan x}{3} - \frac{1}{3} \int x + \frac{-x}{1+x^2} \, dx$$

$$= \frac{x^3 \arctan x}{3} - \frac{x^2}{6} + \frac{\log(1+x^2)}{6}.$$

(viii)

$$\int \frac{x \, dx}{\sqrt{x^2 - 2x + 2}} = \int \frac{x \, dx}{\sqrt{(x-1)^2 + 1}}$$

$$= \int \frac{(x-1) \, dx}{\sqrt{1 + (x-1)^2}} + \int \frac{dx}{\sqrt{1 + (x-1)^2}}$$

$$= 2\sqrt{1 + (x-1)^2} + \log\left(x - 1 + \sqrt{1 + (x-1)^2}\right)$$

$$\text{(by Problem 4(ii))}.$$

(ix)

$$\int \sec^3 x \tan x \, dx = \int (\sec x \tan x) \sec^2 x \, dx$$

$$= \frac{\sec^3 x}{3}.$$

(x) Let $f(x) = x$, $g(x) = \int \tan^2 x \, dx = \tan x - x$ (Problem 1(v)). Then

$$\int x \tan^2 x \, dx = \int f(x) g'(x) \, dx = x(\tan x - x) - \int \tan x - x \, dx$$

$$= x(\tan x - x) + \log \cos x + \frac{x^2}{2}.$$

9. (i) Let $x = a \tan u$, $dx = a \sec^2 u \, du$. The integral becomes

$$\int \frac{a \sec^2 u \, du}{(a^2 + a^2 \tan^2 u)^2} = \int \frac{\sec^2 u \, du}{a^3 (\sec^2 u)^2}$$

$$= \frac{1}{a^3} \int \frac{du}{\sec^2 u} = \frac{1}{a^3} \int \cos^2 u \, du$$

$$= \frac{1}{a^3} \int \frac{1 + \cos 2u}{2} \, du$$

$$= \frac{u}{2a^3} + \frac{\sin 2u}{4a^3}$$

$$= \frac{1}{2a^3} \arctan \frac{x}{a} + \frac{2}{4a^3} \sin \left(\arctan \frac{x}{a} \right) \cos \left(\arctan \frac{x}{a} \right)$$

$$= \frac{1}{2a^3} \arctan \frac{x}{a} + \frac{1}{2a^3} \frac{x/a}{\sqrt{1 + \left(\frac{x}{a} \right)^2}} \cdot \frac{1}{\sqrt{1 + \left(\frac{x}{a} \right)^2}}$$

$$= \frac{1}{2a^3} \arctan \frac{x}{a} + \frac{1}{2a^2} \frac{x}{(x^2 + a^2)}.$$

(Or one could write $x = au$ and use the reduction formula.)

(ii) Let $u = \sin x$, $x = \arcsin u$, $dx = du/\sqrt{1 - u^2}$. The integral becomes

$$\int \frac{\sqrt{1 - u} \, du}{\sqrt{1 - u^2}} = \int \frac{du}{\sqrt{1 + u}} = 2\sqrt{1 + u} = 2\sqrt{1 + \sin x}.$$

(iii) Let $u = \sqrt{x}$, $x = u^2$, $dx = 2u \, du$. The integral becomes

$$\int 2u \arctan u \, du = u^2 \arctan u - u + \arctan u \qquad \text{by Problem 8(iv)}$$

$$= x \arctan \sqrt{x} - \sqrt{x} + \arctan \sqrt{x}.$$

(iv) Let $u = \sqrt{x+1}$, $x = u^2 - 1$, $dx = 2u\,du$. The integral becomes

$$\int 2u \sin u \, du = 2u(-\cos u) - 2 \int (-\cos u) \, du$$

$$= -2u \cos u + 2 \sin u$$

$$= -2\sqrt{x+1} \cos \sqrt{x+1} + 2 \sin \sqrt{x+1}.$$

(v) Let $u = \sqrt{x^3 - 2}$, $x = (u^2 + 2)^{1/3}$, $dx = 2u\,du/3(u^2 + 2)^{2/3}$. The integral becomes

$$\frac{2}{3} \int \frac{u^2 \, du}{(u^2 + 2)^{1/3}(x^2 + 2)^{2/3}} = \frac{2}{3} \int \frac{u^2 \, du}{u^2 + 2} = \frac{2}{3} \int 1 - \frac{2}{u^2 + 2} \, du$$

$$= \frac{2u}{3} - \frac{2\sqrt{2}}{3} \arctan \frac{u}{\sqrt{2}}$$

$$= \frac{2\sqrt{x^3 - 2}}{3} - \frac{2\sqrt{2}}{3} \arctan \sqrt{\frac{x^3 - 2}{2}}.$$

(vi) Let

$$u = x + \sqrt{x^2 - 1}$$

$$u - x = \sqrt{x^2 - 1}$$

$$u^2 - 2u + x^2 = x^2 - 1$$

so that

$$x = \frac{u^2 + 1}{2u}, \qquad dx = \left(\frac{1}{2} - \frac{1}{2u^2} \right) du.$$

The integral becomes

$$\frac{1}{2} \int \log u \, du - \frac{1}{2} \int \frac{1}{u^2} \log u \, du$$

$$= \frac{1}{2}(u \log u - 1) - \frac{1}{2} \left[-\frac{1}{u} \log u + \int \frac{1}{u} \cdot \frac{1}{u} \, du \right]$$

$$= \frac{1}{2}(u \log u - 1) + \frac{1}{2u} \log u + \frac{1}{2u}$$

$$= \frac{1}{2} \left[(x + \sqrt{x^2 - 1}) \log(x + \sqrt{x^2 - 1}) \right.$$

$$\left. + \frac{\log(x + \sqrt{x^2 - 1})}{x + \sqrt{x^2 - 1}} + \frac{1}{x + \sqrt{x^2 - 1}} \right].$$

(vii) Let

$$u = x + \sqrt{x}$$
$$u - x = \sqrt{x}$$
$$u^2 - 2ux + x^2 = x$$
$$x^2 - (2u + 1)x + u^2 = 0$$

so that

$$x = \frac{2u + 1 + \sqrt{(2u + 1)^2 - 4u^2}}{2}$$

$$= \frac{2u + 1 + \sqrt{4u + 1}}{2}$$

$$dx = 1 + \frac{1}{\sqrt{4u + 1}}\, du.$$

The integral becomes

$$\int \log u\, du + \int \frac{1}{\sqrt{4u + 1}} \log u\, du = u \log u - 1 + \frac{\sqrt{4u + 1}}{2} - \int \frac{\sqrt{4u + 1}}{2u}\, du$$

$$= u \log u - 1 + \frac{\sqrt{4u + 1}}{2} - I_1.$$

Now let

$$v = \sqrt{4u + 1}$$

$$u = \frac{v^2 - 1}{4}, \qquad du = \frac{v}{2}\, dv.$$

The integral I_1 becomes

$$\int \frac{v^2\, dv}{v^2 - 1} = \int 1 + \frac{1}{v^2 - 1}\, dv = \int 1 + \frac{1/2}{v - 1} - \frac{1/2}{v + 1}\, dv$$

$$= v + \frac{1}{2} \log(v - 1) - \frac{1}{2} \log(v + 1)$$

$$= \sqrt{4u + 1} + \frac{1}{2} \log(\sqrt{4u + 1} - 1) - \frac{1}{2} \log(\sqrt{4u + 1} + 1).$$

So the answer is

$$(x + \sqrt{x}) \log(x + \sqrt{x}) - 1 - \frac{1}{2}\sqrt{4x + 4\sqrt{x} + 1}$$

$$- \frac{1}{2} \log\left(\sqrt{4x + 4\sqrt{x} + 1} - 1\right) + \frac{1}{2} \log\left(\sqrt{4x + 4\sqrt{x} + 1} + 1\right).$$

(viii) Let $u = x^{1/5}$, $x = u^5$, $dx = 5u^4\, du$. The integral becomes

$$\int \frac{5u^4\, du}{u^5 - u^3} = 5 \int \frac{u\, du}{u^2 - 1} = \frac{5}{2} \log(u^2 - 1) = \frac{5}{2} \log(x^{2/5} - 1).$$

(ix) Let $u = \arcsin x$, $x = \sin u$, $dx = \cos u\, du$. The integral becomes

$$\int u^2 \cos u\, du = u^2 \sin u - \int 2u \sin u\, du$$

$$= u^2 \sin u - \left[2u(-\cos u) + \int 2 \cos u\, du \right]$$

$$= u^2 \sin u + 2u \cos u - 2 \sin u\, du$$

$$= (\arcsin x)^2 x + 2(\arcsin x)\sqrt{1 - x^2} - 2x.$$

(x) Let $u = x^2$, $x = u^{1/2}$, $dx = du/2u^{1/2}$. The integral becomes

$$\int \frac{u^{5/2} \arctan u}{2u^{1/2}}\, du = \frac{1}{2} \int u^2 \arctan u\, du$$

$$= \frac{u^3 \arctan u}{6} - \frac{u^2}{12} + \frac{\log(1 + u^2)}{12} \qquad \text{by Problem 8(vii)}$$

$$= \frac{x^6 \arctan x^2}{6} - \frac{x^4}{12} + \frac{\log(1 + x^4)}{12}.$$

10. (iv) Let $x = \cosh u$, $dx = \sinh u\, du$. The integral in Problem 4(iv) becomes

$$\int \frac{\sinh u\, du}{\cosh u \sinh u} = \int \frac{1}{\cosh u}\, du$$

$$= \int \frac{2}{e^u + e^{-u}}\, du = \int \frac{2e^u}{1 + e^{2u}}\, du$$

$$= 2 \arctan e^u$$

$$= 2 \arctan\left(x + \sqrt{x^2 - 1}\right),$$

since $u = \cosh^{-1} x = \log\left(x + \sqrt{x^2 - 1}\right)$, as found in Problem 18-9.

Comparing with Problem 4(iv) we cannot conclude that

$$2 \arctan\left(x + \sqrt{x^2 - 1}\right) = \arctan\left(\sqrt{x^2 - 1}\right),$$

but only that these two expressions differ by a constant. As a matter of fact, we can only conclude that there are two constants c_1 and c_2 with

$$2 \arctan\left(x + \sqrt{x^2 - 1}\right) = \arctan\left(\sqrt{x^2 - 1}\right) + c_1 \qquad \text{for } x \geq 1,$$

$$2 \arctan\left(x + \sqrt{x^2 - 1}\right) = \arctan\left(\sqrt{x^2 - 1}\right) + c_2 \qquad \text{for } x \leq -1.$$

By setting $x = 1$ and -1 it is easy to see that $c_1 = \pi/2$ and $c_2 = -\pi/2$.

(vi) Let $x = \sinh u$, $dx = \cosh u$, du. The integral becomes

$$\int \frac{\cosh u \, du}{\sinh u \cosh u} = \int \frac{du}{\sinh u}$$

$$= \int \frac{2}{e^u - e^{-u}} \, du = \int \frac{2e^u \, du}{e^{2u} - 1}$$

$$= \int \frac{-e^u}{e^u + 1} + \frac{e^u}{e^u - 1} \, du$$

$$= -\log(e^u + 1) + \log(e^u - 1)$$

$$= \log\left(\frac{e^u - 1}{e^u + 1}\right) = \log\left(\frac{\sqrt{x^2 + 1} + x - 1}{\sqrt{x^2 + 1} + x + 1}\right).$$

(ix) Let $x = \sinh u$, $dx = \cosh u \, du$. The integral becomes

$$\int \cosh^2 u \, du = \int \frac{e^{2u}}{4} + \frac{1}{2} + \frac{e^{-2u}}{4} \, du$$

$$= \frac{e^{2u}}{8} + \frac{u}{3} - \frac{e^{-2u}}{8}$$

$$= \frac{\left(x + \sqrt{1 + x^2}\right)^2}{8} + \frac{\log\left(x + \sqrt{1 + x^2}\right)}{2} - \frac{1}{8\left(x + \sqrt{1 + x^2}\right)^2}.$$

(x) Let $x = \cosh u$, $dx = \sinh u \, du$. The integral becomes

$$\int \sinh^2 u \, du = \int \frac{e^{2u}}{4} - \frac{1}{2} + \frac{e^{-2u}}{4} \, du$$

$$= \frac{e^{2u}}{8} - \frac{u}{2} - \frac{e^{-2u}}{8}$$

$$= \frac{\left(x + \sqrt{x^2 - 1}\right)^2}{8} - \frac{\log\left(x + \sqrt{x^2 - 1}\right)}{2} - \frac{1}{8\left(x + \sqrt{x^2 - 1}\right)^2}.$$

11. (i)

$$\int \frac{1}{1 + \dfrac{2t}{1 + t^2}} \cdot \frac{2}{1 + t^2} \, dt = \int \frac{2 \, dt}{1 + 2t + t^2} = \int \frac{2 \, dt}{(1 + t)^2} = \frac{-2}{1 + t} = \frac{-2}{1 + \tan \dfrac{x}{2}}.$$

Comparing with the formula

$$\int \frac{1}{1 + \sin x} \, dx = \int \frac{1 - \sin x}{1 - \sin^2 x} \, dx = \int \sec^2 x - \sec x \tan x \, dx = \tan x - \sec x,$$

we can conclude that

$$\frac{-2}{1 + \tan \dfrac{x}{2}} = \tan x - \sec x - 1.$$

This can be checked most easily by expressing everything in terms of t:

$$\frac{-2}{1+t} = \frac{2t}{1-t^2} - \frac{1+t^2}{1-t^2} - 1.$$

(ii) Let $t = \tan x$, $dx = 1/(1+t^2)\,dt$. Then $\sin^2 x$ can be expressed in terms of t as

$$\sin^2 x = 1 - \cos^2 x = 1 - \frac{1}{\sec^2 x}$$

$$= 1 - \frac{1}{\tan^2 x + 1}$$

$$= 1 - \frac{1}{1+t^2} = \frac{t^2}{1+t^2}.$$

So the integral becomes

$$\int \frac{1}{1 + \dfrac{t^2}{1+t^2}} \cdot \frac{1}{1+t^2}\,dt = \int \frac{1}{2t^2 + 1}\,dt$$

$$= \frac{\arctan \sqrt{2}t}{\sqrt{2}}$$

$$= \frac{\arctan\left(\sqrt{2}\tan x\right)}{\sqrt{2}}.$$

(iii)

$$\int \frac{1}{\dfrac{2at}{1+t^2} + \dfrac{b-bt^2}{1+t^2}} \cdot \frac{2}{1+t^2}\,dt = \int \frac{2}{2at + b - bt^2}\,dt.$$

If $b > 0$, this can be written

$$\int \frac{-2\,dt}{bt^2 - 2at - b}$$

$$= \int \frac{-2\,dt}{\left(\sqrt{b}\,t - \dfrac{a}{\sqrt{b}}\right)^2 - \dfrac{a^2 + b^2}{b}}$$

$$= \int \left[\frac{\dfrac{\sqrt{b}}{\sqrt{a^2 + b^2}}}{\left(\sqrt{b}\,t - \dfrac{a}{\sqrt{b}} + \dfrac{\sqrt{a^2 + b^2}}{\sqrt{b}}\right)} - \frac{\dfrac{\sqrt{b}}{\sqrt{a^2 + b^2}}}{\left(\sqrt{b}\,t - \dfrac{a}{\sqrt{b}} - \dfrac{\sqrt{a^2 + b^2}}{\sqrt{b}}\right)} \right] dt$$

$$= \frac{1}{\sqrt{a^2 + b^2}} \left[\log \left(\sqrt{b}\,t - \frac{a}{\sqrt{b}} + \frac{\sqrt{a^2 + b^2}}{\sqrt{b}} \right) \right.$$

$$\left. - \log \left(\sqrt{b}\,t - \frac{a}{\sqrt{b}} - \frac{\sqrt{a^2 + b^2}}{\sqrt{b}} \right) \right].$$

If $b < 0$, the integral can be written

$$\int \frac{2\,dt}{-bt^2 + 2at + b} = \int \frac{2\,dt}{\left(\sqrt{-b}\,t + \dfrac{a}{\sqrt{-b}} \right)^2 + \dfrac{a^2 + b^2}{b}}$$

$$= \frac{1}{\sqrt{a^2 + b^2}} \left[\log \left(\sqrt{-b}\,t + \frac{a}{\sqrt{-b}} - \frac{\sqrt{a^2 + b^2}}{\sqrt{-b}} \right) \right.$$

$$\left. - \log \left(\sqrt{-b}\,t + \frac{a}{\sqrt{-b}} + \frac{\sqrt{a^2 + b^2}}{\sqrt{b}} \right) \right].$$

It is also possible to write

$$\int \frac{dx}{a \sin x + b \cos x} = \int \frac{dx}{A \sin(x + b)}$$

$$= -\frac{1}{A} \log(\csc(x + B) + \cot(x + B)),$$

where

$$A = \sqrt{a^2 + b^2}$$

$$\sin B = \frac{b}{\sqrt{a^2 + b^2}}$$

$$\cos B = \frac{a}{\sqrt{a^2 + b^2}}.$$

(iv)

$$\int \frac{4t^2}{(1 + t^2)^2} \cdot \frac{2}{1 + t^2}\,dt$$

$$= 8 \int \frac{1}{(1 + t^2)^2} = \frac{1}{(1 + t^2)^3}\,dt$$

$$= -8 \left[\frac{1}{4} \frac{t}{(1 + t^2)^2} + \frac{3}{4} \int \frac{1}{(1 + t^2)^2}\,dt \right] + 8 \int \frac{1}{(1 + t^2)^2}\,dt$$

$$= \frac{-2t}{(1 + t^2)^2} + 2 \left[\frac{1}{2} \frac{t}{1 + t^2} + \frac{1}{2} \int \frac{1}{1 + t^2}\,dt \right]$$

$$= \frac{-2t}{(1 + t^2)^2} + \frac{t}{1 + t^2} + \arctan t$$

$$= \frac{-2\tan x/2}{\sec^4 x/2} - \frac{\tan x/2}{\sec^2 x/2} + \frac{x}{2}$$

$$= -2\sin x/2\cos^3 x/2 - \sin x/2\cos x/2 + x/2$$

$$= -2\sin x/2\cos x/2\,(1 - \sin^2 x/2) - \sin x/2\cos x/2 + x/2$$

$$= -\sin x\left[1 - \left(\frac{1 - \cos x}{2}\right)\right] - \frac{\sin x}{2} + \frac{x}{2}$$

$$= -\sin x\left[\frac{1 + \cos x}{2} - \frac{1}{2}\right] + \frac{x}{2}$$

$$= \frac{-\sin 2x}{4} + \frac{x}{2}.$$

(v)

$$\int \frac{1}{3 + \dfrac{10t}{1 + t^2}} \cdot \frac{2}{1 + t^2}\,dt = \int \frac{2\,dt}{3t^2 + 10t + 3}$$

$$= \int \frac{3/4}{2t + 1} - \frac{1/4}{t + 3}\,dt$$

$$= \frac{1}{4}\log(3t + 1) - \frac{1}{4}\log(t + 3)$$

$$= \frac{1}{4}\log\left(3\tan\frac{x}{2} + 1\right) - \frac{1}{4}\log\left(\tan\frac{x}{2} + 3\right).$$

12. (a) The given formula shows that

$$\int \sec x\,dx = \frac{1}{2}\int \frac{\cos x}{1 + \sin x}\,dx + \frac{1}{2}\int \frac{\cos x}{1 - \sin x}\,dx$$

$$= \frac{1}{2}\log(1 + \sin x) - \frac{1}{2}\log(1 - \sin x)$$

$$= \log\sqrt{1 + \sin x} - \log\sqrt{1 - \sin x}$$

$$= \log\sqrt{\frac{1 + \sin x}{1 - \sin x}}$$

$$= \log\sqrt{\frac{(1 + \sin x)^2}{1 - \sin^2 x}}$$

$$= \log\left(\frac{1 + \sin x}{\cos x}\right)$$

$$= \log(\sec x + \tan x).$$

(b) With the substitution $t = \tan x/2$ the integral $\int \sec x \, dx$ becomes

$$\int \frac{1+t^2}{1-t^2} \cdot \frac{2}{1+t^2} \, dt = \int \frac{1}{1+t} + \frac{1}{1-t} \, dt$$
$$= \log(1+t) - \log(1-t)$$
$$= \log\left(\frac{1+t}{1-t}\right).$$

Now

$$\sec x + \tan x = \frac{1}{\cos 2(x/2)} + \tan 2(x/2)$$

$$= \frac{1}{\cos^2 x/2 - \sin^2 x/2} + \frac{2 \tan x/2}{1 - \tan^2 x/2}$$

$$= \frac{1}{2 \cos^2 x/2 - 1} + \frac{2 \tan x/2}{1 - \tan^2 x/2}$$

$$= \frac{1}{\dfrac{2}{1 + \tan^2 x/2} - 1} + \frac{2 \tan x/2}{1 - \tan^2 x/2}$$

$$= \frac{1 + \tan^2 x/2 + 2 \tan x/2}{1 - \tan^2 x/2}$$

$$= \frac{1 + 2t + t^2}{t - t^2} = \frac{(1+t)^2}{1-t^2} = \frac{1+t}{1-t}.$$

14. We have

$$\int_0^\pi f''(x) \sin x \, dx = f'(x) \sin x \Big|_0^\pi - \int_0^\pi f'(x) \cos x \, dx$$

$$= 0 - \left[f(x) \cos x \Big|_0^\pi + \int_0^\pi f(x) \sin x \, dx \right]$$

$$= f(\pi) - f(0) - \int_0^\pi f(x) \sin x \, dx.$$

So

$$2 = \int_0^\pi [f(x) + f''(x)] \sin x \, dx = f(\pi) - f(0) = 1 - f(0),$$

hence $f(0) = -1$.

15. (b) We have

$$\int f^{-1}(x) \, dx = \int 1 \cdot f^{-1}(x) \, dx = x f^{-1}(x) - \int x (f^{-1})'(x) \, dx$$

$$= xf^{-1}(x) - \int \frac{x}{f'(f^{-1}(x))} \, dx.$$

If $F = \int f(x) \, dx$, the substitution $u = f^{-1}(x)$, $x = f(u)$, $dx = f'(u) \, du$ changes the new integral to

$$\int \frac{f(u)}{f'(u)} f'(u) \, du = F(u) = F(f^{-1}(x)),$$

so

$$\int f^{-1}(x) \, dx = xf^{-1}(x) - F(f^{-1}(x)).$$

17.

$$\int \log(\log x) \, dx = \int 1 \cdot \log(\log x) \, dx$$

$$= x \log(\log x) - \int x \cdot \frac{1}{\log x} \cdot \frac{1}{x} \, dx$$

$$= x \log(\log x) - \int \frac{1}{\log x} \, dx.$$

18.

$$\int x^2 e^{-x^2} \, dx = \int x(x e^{-x^2}) \, dx$$

$$= \frac{-e^{-x^2}}{2} x + \frac{1}{2} \int e^{-x^2} \, dx.$$

19. (Use the substitution $u = e^x$.) The function $g(x) = 1/(x^5 + x + 1)$ has an elementary primitive G, since it is a rational function. Then $G \circ \exp$ is a primitive of f.

21. (a)

$$\int x^n e^x \, dx = x^n e^x - n \int x^{n-1} e^x \, dx.$$

(b)

$$\int (\log x)^n = x(\log x)^n - n \int x(\log x)^{n-1} \cdot \frac{1}{x} \, dx$$

$$= x(\log x)^n - n \int (\log x)^{n-1} \, dx.$$

22. By Problem 4(x),

$$\int_1^{\cosh x} \sqrt{t^2 - 1}\, dt = \frac{1}{2} \cosh x \sqrt{\cosh^2 x - 1} - \frac{1}{2} \log\left(\cosh x + \sqrt{\cosh^2 - 1}\right)$$

$$= \frac{\cosh x \sinh x}{2} - \frac{1}{2} \log(\cosh x + \sinh x)$$

$$= \frac{\cosh x \sinh x}{2} - \frac{x}{2}.$$

23. By Theorem 2, with $g(x) = a + b - x$,

$$\int_a^b f(a + b - x)\, dx = -\int_a^b f(g(x)) \cdot g'(x)\, dx$$

$$= -\int_{g(a)}^{g(b)} f(x)\, dx = -\int_b^a f(x)\, dx = \int_a^b f(x)\, dx.$$

24. By Theorem 2, with $g(x) = x/r$,

$$\int_{-r}^r \sqrt{r^2 - x^2}\, dx = r^2 \int_{-r}^r \frac{1}{r}\sqrt{1 - \left(\frac{x}{r}\right)^2}\, dx$$

$$= r^2 \int_{-r}^r \sqrt{1 - [g(x)]^2}\, g'(x)\, dx$$

$$= r^2 \int_{-1}^1 \sqrt{1 - x^2}\, dx$$

$$= \frac{\pi r^2}{2}.$$

25. (a) If $|x| \geq h$, then $|x/h| \geq 1$, so

$$\phi_h(x) = \frac{1}{h}\phi(x/h) = 0.$$

Moreover, using the substitution $g(x) = x/h$ we have

$$\int_{-h}^h \phi_h(x)\, dx = \int_{-h}^h \phi(x/h)\frac{1}{h}\, dx$$

$$= \int_{-1}^1 \phi(u)\, du = 1.$$

(b) We have

$$\lim_{h \to 0^+} \int_{-1}^1 \phi_h f = \lim_{h \to 0^+} \int_{-h}^h \phi_h f \qquad \text{since } \phi_h(x) = 0 \text{ for } |x| \geq h.$$

Since f is continuous at 0, for any $\varepsilon > 0$ there is $\delta > 0$ such that $|f(0) - f(x)| < \varepsilon$ for $|x| < \delta$. Then for $0 < h < \delta$ we have

$$\left| f(0) - \int_{-h}^{h} \phi_h f \right| = \left| \int_{-h}^{h} \phi_h f(0) - \int_{-h}^{h} \phi_h f \right| \leq \int_{-h}^{h} \phi_h(x) \, |f(x) - f(0)| \, dx$$

$$\leq \varepsilon \int_{-h}^{h} \phi_h(x) \, dx = \varepsilon.$$

[If f is continuous on an interval around 0 then there is a simpler argument, using the Mean Value Theorem for Integrals (Problem 13-23).]

(c) We have

$$\int \frac{h}{h^2 + x^2} \, dx = \frac{1}{h} \int \frac{1}{1 + \left(\dfrac{x}{h}\right)^2} \, dx$$

$$= \arctan \frac{x}{h},$$

so

$$\lim_{h \to 0^+} \int_{-1}^{1} \frac{h}{h^2 + x^2} \, dx = \lim_{h \to 0^+} \arctan \frac{x}{h} \Big|_{-1}^{1}$$

$$= \frac{\pi}{2} + \frac{\pi}{2} = \pi.$$

(d) Let $|f(x)| \leq M$ on $[-1, 1]$. Notice that if $0 < h < \delta$, then for all $|x| \geq d = \sqrt[4]{\delta}$ we have

$$\frac{h}{h^2 + x^2} < \frac{\delta}{\sqrt{\delta}} = \sqrt{\delta}.$$

So for $0 < h < \delta$ we have

(1) $$\left| \int_{-1}^{1} \frac{h}{h^2 + x^2} f(x) \, dx - \int_{-d}^{d} \frac{h}{h^2 + x^2} f(x) \, dx \right|$$

$$\leq \int_{-1}^{-d} \sqrt{\delta} \, M + \int_{d}^{1} \sqrt{\delta} \, M < 2\sqrt{\delta} \, M.$$

In particular, choosing the constant function $f(0)$ we have

(2) $$\left| \pi f(0) - \int_{-d}^{d} \frac{h}{h^2 + x^2} f(0) \, dx \right| < 2\sqrt{\delta} f(0).$$

Given $\varepsilon > 0$, choose $\delta > 0$ so that $|f(0) - f(x)| < \varepsilon$ for $|x| < \sqrt[4]{\delta}$. Then

(3)
$$\left| \int_{-d}^{d} \frac{h}{h^2 + x^2} f(0)\, dx - \int_{-d}^{d} \frac{h}{h^2 + x^2} f(x)\, dx \right|$$

$$\leq \int_{-d}^{d} \frac{h}{h^2 + x^2} |f(0) - f(x)|\, dx$$

$$\leq \varepsilon \int_{-d}^{d} \frac{h}{h^2 + x^2}\, dx < \varepsilon \int_{-1}^{1} \frac{h}{h^2 + x^2}\, dx = \pi \varepsilon.$$

It follows form (1), (2) and (3) that for $0 < h < \delta$ we have

$$\left| \pi f(0) - \int_{-1}^{1} \frac{h}{h^2 + x^2} f(x)\, dx \right| < 2\sqrt{\delta} M + 2\sqrt{\delta} f(0) + \pi \varepsilon,$$

$$< (\pi + 1)\varepsilon \qquad \text{for small enough } \delta.$$

26. (i) The whole circle of radius $a/2$ is traversed as θ goes from 0 to π. So

$$\text{area} = \frac{1}{2} \int_{0}^{\pi} a^2 \sin^2 \theta\, d\theta = \frac{a^2}{4} \int_{0}^{\pi} (1 - \cos 2\theta)\, d\theta$$

$$= \frac{a^2}{4} \left[\pi - \frac{\sin 2\theta}{2} \Big|_{0}^{\pi} \right]$$

$$= \frac{\pi a^2}{4}.$$

(ii)

$$\text{area} = \frac{1}{2} \int_{0}^{2\pi} (1 + \cos \theta)^2\, d\theta = \frac{1}{2} \int_{0}^{2\pi} 4 + 4\cos \theta + \cos^2 \theta\, d\theta$$

$$= \frac{1}{2} \int_{0}^{2\pi} 4 + 4\cos \theta + \frac{1 + \cos 2\theta}{2}\, d\theta$$

$$= \frac{1}{2} \left[8\pi + 4\sin \theta \Big|_{0}^{2\pi} + \pi + \frac{\sin 2\theta}{4} \Big|_{0}^{2\pi} \right]$$

$$= \frac{9\pi}{2}.$$

(iii) This will look something like the graph of $f(\theta) = a \cos 2\theta$ (Problem 3(iii) of Chapter 4, Appendix 3), but there can be only two leaves, since $\cos 2\theta$ must be ≥ 0. Each leaf has area

$$\frac{1}{2} \int_{-\pi/4}^{\pi/4} 2a^2 \cos 2\theta\, d\theta = a^2 \cdot \frac{\sin 2\theta}{4} \Big|_{-\pi/4}^{\pi/4}$$

$$= \frac{a^2}{2}.$$

(iv) Each leaf (two or four, depending on conventions for the sign of r) has area

$$\frac{1}{2}\int_{-\pi/4}^{\pi/4} a^2 \cos^2 2\theta \, d\theta = \frac{a^2}{2}\int_{-\pi/4}^{\pi/4} \frac{1+\cos 2\theta}{2}\, d\theta$$

$$= \frac{a^2}{2}\left[\frac{\pi}{4} + \frac{\sin 2\theta}{4}\Big|_{-\pi/4}^{\pi/4}\right]$$

$$= a^2\left(\frac{\pi}{8} + \frac{1}{4}\right).$$

27. In the integral

$$\int_{x_1}^{x_0} g(x)\, dx$$

make the substitution

$$x = f(\theta)\cos\theta$$
$$dx = f'(\theta)\cos\theta - f(\theta)\sin\theta\, d\theta,$$
$$g(x) = f(\theta)\sin\theta$$

to obtain

$$\int_{x_1}^{x_0} g(x)\, dx = \int_{\theta_1}^{\theta_0} f(\theta) f'(\theta)\sin\theta\cos\theta - f(\theta)^2 \sin^2\theta\, d\theta$$

$$= \frac{f(\theta)^2}{2}\sin\theta\cos\theta\Big|_{\theta_1}^{\theta_0} - \int_{\theta_1}^{\theta_0} \frac{f(\theta)^2}{2}[-\sin^2\theta + \cos^2\theta] - \int_{\theta_1}^{\theta_0} f(\theta)^2 \sin^2\theta\, d\theta$$

$$= \frac{f(\theta)\sin\theta\, f(\theta)\cos\theta}{2}\Big|_{\theta_0}^{\theta_1} - \frac{1}{2}\int_{\theta_1}^{\theta_0} f(\theta)^2\, d\theta$$

$$= \frac{x_0 y_0 - x_1 y_1}{2} - \frac{1}{2}\int_{\theta_1}^{\theta_0} f(\theta)^2\, d\theta,$$

where y_0 and y_1 are the second coordinates of A and B. Hence

$$\int_{x_1}^{x_0} g(x)\, dx = \text{area } \triangle O x_0 A - \text{area } \triangle O x_1 B + \frac{1}{2}\int_{\theta_0}^{\theta_1} f(\theta)^2\, d\theta,$$

as desired.

28. **(a)** For each partition $P = \{t_0, \ldots, t_n\}$ of $[a, b]$, let

$$\bar{P} = \{h^{-1}(t_0), \ldots, h^{-1}(t_n)\} = \{\bar{t}_0 \ldots, \bar{t}_n\}.$$

Then $\bar{P}$ is a partition of $[\bar{a}, \bar{b}]$ and

$$\bar{u}(\bar{t}_i) = u(t_i), \qquad \bar{v}(\bar{t}_i) = v(t_i).$$

So the $\bar{\ell}$ for $\bar{c}$ corresponding to $\bar{P}$ is

$$\bar{l}(\bar{c}, \bar{P}) = \sum_{i=1}^{n} \sqrt{[\bar{u}(\bar{t}_i) - \bar{u}(\bar{t}_{i-1})]^2 + [\bar{v}(\bar{t}_i) - \bar{v}(\bar{t}_{i-1})]^2}$$

$$= \sum_{i=1}^{n} \sqrt{[u(t_i) - u(t_{i-1})]^2 + [v(t_i) - v(t_{i-1})]^2}$$

$$= \ell(c, P).$$

Thus, every $\ell(c, P)$ is $\bar{l}(\bar{c}, \bar{P})$ for some partition $\bar{P}$ of $[\bar{a}, \bar{b}]$ and, conversely, it is easy to see that every $\bar{l}$ for $\bar{c}$ is $\ell(c, P)$ for some partition P. So the length of ℓ on $[a, b]$ and the length of $\bar{\ell}$ on $[\bar{a}, \bar{b}]$ are the sup's of the same set of numbers, and hence are equal.

(b) The length of c on $[a, b]$ is

$$\int_a^b \sqrt{u'(x)^2 + v'(x)^2}\, dx.$$

Letting

$$x = h(y)$$
$$dx = h'(y)\, dy,$$

the integral becomes

$$\int_{h^{-1}(a)}^{h^{-1}(b)} \sqrt{u'(h(y))^2 + v'(h(y))^2} \cdot h'(y)\, dy$$

$$= \int_{h^{-1}(a)}^{h^{-1}(b)} \sqrt{[u'(h(y)) \cdot h'(y)]^2 + [v'(h(y)) \cdot h'(y)]^2}\, dy$$

$$= \int_{h^{-1}(a)}^{h^{-1}(b)} \sqrt{(u \circ h)'(y)^2 + (v \circ h)'(y)^2}\, dy$$

$$= \int_{\bar{a}}^{\bar{b}} \sqrt{\bar{u}'(y)^2 + \bar{v}(y)^2}\, dy$$

$$= \text{length of } \bar{c} \text{ on } [\bar{a}, \bar{b}].$$

29. (i) Since

$$f'(x) = x(x^2 + 2)^{1/2}$$
$$f'(x)^2 = x^2(x^2 + 2)$$
$$1 + f'(x)^2 = 1 + 2x^2 + x^4 = (1 + x^2)^2,$$

we have

$$\int_0^1 \sqrt{1 + f'(x)^2}\, dx = \int_0^1 +x^2\, dx$$
$$= 1 + \tfrac{1}{3}.$$

(ii) Since

$$f'(x) = 3x^2 - \frac{1}{12x^2}$$
$$f'(x)^2 = 9x^4 - \frac{1}{2} + \frac{1}{144x^4}$$
$$1 + f'(x)^2 = 9x^4 + \frac{1}{2} + \frac{1}{144x^4} = \left(3x^2 + \frac{1}{12x^2}\right)^2,$$

we have

$$\int_1^2 \sqrt{1 + f'(x)^2}\, dx = \int_1^2 3x^2 + \frac{1}{12x^2}\, dx$$
$$= 7 + \tfrac{1}{24}.$$

(iii) For $u(t) = a^3 \cos^3 t$, $v(t) = a^3 \sin^3 t$ we have

$$u'(t) = 3a^3 \cos^2 t \cdot (-\sin t)$$
$$v'(t) = 3a^3 \sin^2 t \cos t$$

so

$$u'(t)^2 + v'(t)^2 = 9a^6[\sin^2 t \cos^4 t + \cos^2 t \sin^4 t]$$
$$= 9a^6 \sin^2 t \cos^2 t,$$

so

$$\int_0^{2\pi} \sqrt{u'^2 + v'^2} = 3a^2 \int_0^{2\pi} |\sin t \cos t|\, dt.$$

On each of the four intervals $[0, \pi/2]$, $[\pi/2, \pi]$, $[\pi, 3\pi/2]$, $[3\pi/2, 2\pi]$ we can write

$$|\sin t \cos t| = \sin t \cos t \quad \text{or} \quad |\sin t \cos t| = -\sin t \cos t,$$

so on each interval we are considering

$$\int \sin t \cos t\, dt = \frac{\sin^2 t}{2} \quad \text{or} \quad -\frac{\sin^2 t}{2}.$$

So the integrals are the same on all four intervals, namely

$$\frac{\sin^2 t}{2}\Big|_0^{\pi/2} = -\frac{\sin^2 t}{2}\Big|_{\pi/2}^{\pi} = \frac{\sin^2 t}{2}\Big|_{\pi/2}^{3\pi/2} = -\frac{\sin^2 t}{2}\Big|_{3\pi/2}^{2\pi} = \frac{1}{2}.$$

Hence the total integral is

$$3a^3 \cdot 4 \cdot \frac{1}{2} = 6a^3.$$

(iv) Since

$$f'(x) = -\tan x$$
$$f'(x)^2 = \tan^2 x$$
$$1 + f'(x)^2 = \sec^2 x,$$

and $\sec x \geq 0$ for $0 \leq x \leq \pi/6$, we have

$$\int_0^{\pi/6} \sqrt{1 + f'(x)^2} \, dx = \int_0^{\pi/6} \sec x \, dx = \log(\sec x + \tan x)\Big|_0^{\pi/6}$$
$$= \log \sqrt{3} = \frac{1}{2} \log 3.$$

(v) We have

$$\int_1^e \sqrt{1 + f'(x)^2} \, dx = \int_1^e \sqrt{1 + \frac{1}{x^2}} \, dx = \int_1^e \frac{\sqrt{x^2 + 1}}{x} \, dx.$$

Letting

$$x = \tan u$$
$$dx = \sec^2 u \, du$$

we have

$$\int \frac{\sqrt{x^2 + 1}}{x} \, dx = \int \frac{\sqrt{\tan^2 u + 1} \sec^2 u \, du}{\tan u}$$
$$= \int \frac{\sec^3 u \, du}{\tan u}$$
$$= \int \frac{\sec u (1 + \tan^2 u)}{\tan u} \, du$$
$$= \int \csc u + \sec u \tan u \, du$$
$$= -\log(\csc u + \cot u) + \sec u$$
$$= -\log\left(\frac{\sqrt{1 + x^2}}{x} + \frac{1}{x}\right) + \sqrt{1 + x^2}.$$

So our integral is

$$-\log\left(\frac{1 + \sqrt{1 + x^2}}{x}\right) + \sqrt{1 + x^2} \, \Big|_1^e$$
$$= -\log\left(1 + \sqrt{1 + e^2}\right) - 1 + \log\left(1 + \sqrt{2}\right) + \sqrt{1 + e^2} - \sqrt{2}.$$

(vi) Since

$$f'(x) = \frac{e^x}{\sqrt{1 - e^{2x}}}$$

$$f'(x)^2 = \frac{e^{2x}}{1 - e^{2x}}$$

$$1 + f'(x)^2 = \frac{1}{1 - e^{2x}}$$

we have

$$\int_{-\log 2}^0 \sqrt{1 + f'(x)^2} = \int_{-\log 2}^0 \frac{1}{\sqrt{1 - e^{2x}}}\, dx.$$

Letting

$$y = e^x$$

$$x = \log y$$

$$dx = \frac{dy}{y}$$

we have

$$\int \frac{dx}{\sqrt{1 - e^{2x}}} = \int \frac{dy}{y\sqrt{1 - y^2}}$$

$$= -\log\left(\frac{1}{y} + \frac{\sqrt{1 - y^2}}{y}\right) \qquad \text{by Problem 4(v)}$$

$$= -\log\left(e^{-x} + e^{-x}\sqrt{1 - e^{2x}}\right)$$

$$= -\log\left(e^{-x}\left[1 + \sqrt{1 - e^{2x}}\right]\right)$$

$$= x - \log\left(1 + \sqrt{1 - e^{2x}}\right).$$

So our integral is

$$x - \log\left(1 + \sqrt{1 - e^{2x}}\right)\Big|_{-\log 2}^0 = -\log 2 + \log 2 + \log\left(1 + \sqrt{1 - e^{2\log 2}}\right)$$

$$= \log\left(1 + \sqrt{1 - \tfrac{1}{2^2}}\right)$$

$$= \log\left(1 + \sqrt{\tfrac{3}{4}}\right)$$

$$= \log(1 + \tfrac{1}{2}\sqrt{3}).$$

30. According to the Appendix to Chapter 13, the graph on the interval $[\theta_0, \theta_1]$ has length

$$\int_{\theta_0}^{\theta_1} \sqrt{f^2 + f'^2}.$$

(i) We have

$$f(\theta) = a\cos\theta$$
$$f'(\theta) = -a\sin\theta,$$

so

$$f(\theta)^2 + f'(\theta)^2 = a^2,$$

and the length is

$$\int_0^\pi a = \pi a \qquad (= 2\pi \cdot (a/2)).$$

(ii) The length is

$$\int_0^{2\pi} \sqrt{a^2(1-\cos\theta)^2 + a^2\sin^2\theta}\, d\theta = \int_0^{2\pi} a\sqrt{2-2\cos 2\theta}\, d\theta$$
$$= \int_0^{2\pi} 2a\sqrt{\frac{1-\cos 2\theta}{2}}\, d\theta$$
$$= \int_0^{2\pi} 2a\,|\sin\theta|\, d\theta.$$

Breaking up the interval as in Problem 29(iii), we find that the length is

$$4 \cdot 2a \cdot \int_0^{\pi/2} \sin\theta\, d\theta = 8a.$$

(iii) Since

$$a\sin^2\frac{\theta}{2} = a\frac{1-\cos\theta}{2},$$

by (ii) the length is $4a$.

(iv) The length is

$$\int_0^{2\pi} \sqrt{1+\theta^2}\, d\theta = \frac{1}{2}\left[\theta\sqrt{1+\theta^2} + \log\left(\theta + \sqrt{1+\theta^2}\right)\right]\Big|_0^{2\pi}$$

by Problem 4(ix)

$$= \frac{1}{2}\left[2\pi\sqrt{1+4\pi^2} + \log\left(2\pi + \sqrt{1+4\pi^2}\right)\right].$$

(v) The length is

$$3\int_0^{\pi/3} \sqrt{\sec^2\theta + \sec^2\theta\tan^2\theta}\, d\theta = 3\int_0^{\pi/3} \sec\theta\sqrt{1+\tan^2\theta}\, d\theta$$
$$= 3\int_0^{\pi/3} \sec^2\theta\, d\theta$$

$$= 3 \tan \theta \Big|_0^{\pi/3}$$

$$= 3 \cdot \frac{\sqrt{3}}{2},$$

which is hardly surprising, since the graph is a straight line from $(3, 0)$ to $(3, 3 \tan \frac{\pi}{3})$.

31. (a) The length is

$$\int_0^{2\pi} \sqrt{a^2(1 - \cos t)^2 + a^2 \sin^2 t} \, dt = a \int_0^{2\pi} \sqrt{2 - 2 \cos t} \, dt$$

$$= 2a \int_0^{2\pi} \sqrt{\frac{1 - \cos t}{2}} \, dt$$

$$= 2a \int_0^{2\pi} \sin \frac{t}{2} \, dt$$

$$= (2a) \left(-2 \cos \frac{t}{2} \Big|_0^{2\pi} \right)$$

$$= 8a.$$

(b) In the integral

$$\int_0^{2\pi a} f(x) \, dx = \int_0^{2\pi a} v(u^{-1}(x)) \, dx$$

let

$$t = u^{-1}(x)$$
$$x = u(t) = a(t - \sin t)$$
$$dx = a - a \cos t \, dt;$$

the integral becomes

$$a \int_0^{2\pi} v(t)(1 - \cos t) \, dt = a^2 \int_0^{2\pi} (1 - 2 \cos t + \cos^2 t) \, dt$$

$$= a^2 \int_0^{2\pi} \left(1 - 2 \cos t + \frac{1}{2} + \frac{\cos 2t}{2} \right) dt$$

$$= 3a^2 \pi.$$

32. The formula is true for $n = 1$ by Problem 14-12. Suppose that it is true for n. Let $F(u) = \int_0^u f(t) \, dt$. Then F is a primitive of f with $F(0) = 0$. So

$$\int_0^x \frac{f(u)(x - u)^{n+1}}{(n + 1)!} \, du$$

$$= \frac{F(u)(x - u)^{n+1}}{(n + 1)!} \Big|_{u=0}^{u=x} - \int_0^x \frac{-F(u)(x - u)^n}{n!} \, dx$$

$$= 0 + \int_0^x \left(\int_0^{u_n} \left(\cdots \left(\int_0^{u_1} F(t)\,dt \right) du_1 \right) \cdots \right) du_n$$

$$= \int_0^x \left(\int_0^{u_n} \left(\cdots \left(\int_0^{u_1} \left(\int_0^t f(s)\,ds \right) dt \right) du_1 \right) \cdots \right) du_n,$$

which can also be written as

$$\int_0^x \left(\int_0^{u_{n+1}} \left(\cdots \left(\int_0^{u_1} f(t)\,dt \right) du_1 \right) \cdots \right) du_{n+1}.$$

33.

$$\lim_{\lambda \to \infty} \int_a^b f(t) \sin \lambda t \, dt = \lim_{\lambda \to \infty} \left[\left. \frac{-f(t) \cos \lambda t}{\lambda} \right|_a^b + \int_a^b \frac{f'(t) \cos \lambda t}{\lambda} \, dt \right]$$

$$= 0,$$

since

$$\left| \left. \frac{-f(t) \cos \lambda t}{\lambda} \right|_a^b \right| \le \frac{1}{\lambda} (|f(b)| + |f(a)|),$$

$$\left| \int_a^b \frac{f'(t) \cos \lambda t}{\lambda} \, dt \right| \le \frac{1}{\lambda} \int_a^b |f'(t)| \, dt.$$

34. (a) Simply replace ϕ by $-\phi$; multiplying the resulting formula for $-\phi$ by -1 we get the formula for ϕ, with the same ξ.

(b) The function $\psi = \phi - \phi(b)$ satisfies $\psi(b) = 0$. The formula for ψ gives

$$\int_a^b f(x)[\phi(x) - \phi(b)] \, dx = [\phi(a) - \phi(b)] \int_0^\xi f(x) \, dx + 0 \cdot \int_\xi^b f(x) \, dx.$$

So

$$\int_a^b f(x)\phi(x)\,dx = \phi(a) \int_a^\xi f(x)\,dx + \phi(b) \int_a^b f(x)\,dx - \phi(b) \int_a^\xi f(x)\,dx$$

$$= \phi(a) \int_a^\xi f(x)\,dx + \phi(b) \int_\xi^b f(x)\,dx.$$

(c) If $F(x) = \int_a^x f$, then

$$\int_a^b f(x)\phi(x)\,dx = \left. F(x)\phi(x) \right|_a^b - \int_a^b F(x)\phi'(x)\,dx$$

$$= \left. F(x)\phi(x) \right|_a^b - F(\xi) \int_a^b \phi'(x)\,dx, \qquad \text{by Problem 13-23}$$

$$= 0 - \left(\int_a^\xi f \right) \cdot [-\phi(a)]$$

$$= \phi(a) \int_a^\xi f(x)\, dx.$$

(d) If $\phi(a) = \phi(b) = 0$, but $\phi > 0$ on (a, b), and $f > 0$ on (a, b), then we clearly cannot have

$$\int_a^b f\phi = \phi(a) \int_a^b f + \phi(b) \int_\xi^b f.$$

35. (a) We have

$$a_1 b_1 + \cdots + a_n b_n$$
$$= b_1 s_1 + b_2(s_2 - s_1) + b_3(s_3 - s_1) + \cdots$$
$$+ b_{n-1}(s_{n-1} - s_{n-2}) + b_n(s_n - s_{n-1})$$
$$= s_1(b_1 - b_2) + s_2(b_2 - b_3) + \cdots + s_{n-1}(b_{n-1} - b_n) + s_n b_n.$$

(b) Since $\{b_k\}$ is nonincreasing we have $b_k - b_{k-1} \geq 0$ for each k. Also, $m \leq s_k \leq M$ for each k. So

$$m(b_1 - b_2) + m(b_2 - b_3) + \cdots + m(b_{n-1} - b_n) + m b_n$$
$$\leq s_1(b_1 - b_2) + s_2(b_2 - b_3) + \cdots + s_n(b_{n-1} - b_n) + s_n b_n$$
$$\leq M(b_1 - b_2) + M(b_2 - b_3) + \cdots + M(b_{n-1} - b_n) + M b_n,$$

or

$$m b_1 \leq a_1 b_1 + \cdots + a_n b_n \leq M b_1.$$

Applying this result to $a_k, a_{k+1}, \ldots, a_n$, and $b_k, b_{k+1}, \ldots, b_n$, we get $b_k m \leq a_k b_k + \cdots + a_n b_n \leq b_k M$.

(c) If we set

$$a_i = f(x_i)(t_i - t_{i-1})$$

and let

$$m = \text{smallest of the } \sum_{i=1}^k f(x_i)(t_i - t_{i-1})$$

$$M = \text{largest of the } \sum_{i=1}^k f(x_i)(t_i - t_{i-1})$$

then $m \leq a_1 + \cdots + a_k \leq M$ for all k. Letting $b_k = \phi(x_k)$ in part (b), we find that

$$\sum_{i=1}^n f(x_i)\phi(x_i)(t_i - t_{i-1})$$

lies between the smallest and the largest of the sums

$$\phi(x_1) \sum_{i=1}^{k} f(x_i)(t_i - t_{i-1}).$$

Since we can approximate $\int_a^b f(x_i)\phi(x)\, dx$ by sums $\sum_{i=1}^{n} f(x_i)\phi(x_i)(t_i - t_{i-1})$, and $\int_a^x f(t)\, dt$ by sums like $\sum_{i=1}^{k} f(x_i)(t_i - t_{i-1})$, the final result should follow from the above. However, some care is required for the argument:

Given $\varepsilon > 0$ we can choose $\delta > 0$ so that whenever all $t_i - t_{i-1} < \delta$ we have

(1)
$$\left| \int_a^b f(x)\, dx - \sum_{i=1}^{n} f(t_{i-1})(t_i - t_{i-1}) \right| < \varepsilon.$$

We claim that for any $\varepsilon' > \varepsilon$ it also follows that for each k

$$\left| \int_a^{t_k} f(x)\, dx - \sum_{i=1}^{k} f(t_{i-1})(t_i - t_{i-1}) \right| < \varepsilon'.$$

The idea is that if we had

(2)
$$\left| \int_a^{t_k} f(x)\, dx - \sum_{i=1}^{k} f(t_{i-1})(t_i - t_{i-1}) \right| \geq \varepsilon',$$

then by choosing some $p > n$ and new $t_{k+1} < t_{k+2} < \cdots < t_p = b$ we could make the sums on $[t_k, b]$ so close to $\int_{t_k}^b f(x)\, dx$ that inequality (2) would contradict (1). More precisely, choose $t_{k+1}, \ldots, t_p$, still with $t_i - t_{i-1} < \delta$, so that

(3)
$$\left| \int_{t_k}^b f(x)\, dx - \sum_{i=k+1}^{p} f(t_{i-1})(t_i - t_{i-1}) \right| < \varepsilon' - \varepsilon > 0.$$

Then

$$\left| \int_a^b f(x)\, dx - \sum_{i=1}^{p} f(t_{i-1})(t_i - t_{i-1}) \right|$$

$$= \left| \left[\int_a^{t_k} f(x)\, dx - \sum_{i=1}^{k} f(t_{i-1})(t_i - t_{i-1}) \right] \right.$$

$$\left. - \left[\sum_{i=k+1}^{p} f(t_{i-1})(t_i - t_{i-1}) - \int_{t_k}^b f(x)\, dx \right] \right|$$

$$\geq \left| \int_a^{t_k} f(x)\,dx - \sum_{i=1}^{k} f(t_{i-1})(t_i - t_{i-1}) \right|$$

$$- \left| \int_{t_k}^{b} f(x)\,dx - \sum_{i=k+1}^{p} f(t_{i-1})(t_i - t_{i-1}) \right|$$

$$\geq \varepsilon' - (\varepsilon' - \varepsilon) = \varepsilon,$$

contradicting the fact that (1) is supposed to hold whenever all $t_i - t_{i-1} < \delta$.

If we now choose the t_i so that for some t_k and t_l the integrals $\int_a^{t_k} f(x)\,dx$ and $\int_a^{t_l} f(x)\,dx$ are the minimum and maximum of $\int_a^x f(t)\,dt$ on $[a, b]$, then the smallest and largest of the sums

$$\phi(a) \sum_{i=1}^{k} f(t_{i-1})(t_i - t_{i-1})$$

includes two sums within $\phi(a)\varepsilon'$ of the minimum and maximum of $\int_a^x f(t)\,dt$. The remainder of the argument is straightforward.

36. **(a)** Using the substitution

$$y = \frac{1}{x}$$

$$x = \frac{1}{y}$$

$$dx = -\frac{1}{y^2}\,dy$$

the integrals become

(i) $\displaystyle \int_1^\infty \frac{1}{y^2} \sin\left(y + \frac{1}{y}\right) dy$ (ii) $\displaystyle \int_1^\infty \frac{1}{y^2} \sin^2\left(y + \frac{1}{y}\right) dy.$

The second integral is the easiest, since $\int_1^N \sin^2(y + 1/y)/y^2\,dy$ is an increasing function $\leq \int_1^N 1/y^2\,dy$, which is bounded. For the first we have to argue slightly differently: Since

$$\left| \int_M^N \frac{1}{y^2} \sin\left(y + \frac{1}{y}\right) dy \right| \leq \int_M^N \frac{1}{y^2} \left| \sin\left(y + \frac{1}{y}\right) \right| dy$$

$$\leq \int_M^N \frac{1}{y^2}\,dy \leq \frac{1}{M}, \qquad \text{for all } N \geq M,$$

the value of the integral from 1 to M is within $1/M$ of all later values, so the limit must exist. [For a precise proof, prove an analogue of Theorem 3 in Chapter 22: If $\displaystyle \lim_{M,N\to\infty} |F(M) - F(N)| = 0$, then $\displaystyle \lim_{N\to\infty} F(N)$ exists.]

(b) The substitution $y = 1/x$ yields

$$\text{(i)} \quad \int_0^1 \frac{1}{y^2} \sin y \, dy \qquad \text{(ii)} \quad \int_0^1 \frac{1}{y^2} \sin^2 \, dy.$$

Since

$$\lim_{y \to 0} \frac{\sin y}{y} = 1,$$

the second integral involves a bounded function and converges, while the first is essentially like $\int_0^1 1/y \, dy$ and does not converge.

37. (a) We have

$$\lim_{\varepsilon \to 0} \int_\varepsilon^1 \log x \, dx = \lim_{\varepsilon \to 0} \left(x \log x - x \Big|_\varepsilon^1 \right)$$
$$= -1,$$

since $\lim\limits_{x \to 0} x \log x = 0$ by Problem 12(d) of Chapter 18.

(b) To investigate the behavior near 0, write

$$\log(\sin x) = \log \left(\frac{\sin x}{x} \cdot x \right)$$
$$= \log \left(\frac{\sin x}{x} \right) + \log x.$$

Since $(\sin x)/x$ is close to 1, this is close to $\log x$, and part (a) shows that this causes no problem near 0.

The behavior near π is essentially the same, since $\sin(\pi - x) = \sin x$.

(c) The substitution $x = 2u$, $dx = 2 \, du$ gives

$$\int_0^\pi \log(\sin x) \, dx = 2 \int_0^{\pi/2} \log(\sin 2u) \, du$$
$$= 2 \int_0^{\pi/2} \log(2 \sin u \cos u) \, du$$
$$= 2 \int_0^{\pi/2} \log 2 + \log(\sin u) + \log(\cos u) \, du.$$

(d) Since the substitution $u = \pi - x$ gives

$$\int_0^{\pi/2} \log(\sin x) \, dx = \int_{\pi/2}^\pi \log(\sin(\pi - u)) \, du$$
$$= \int_{\pi/2}^\pi \log(\sin u) \, du,$$

and thus

$$\int_0^{\pi} \log(\sin x)\, dx = 2 \int_0^{\pi/2} \log(\sin x)\, dx,$$

the result of part (c) becomes

$$\int_0^{\pi/2} \log(\cos x) = -\frac{\pi \log 2}{2}.$$

(e) The substitution $x = \pi/2 - u$ gives

$$\int_0^{\pi/2} \log(\cos x)\, dx = \int_0^{\pi/2} \log(\cos(\pi/2 - u))\, du$$

$$= \int_0^{\pi/2} \log(\sin u)\, du,$$

so

$$\int_0^{\pi} \log(\sin x)\, dx = 2 \int_0^{\pi/2} \log(\cos x)\, dx$$

$$= -\pi \log 2.$$

38. For each N we have

$$\int_a^N u'(x)v(x)\, dx = u(x)v(x)\Big|_a^N - \int_a^N u(x)v'(x)\, dx.$$

The desired equation follows by taking limits (and shows that if any two of the three symbols involved exist, the third does also).

39. (a) The integral

$$\int_1^{\infty} e^{-t} t^{x-1}\, dt$$

certainly exists, because $\int_1^{\infty} t^{-2}\, dt$ exists (Problem 14-25), and for sufficiently large t we have $e^{-t} t^{x-1} < t^{-2}$ (by Theorem 18-6). On the other hand, if $t > 0$, then $e^{-t} t^{x-1} < t^{x-1}$; since the integral $\int_0^1 t^{x-1}\, dt$ exists for $x > 0$ (Problem 14-28), it follows that $\int_0^1 e^{-t} t^{x-1}\, dt$ exists for $x > 0$ (it is an improper integral if $x < 1$).

(b)

$$\Gamma(x+1) = \int_0^{\infty} e^{-t} t^x\, dt$$

$$= -e^{-t} t^x \Big|_{t=0}^{t=\infty} + \int_0^{\infty} x e^{-t} t^{x-1}\, dt$$

$$= 0 + x \int_0^{\infty} e^{-t} t^{x-1}\, dt = x\Gamma(x).$$

(If $x < 1$, then we are also using a second version of integration by parts to take care of the integral from 0 to 1.)

(c)

$$\Gamma(1) = \int_0^\infty e^{-t}\, dt = -e^{-t}\Big|_0^\infty = 1.$$

This proves that $\Gamma(1) = (1-1)!$. If $\Gamma(n) = (n-1)!$, then $\Gamma(n+1) = n\Gamma(n) = n \cdot (n-1)! = n!$, so the formula is true for all n, by induction.

40. (a)

$$\int_0^{\pi/2} \sin^n x\, dx = -\frac{1}{n}\sin^{n-1}x \cos x\Big|_0^{\pi/2} + \frac{n-1}{n}\int_0^{\pi/2}\sin^{n-2}x\, dx$$

$$= \frac{n-1}{n}\int_0^{\pi/2}\sin^{n-2}x\, dx.$$

(b)

$$\int_0^{\pi/2}\sin^{2n+1}x\, dx = \frac{2n}{2n+1}\cdot\int_0^{\pi/2}\sin^{2n-1}x\, dx$$

$$= \frac{2n}{2n+1}\cdot\frac{2n-2}{2n-1}\cdot\int_0^{\pi/2}\sin^{2n-3}x\, dx$$

$$= \cdots = \frac{2n}{2n+1}\cdot\frac{2n-2}{2n-1}\cdots\frac{2}{3}\int_0^{\pi/2}\sin^1 x\, dx$$

$$= \frac{2n}{2n+1}\cdot\frac{2n-2}{2n-1}\cdots\frac{2}{3}.$$

(A proof by induction is lurking in the wings.) Similarly,

$$\int_0^{\pi/2}\sin^{2n}x\, dx = \frac{2n-1}{2n}\cdots\frac{1}{2}\int_0^{\pi/2}\sin x\, dx$$

$$= \frac{2n-1}{2n}\cdots\frac{1}{2}\cdot\frac{\pi}{2}.$$

(c)

$$0 < \int_0^{\pi/2}\sin^{2n+1}x\, dx \le \int_0^{\pi/2}\sin^{2n}x\, dx \le \int_0^{\pi/2}\sin^{2n-1}x\, dx$$

so

$$1 \leq \frac{\displaystyle\int_0^{\pi/2} \sin^{2n} x \, dx}{\displaystyle\int_0^{\pi/2} \sin^{2n+1} x \, dx} \leq \frac{\displaystyle\int_0^{\pi/2} \sin^{2n-1} x \, dx}{\displaystyle\int_0^{\pi/2} \sin^{2n+1} x \, dx}$$

$$= \frac{\displaystyle\int_0^{\pi/2} \sin^{2n-1} x \, dx}{\dfrac{2n}{2n+1} \displaystyle\int_0^{\pi/2} \sin^{2n-1} x \, dx}$$

$$= 1 + \frac{1}{2n}.$$

(d) If n is large, then $\dfrac{\sqrt{\pi}}{\sqrt{2}}$ is close to

$$\sqrt{\frac{2\cdot 2\cdot 4\cdot 4\cdots 2n\cdot 2n}{1\cdot 3\cdot 3\cdot 5\cdots(2n-1)(2n+1)}} = \frac{2\cdot 4\cdots 2n}{\sqrt{2n+1}\cdot 1\cdot 3\cdot 5\cdots(2n-1)}$$

$$= \frac{\sqrt{\dfrac{2n}{2n+1}}}{\sqrt{2}} \frac{1}{\sqrt{n}} \frac{2\cdots 2n}{1\cdot 3\cdots(2n-1)}.$$

Since $\sqrt{(2n)/(2n+1)}$ is close to 1 for large n, the result follows. [Wallis' procedure was quite different. He worked with the integral

$$\int_0^1 (1 - x^2)^n \, dx$$

(which appears in Problem 41), hoping to recover, from the values obtained for natural numbers n, a formula for

$$\frac{\pi}{4} = \int_0^1 (1 - x^2)^{1/2} \, dx.$$

Wallis first obtained the formula

$$\int_0^1 (1 - x^2)^n \, dx = \frac{2}{3}\cdot\frac{4}{5}\cdots\frac{2n}{2n+1}$$

$$= \frac{(2\cdot 4\cdots 2n)^2}{2\cdot 3\cdot 4\cdots 2n(2n+1)} = \frac{2^n}{2n+1}\frac{(n!)^2}{(2n)!},$$

(by what method I am not certain). He then reasoned that $\pi/4$ should be

$$\int_0^1 (1 - x^2)^{1/2} \, dx = \frac{2^1}{2}\frac{\left(\frac{1}{2}!\right)^2}{1!} = \left(\tfrac{1}{2}!\right)^2.$$

If we interpret $\frac{1}{2}!$ to mean $\Gamma(1/2)$, this agrees with Problem 44, but Wallis did not know of the gamma function (which was invented by Euler, guided principally by Wallis' work). Since $(2n)!/(n!)^2$ is the binomial coefficient $\binom{2n}{n}$, Wallis hoped to find $\frac{1}{2}!$ by finding $\binom{p+q}{p}$ for $p = q = 1/2$. Now

$$\binom{p+q}{p} = \frac{(p+q)(p+q-1)\cdots(p+1)}{q!}$$

and this makes sense even if p is not a natural number. Wallis therefore decided that

$$\binom{\frac{1}{2}+q}{\frac{1}{2}} = \frac{(\frac{1}{2}+q)\cdots(\frac{3}{2})}{q!}.$$

With this interpretation of $\binom{p+q}{p}$ for $p = 1/2$, it is still true that

$$\binom{p+q+1}{p} = \frac{p+q+1}{q+1}\binom{p+q}{p}.$$

Denoting $\binom{\frac{1}{2}+q}{\frac{1}{2}}$ by $W(q)$ this equation can be written

$$W(q+1) = \frac{\frac{1}{2}+q+1}{q+1}W(q) = \frac{2q+3}{2q+2}W(q),$$

which leads to the table

q	1	2	3	
$W(q)$	$\frac{3}{2}$	$\frac{3}{2}\cdot\frac{5}{4}$	$\frac{3}{2}\cdot\frac{5}{4}\cdot\frac{7}{6}$	$\cdots$

But, since $W(\frac{1}{2})$ should be $4/\pi$, Wallis also constructs the table

q	$\frac{1}{2}$	$\frac{3}{2}$	$\frac{5}{2}$	
$W(q)$	$\frac{4}{\pi}$	$\frac{4}{\pi}\cdot\frac{4}{3}$	$\frac{4}{\pi}\cdot\frac{4}{3}\cdot\frac{6}{5}$	$\cdots$

Next Wallis notes that if a_1, a_2, a_3, a_4 are 4 successive values $W(q)$, $W(q+1)$, $W(q+2)$, $W(q+3)$, appearing in either of these tables, then

$$\frac{a_2}{a_1} > \frac{a_3}{a_2} > \frac{a_4}{a_3} \quad \left(\text{because } \frac{2q+3}{2q+2} > \frac{2q+5}{2q+4}\cdot\frac{2q+7}{2q+6}\right),$$

which implies that

$$\sqrt{\frac{a_3}{a_1}} > \frac{a_3}{a_2} > \sqrt{\frac{a_4}{a_2}}.$$

Wallis then argues that this should still be true when a_1, a_2, a_3, a_4 are four successive values in a combined table where q is given *both* integer and half-integer values! Thus, taking as the four successive values $W(n+\frac{1}{2})$, $W(n)$, $W(n+\frac{3}{2})$, $W(n+1)$, he obtains

$$\sqrt{\dfrac{\dfrac{4}{\pi} \cdot \dfrac{4}{3} \cdot \dfrac{6}{5} \cdots \dfrac{2n+4}{2n+3}}{\dfrac{4}{\pi} \cdot \dfrac{4}{3} \cdot \dfrac{6}{5} \cdots \dfrac{2n+2}{2n+1}}} > \dfrac{\dfrac{4}{\pi} \cdot \dfrac{4}{3} \cdot \dfrac{6}{5} \cdots \dfrac{2n+2}{2n+1}}{\dfrac{3}{2} \cdot \dfrac{5}{4} \cdot \dfrac{7}{6} \cdots \dfrac{2n+1}{2n}} > \sqrt{\dfrac{\dfrac{3}{2} \cdot \dfrac{5}{4} \cdots \dfrac{2n+3}{2n+2}}{\dfrac{3}{2} \cdot \dfrac{5}{4} \cdot \dfrac{7}{6} \cdots \dfrac{2n+1}{2n}}}$$

which yields simply

$$\sqrt{\frac{2n+4}{2n+3}} > \frac{4}{\pi} \cdot \left[\frac{2 \cdot 4 \cdot 4 \cdot 6 \cdot 6 \cdots (2n)(2n)(2n+2)}{3 \cdot 3 \cdot 5 \cdot 5 \cdots (2n+1)(2n+1)} \right] > \sqrt{\frac{2n+3}{2n+2}},$$

from which Wallis' product follows immediately.]

41. (a) Let $x = \cos u$, $dx = -\sin u \, du$. Then

$$\int_0^1 (1 - x^2)^n \, dx = \int_{-\pi/2}^0 (\sin^{2n} u)(-\sin u) \, du = \int_0^{\pi/2} \sin^{2n+1} u \, du$$

$$= \frac{2}{3} \cdot \frac{4}{5} \cdots \frac{2n}{2n+1} \qquad \text{by Problem 40.}$$

Now let $x = \cot u$, $dx = -\csc^2 u \, du$. Then

$$\int_0^\infty \frac{1}{(1+x^2)^n} \, dx = \int_{\pi/2}^0 (\sin^{2n} u) \left(\frac{-1}{\sin^2 u} \right) \, du$$

$$= \int_0^{\pi/2} \sin^{2(n-1)} u \, du$$

$$= \frac{\pi}{2} \cdot \frac{1}{2} \cdot \frac{3}{4} \cdot \frac{5}{6} \cdots \frac{2n-3}{2n-2} \qquad \text{by Problem 40.}$$

(b) If $f(y) = 1 - y$ and $g(y) = e^{-y}$, then $f(0) = g(0)$ and

$$f'(y) = -1 \le -e^{-y} \qquad \text{for } y \ge 0,$$

so $f(y) \le g(y)$ for $y \ge 0$, i.e., $1 - y \le e^{-y}$ for $y \ge 0$. So, in particular, $1 - x^2 \le e^{-x^2}$ (for all x).

The second inequality follows from the inequality $1 + y \le e^y$, which can be proved similarly (and has already appeared in Problem 18-29).

(c)

$$\int_0^1 (1 - x^2)^n \, dx \le \int_0^1 e^{-nx^2} \, dx \le \int_0^\infty e^{-nx^2} \, dx \le \int_0^\infty \frac{1}{(1+x^2)^n} \, dx,$$

so

$$\frac{2}{3} \cdots \frac{2n}{2n+1} \le \int_0^1 e^{-nx^2} \, dx \le \int_0^\infty e^{-nx^2} \, dx \le \frac{\pi}{2} \cdot \frac{1}{2} \cdots \frac{2n-3}{2n-2}.$$

Using the substitution $y = \sqrt{n}\,x$, $dx = 1/\sqrt{n}\,dy$, we obtain

$$\int_0^1 e^{-nx^2}\,dx = \frac{1}{\sqrt{n}} \int_0^{\sqrt{n}} e^{-y^2}\,dy,$$

$$\int_0^\infty e^{-nx^2}\,dx = \frac{1}{\sqrt{n}} \int_0^\infty e^{-y^2}\,dy,$$

from which the desired inequalities follow.

(d) It follows from Problem 40(d) that by choosing a sufficiently large, the numbers

$$\frac{\pi}{2}\sqrt{n}\,\frac{1}{2} \cdot \frac{3}{4} \cdots \frac{2n-3}{2n-2} = \frac{\pi}{2}\sqrt{\frac{n}{n-1}}\left[\sqrt{n-1}\,\frac{1}{2} \cdot \frac{3}{4} \cdots \frac{2n-3}{2n-2}\right]$$

and

$$\sqrt{n}\,\frac{2}{3} \cdot \frac{4}{5} \cdots \frac{2n}{2n+1} = \frac{n}{2n+1}\left[\frac{1}{\sqrt{n}}\,\frac{2}{3} \cdots \frac{2n}{2n-1}\right]$$

can be made as close as desired to

$$\frac{\pi}{2} \cdot 1 \cdot \frac{1}{\sqrt{\pi}} = \frac{\sqrt{\pi}}{2}$$

and

$$\frac{1}{2} \cdot \sqrt{\pi} = \frac{\sqrt{\pi}}{2}.$$

42. (a)

$$\int_a^b \sin x \cdot \frac{1}{x}\,dx = -\cos x \cdot \frac{1}{x}\Big|_a^b - \int_a^b -\cos x \cdot -\frac{1}{x^2}\,dx$$

$$= \frac{\cos a}{a} - \frac{\cos b}{b} - \int_a^b \frac{\cos x}{x^2}\,dx.$$

In particular,

$$\int_1^\infty \frac{\sin x}{x}\,dx = \frac{\cos 1}{1} - \int_1^\infty \frac{\cos x}{x^2}\,dx;$$

the latter integral exists because the integral

$$\int_1^\infty \left|\frac{\cos x}{x^2}\right|\,dx \leq \int_1^\infty \frac{1}{x^2}\,dx$$

exists (compare Theorem 23-4).

On the other hand, the integral

$$\int_0^1 \frac{\sin x}{x}\,dx$$

exists and equals $\int_0^1 f(x)\,dx$, where f is the continuous function with

$$f(x) = \begin{cases} 1, & x = 0 \\ \dfrac{\sin x}{x}, & x \neq 0. \end{cases}$$

(b) According to Problem 15-33,

$$\int_0^\pi \frac{\sin(n+\frac{1}{2})t}{\sin t/2}\,dt = \int_0^\pi (1 + 2\cos t + \cdots + 2\cos nt)\,dt$$
$$= \pi.$$

(c) The hint is the whole answer, since the function

$$f(t) = \begin{cases} 0, & t = 0 \\ \dfrac{2}{t} - \dfrac{1}{\sin t/2}, & t \neq 0 \end{cases}$$

is integrable on $[0, \pi]$.

(d) From parts (b) and (c) we have

$$\lim_{\lambda \to \infty} \int_0^\pi \frac{2\sin(\lambda + \frac{1}{2})t}{t}\,dt = \lim_{\lambda \to \infty} \int_0^\pi \frac{\sin(\lambda + \frac{1}{2})t}{\sin t/2}\,dt = \pi.$$

Using the substitution $u = (\lambda + \frac{1}{2})t$, we have

$$\lim_{\lambda \to \infty} \int_0^\pi \frac{2\sin(\lambda + \frac{1}{2})t}{t}\,dt = \lim_{\lambda \to \infty} \int_0^{(\lambda + 1/2)\pi} 2\sin u \cdot \frac{(\lambda + \frac{1}{2})}{u} \cdot \frac{du}{\lambda + \frac{1}{2}}$$
$$= 2\int_0^\infty \frac{\sin u}{u}\,du.$$

43. We have

$$\int_0^\infty \sin^2 x \cdot \frac{1}{x^2}\,dx = \sin^2 x \left(-\frac{1}{x}\right)\Big|_0^\infty + \int_0^\infty \frac{1}{x}\, 2\sin x \cos x\,dx$$
$$= \int_0^\infty \frac{\sin 2x}{x}\,dx.$$

Setting $u = 2x$, $du = 2\,dx$, so that $du/u = dx/x$, this becomes

$$\int_0^\infty \frac{\sin u}{u}\,du = \frac{\pi}{2}.$$

44. (a) Let $u = t^x$, $du = xt^{x-1} dt$. Then

$$\Gamma(x) = \int_0^\infty e^{-t} t^{x-1} dt = \int_0^\infty e^{-u^{1/x}} \frac{du}{x}$$
$$= \frac{1}{x} \int_0^\infty e^{-u^{1/x}} du.$$

(b)

$$\Gamma(\tfrac{1}{2}) = 2 \int_0^\infty e^{-u^2} du$$
$$= \sqrt{\pi} \qquad \text{by Problem 40.}$$

45. (a) The substitution $u = \alpha x$, $du = \alpha\, dx$ gives

$$\int_\varepsilon^N \frac{f(\alpha x)}{x} dx = \int_{\alpha\varepsilon}^{\alpha N} \frac{f(u)}{u} du.$$

Similarly, the substitution $u = \beta x$, $du = \beta\, dx$ gives

$$\int_\varepsilon^N \frac{f(\beta x)}{x} dx = \int_{\beta\varepsilon}^{\beta N} \frac{f(u)}{u} du.$$

So

$$\int_\varepsilon^N \frac{f(\alpha x) - f(\beta x)}{x} dx = \int_{\alpha\varepsilon}^{\alpha N} \frac{f(u)}{u} du - \int_{\beta\varepsilon}^{\beta N} \frac{f(u)}{u} du$$
$$= \int_{\alpha\varepsilon}^{\beta\varepsilon} \frac{f(u)}{u} du - \int_{\alpha N}^{\beta N} \frac{f(u)}{u} du.$$

As $\varepsilon \to 0$ and $N \to \infty$, this approaches

$$\int_{\alpha\varepsilon}^{\beta\varepsilon} \frac{A}{u} du - \int_{\alpha N}^{\beta N} \frac{B}{u} du = (A - B) \log \frac{\beta}{\alpha}.$$

(b) In this case the same substitutions give

$$\int_\varepsilon^\infty \frac{f(\alpha x)}{x} dx = \int_{\alpha\varepsilon}^\infty \frac{f(u)}{u} du, \qquad \int_\varepsilon^\infty \frac{f(\beta x)}{x} dx = \int_{\beta\varepsilon}^\infty \frac{f(u)}{u} du,$$

so

$$\int_\varepsilon^\infty \frac{f(\alpha x) - f(\beta x)}{x} dx = \int_{\alpha e}^{\beta\varepsilon} \frac{f(u)}{u} du \to A \log \frac{\beta}{\alpha}.$$

(c) (i) Since

$$\int_0^\infty \frac{e^{-x}}{x} dx$$

converges and $\lim\limits_{x \to 0} e^{-x} = 1$, we have

$$\int_0^\infty \frac{e^{-\alpha x} - e^{-\beta x}}{x}\, dx = \log \frac{\beta}{\alpha}.$$

(iii) For $a > 0$, the integral

$$\int_a^\infty \frac{\cos x}{x}\, dx$$

exists (same reasoning as in Problem 42(a)), and $\lim\limits_{x \to 0} \cos x = 1$, so

$$\int_0^\infty \frac{\cos(\alpha x) - \cos(\beta x)}{x}\, dx = \log \frac{\beta}{\alpha}.$$

46. (a) Choosing $n = 1$ in Problem 11-43, with $x_1 = t_{i-1}$ and $x_2 = t_i$, so that $Q(x) = (x - t_{i-1})(x - t_i)$, it follows that for each x in $[t_{i-1}, t_i]$ we have

$$f(x) - P_i(x) = (x - t_{i-1})(x - t_i) \cdot \frac{f''(c)}{2}$$

for some c in $[t_{i-1}, t_i]$. So

$$\frac{n_i}{2}(x - t_{i-1})(x - t_i) \geq f(x) - P_i(x) \geq \frac{N_i}{2}(x - t_{i-1})(x - t_i)$$

(the inequalities are reversed because $(x - t_{i-1})(x - t_i) \leq 0$ on $[t_{i-1}, t_i]$).

(b)

$$
\begin{aligned}
I &= \int_{t_{i-1}}^{t_i} x^2 - (t_i + t_{i-1})x + t_{i-1}t_i \, dx \\[2mm]
&= \frac{x^3}{3}\Big|_{t_{i-1}}^{t_i} - (t_i + t_{i-1}) \cdot \frac{x^2}{2}\Big|_{t_{i-1}}^{t_i} + (t_i - t_{i-1})(t_{i-1}t_i) \\[2mm]
&= \frac{t_i{}^3}{3} - \frac{t_{i-1}{}^3}{3} - \frac{t_i{}^2(t_i + t_{i-1})}{2} + \frac{t_{i-1}{}^2(t_i + t_{i-1})}{2} + t_i{}^2 t_{i-1} - t_{i-1}{}^2 t_i \\[2mm]
&= \frac{t_{i-1}{}^3}{6} - \frac{t_i{}^3}{6} - \frac{3t_{i-1}{}^2 t_i}{6} + \frac{3t_i{}^2 t_{i-1}}{6} \\[2mm]
&= \frac{(t_{i-1} - t_i)^3}{6} \\[2mm]
&= -\frac{h^3}{6}.
\end{aligned}
$$

(c) Summing the equations in (b) for $i = 1, \ldots, n$ and using $h = (b - a)/n$, we get

$$\frac{(b - a)^3}{12n^2} \cdot \frac{\displaystyle\sum_{i=1}^n n_i}{n} \leq \int_a^b f - \Sigma_n \leq \frac{(b - a)^3}{12n^2} \cdot \frac{\displaystyle\sum_{i=1}^n N_i}{n}.$$

Now the minimum m of f'' on $[a, b]$ is $\leq n_i$, for each i, so

$$m \leq \frac{\sum\limits_{i=1}^{n} n_i}{n}.$$

And similarly the maximum M of f'' on $[a, b]$ satisfies

$$\frac{\sum\limits_{i=1}^{n} N_i}{n} \leq M,$$

so we obtain

$$\frac{(b-a)^3}{12n^2}m \leq \int_a^b f - \Sigma_n \leq \frac{(b-a)^3}{12n^2}M,$$

from which the desired result follows.

47. (a) Using Problem 3-6, we can explicitly write P as

$$P(x) = \frac{f(0)(x-1)(x-2)}{2} - f(1)x(x-2) + \frac{f(2)x(x-1)}{2}$$

$$= x^2\left[\frac{f(0)}{2} - f(1) + \frac{f(2)}{2}\right] + x\left[-\frac{3}{2}f(0) + 2f(1) - \frac{f(2)}{2}\right] + f(0).$$

So

$$\int_0^2 P = \frac{8}{3}\left[\frac{f(0)}{2} - f(1) + \frac{f(2)}{2}\right] + 2\left[-\frac{3}{2}f(0) + 2f(1) - \frac{f(2)}{2}\right] + 2f(0)$$

$$= \frac{1}{3}[f(0) + 4f(1) + f(2)].$$

(b) If

$$\bar{P}(x) = P\left(a + \frac{(b-a)x}{2}\right)$$

$$\bar{f}(x) = f\left(a + \frac{(b-a)x}{2}\right),$$

then since P agrees with f at a, $(a+b)/2$ and b, it is easy to see that $\bar{P}$ agrees with $\bar{f}$ at 0, 1 and 2. So the substitution

$$u = a + \frac{(b-a)x}{2}$$

$$du = \frac{b-a}{2}dx$$

gives

$$\int_a^b P(u)\,du = \frac{b-a}{2}\int_0^2 \bar{P}(x)\,dx$$

$$= \frac{b-a}{2}\cdot\frac{1}{3}[\bar{f}(0) + 4\bar{f}(1) + \bar{f}(2)] \qquad \text{by part (a)}$$

$$= \frac{b-a}{6}\left[f(a) + 4f\left(\frac{a+b}{2}\right) + f(b)\right].$$

(c) According to Problem 11-43, for each x in $[a, b]$ there is a number c in (a, b) with

$$f(x) - P(x) = (x - a)(x - b)\left(x - \frac{a+b}{2}\right)\cdot\frac{f''(c)}{6}$$

$$= (x - a)(x - b)\left(x - \frac{a+b}{2}\right)\cdot C$$

for some constant C. So

$$\int_a^b f - P = C\cdot\int_a^b (x - a)(x - b)\left(x - \frac{a+b}{2}\right)dx.$$

This latter integral has the value 0. An easy way to see this is to check it for $a = 0$, $b = 2$ and then use the substitution of part (b) to express the general case in terms of this one. Another way is by using the substitution

$$u = x - \frac{a+b}{2}, \qquad x = u + \frac{a+b}{2}$$

$$du = dx$$

to make the integral more symmetric. Letting

$$h = \frac{b-a}{2}$$

we have

$$\int_a^b (x - a)(x - b)\left(x - \frac{a+b}{2}\right)dx = \int_{-h}^h (u + h)(u - h)u\,du$$

$$= \int_{-h}^h u^3 - uh^2\,du = 0,$$

since we are integrating an odd function.

48. **(a)** Writing $Q(x)$ as in the hint, so that the first three equations automatically hold, we have

$$Q'\left(\frac{a+b}{2}\right) = P'\left(\frac{a+b}{2}\right) + A\left(\frac{a+b}{2}-a\right)\left(\frac{a+b}{2}-b\right)$$

$$= P'\left(\frac{a+b}{2}\right) - \frac{A(b-a)^2}{4}.$$

Since $b - a \neq 0$, we can then choose A so that $Q'(\frac{a+b}{2})$ has any desired value.

(b) If x is a, b or $\frac{a+b}{2}$ there is nothing to prove. Otherwise, consider the function

$$F(t) = (x-a)\left(x-\frac{a+b}{2}\right)^2 (x-b)[f(t) - Q(t)]$$

$$- (t-a)\left(t-\frac{a+b}{2}\right)^2 (t-b)[f(x) - Q(x)].$$

The F is 0 at a, b, $\frac{a+b}{2}$ and x. To be specific, say $\frac{a+b}{2} < x < b$. Then F' is 0 at points ξ_1, ξ_2, ξ_3 with

$$a < \xi_1 < \frac{a+b}{2} < \xi_2 < x < \xi_3 < b.$$

But it is easy to see that we also have

$$F'\left(\frac{a+b}{2}\right) = 0.$$

So F' is 0 at 4 points in (a, b), and consequently, as in Problem 11-42, $F^{(4)}$ is 0 at some point ξ, that is

$$0 = F^{(4)}(\xi) = (x-a)\left(x-\frac{a+b}{2}\right)^2 (x-b)f^{(4)}(\xi) - 4![f(x) - Q(x)],$$

as required.

(c) If m and M are the minimum and maximum of $f^{(4)}$ on $[a, b]$, it follows that

$$\frac{m_i}{4!}(x-a)\left(x-\frac{a+b}{2}\right)^2 (x-b) \leq f(x)-Q(x) \leq \frac{M_i}{4!}(x-a)\left(x-\frac{a+b}{2}\right)^2 (x-b)$$

for all x in $[a, b]$ (note that the expression $(x-a)(x-\frac{a+b}{2})(x-b)$ is ≥ 0 on $[a, b]$). It follows that

$$\int_a^b f - Q = \frac{f^{(4)}(c)}{24}\int_a^b (x-a)\left(x-\frac{a+b}{2}\right)^2 (x-b)\,dx$$

for some c in $[a, b]$. To evaluate the integral on the right we use the same substitution as in Problem 47(c), to obtain

$$\int_{-h}^{h} (u+h)(u-h)u^2 \, du = \int_{-h}^{h} u^4 - u^2 h^2 \, du$$

$$= \frac{2h^5}{5} - \frac{2h^3}{3}$$

$$= -\frac{4}{15} h^5 = -\frac{4}{15} \left(\frac{b-a}{2} \right)^5$$

$$= -\frac{(b-a)^5}{120}.$$

(d) By part (c) we have, noting that $t_{2i} - t_{2i-2} = (b-a)/n$,

$$\int_{t_{2i-1}}^{t_{2i}} f = \frac{b-a}{6n}[f(t_{2i-1}) + 4f(t_{2i-1}) + f(t_{2i})] - \frac{(b-a)^5}{2880n^5} f^{(4)}(c_i)$$

for some c_i in (t_{2i-2}, t_{2i}). When we sum for $i = 1, \ldots, n$, each t_{2i} occurs twice, once in the above expression, and once in the same expression for $i + 1$; the only exceptions are $f(t_0) = f(a)$ and $f(t_{2n}) = f(b)$, which occur just once. Moreover, if $m \leq f^{(4)} \leq M$ on $[a, b]$, then

$$nm \leq \sum_{i=1}^{n} f^{(4)}(c_i) \leq nM,$$

so

$$\sum_{i=1}^{n} f^{(4)}(c_i) = nf^{(4)}(\bar{c})$$

for some $\bar{c}$ in $[a, b]$. Thus,

$$\int_{a}^{b} f = \frac{b-a}{6n} \left(f(a) + 4 \sum_{i=1}^{n} f(t_{2i-1}) + 2 \sum_{i=1}^{n-1} f(t_{2i}) + f(b) \right) - \frac{(b-a)^5}{2880n^4} f^{(4)}(\bar{c}).$$

1. (a) The graphs intersect at $(0, 0)$ and $(1, 1)$, so the volume is

$$\pi \int_0^1 x^2 - x^4 \, dx = \pi \left(\frac{x^3}{3} - \frac{x^5}{5} \Big|_0^1 \right)$$
$$= \pi(\tfrac{1}{3} - \tfrac{1}{5}).$$

(b) The shell method gives

$$2\pi \int_0^1 x \cdot (x - x^2) \, dx = 2\pi \left(\frac{x^3}{3} - \frac{x^4}{4} \Big|_0^1 \right)$$
$$= 2\pi(\tfrac{1}{3} - \tfrac{1}{4}).$$

2. Rotating the graph of

$$f(x) = \sqrt{r^2 - x^2} \qquad -r \le x \le r,$$

we get

$$\pi \int_{-r}^r r^2 - x^2 \, dx = \pi \left(r^2 x - \frac{x^3}{3} \Big|_{-r}^r \right)$$
$$= 2\pi \left(r^3 - \frac{r^3}{3} \right)$$
$$= \tfrac{4}{3}\pi r^3.$$

3. Rotating the graph of

$$f(x) = b\sqrt{1 - \frac{x^2}{a^2}} \qquad -a \le x \le a,$$

we get

$$\pi \int_{-a}^a b^2 \left(1 - \frac{x^2}{a^2} \right) dx = b^2 \pi \left(x - \frac{x^3}{3a^2} \Big|_{-a}^a \right)$$
$$= 2b^2 \pi \left(a - \frac{a}{3} \right)$$
$$= \tfrac{4}{3}\pi a b^2.$$

4. The shell method gives

$$2\pi \int_{a-b}^{a+b} x\sqrt{b^2 - (x - a)^2} \, dx.$$

Letting

$$x - a = bu$$
$$dx = b\,du$$

we get

$$2\pi \int_{-1}^{1} (a + bu)b^2\sqrt{1 - u^2}\,du$$

$$= 2\pi ab^2 \int_{-1}^{1} \sqrt{1 - u^2}\,du + 2\pi ab^3 \int_{-1}^{1} u\sqrt{1 - u^2}\,du$$

$$= 2\pi ab^2 \left(\frac{\arcsin u}{2} + \frac{u\sqrt{1 - u^2}}{2} \bigg|_{-1}^{1} \right) + 0 \qquad \text{by Problem 19-4(viii)}$$

$$= 2\pi ab^2 \left(\frac{\pi}{2} \right)$$

$$= \pi^2 ab^2$$

$$= \pi a \cdot (\pi b^2).$$

(Thus, the volume is the area of the circle of radius b times the length πa of the circle that its center revolves around. This is a special case of "Pappus' rule".)

5. Using the shell method the volume is

$$2 \cdot 2\pi \cdot \int_{a}^{2a} x\sqrt{4a^2 - x^2}\,dx$$

(the extra factor of 2 comes about because the shell method gives only the part with $y \geq 0$). Letting

$$x = 2au$$
$$dx = 2a\,du$$

this becomes

$$4\pi \int_{1/2}^{1} 8a^3 u\sqrt{1 - u^2}\,du = 32\pi a^3 \cdot \left(\frac{-(1 - u^2)^{3/2}}{3} \bigg|_{1/2}^{1} \right)$$

$$= \frac{32\pi a^3}{3} \cdot \left(\frac{3}{4} \right)^{3/2}$$

$$= \frac{4}{3}(2a)^3\pi \cdot \frac{3\sqrt{3}}{8}$$

(as compared to $\dfrac{4}{3}(2a)^3\pi$, the volume of the entire sphere).

6. (a) The volume is

$$2\pi \int_0^{\pi/2} x \cos x \, dx = 2\pi \left[x \sin x \Big|_0^{\pi/2} - \int_0^{\pi/2} \sin x \, dx \right]$$

$$= 2\pi \left[\frac{\pi}{2} - 1 \right]$$

$$= \pi(\pi - 2).$$

(b) The volume is also

$$\pi \int_0^1 (\arccos)^2.$$

7. Actually, instead of using the formula for $\int f^{-1}$, it is simplest to go through the steps by which this formula was derived: In the integral

$$\int_{f(a)}^{f(b)} y f^{-1}(y) \, dy$$

let

$$x = f^{-1}(y)$$
$$y = f(x)$$
$$dy = f'(x) \, dx.$$

The integral becomes

$$\int_a^b x f(x) f'(x) \, dx = \frac{1}{2} \int_a^b x (f^2)'(x) \, dx$$

$$= \frac{1}{2} \left[x f(x)^2 \Big|_a^b - \int_a^b f(x)^2 \, dx \right]$$

$$= \frac{1}{2} \left[b f(b)^2 - a f(a)^2 - \int_a^b f(x)^2 \, dx \right],$$

as required.

8. (a) If the diameter AB lies on the horizontal axis with A at $(-a, 0)$ and B at $(0, a)$, then the square intersected by the plane through $(x, 0)$ has sides of length $2\sqrt{a^2 - x^2}$, so

$$4(a^2 - x^2)(t_i - t_{i-1})$$

is the volume of a "slab", and the sum of these approaches

$$4 \int_{-a}^{a} (a^2 - x^2)\, dx = 4 \left[a^2 x - \frac{x^3}{3} \Big|_{-a}^{a} \right]$$

$$= 4 \left[2a^3 - \frac{2a^3}{3} \right]$$

$$= \frac{16a^3}{3}$$

(as compared to $\dfrac{2\pi a^3}{3}$, the volume of the top part of the sphere of radius a).

(b) Now the triangle intersected by the plane through $(x, 0)$ has area

$$\frac{\sqrt{3}}{4} \cdot 4(a^2 - x^2)$$

so the volume is

$$\frac{\sqrt{3}}{4} \cdot \frac{16a^3}{3} = \frac{4\sqrt{3}a^3}{3}.$$

9. A plane parallel to the base at distance x from the vertex has area

$$\left(\frac{x}{h} \right)^2 A$$

so the volume is

$$\int_{0}^{h} \frac{x^2 A}{h^2}\, dx = \frac{1}{3} h A.$$

10. If (x, y, z) are the coordinates of a point P, then P is inside the first cylinder of radius a if and only if

$$x^2 + z^2 \le a^2$$

and inside the second if and only if

$$y^2 + z^2 \le a^2.$$

For points with $z = b$ (i.e, the horizontal plane at distance b above the plane with the axes) we must have

$$x^2 \le a^2 - b^2, \quad y^2 \le a^2 - b^2,$$

so we have a square with sides of length $\sqrt{a^2 - b^2}$, and area $a^2 - b^2$. So the volume of the intersection is

$$\int_{-a}^{a} a^2 - z^2 \, dz = a^2 z - \frac{z^3}{3} \Big|_{-a}^{a}$$

$$= \frac{4a^3}{3}.$$

11. (a) Using the formula

$$2\pi \int_a^b f(x)\sqrt{1 + f'(x)^2}\, dx,$$

we have

$$f(x) = \sqrt{r^2 - x^2}$$

$$f'(x) = \frac{-x}{\sqrt{r^2 - x^2}}$$

$$1 + f'(x)^2 = \frac{r^2}{r^2 - x^2}$$

so the surface area is

$$2\pi \int_{-r}^{r} \sqrt{r^2 - x^2} \cdot \frac{r}{\sqrt{r^2 - x^2}}\, dx = 2\pi \int_{-r}^{r} r\, dx$$

$$= 4\pi r^2.$$

(b) The area of the portion is

$$2\pi \int_a^{a+h} r\, dx = 2\pi r h.$$

12. (a) The ellipse is the graph of

$$f(x) = b\sqrt{1 - \frac{x^2}{a^2}} = \frac{b}{a}\sqrt{a^2 - x^2}.$$

It is convenient to set

$$\mu = \frac{b}{a}.$$

Then

$$f(x) = \mu\sqrt{a^2 - x^2}$$

$$f'(x) = \frac{-\mu x}{\sqrt{a^2 - x^2}}$$

$$f'(x)^2 = \frac{\mu^2 x^2}{a^2 - x^2}$$

$$1 + f'(x)^2 = \frac{a^2 + (\mu^2 - 1)x^2}{a^2 - x^2}$$

$$f(x)\sqrt{1 + f'(x)^2} = \mu\sqrt{a^2 + (\mu^2 - 1)x^2}.$$

So the area is

$$A = 2\pi\mu \int_{-a}^{a} \sqrt{a^2 + (\mu^2 - 1)x^2}\, dx.$$

Case 1: $a < b$, so $\mu > 1$. We use the substitution

$$y = \frac{\sqrt{\mu^2 - 1}}{a} x$$

$$x = \frac{ay}{\sqrt{\mu^2 - 1}}$$

$$dx = \frac{a\,dy}{\sqrt{\mu^2 - 1}}.$$

Then

$$A = \frac{2\pi a\mu}{\sqrt{\mu^2 - 1}} \int_{-\sqrt{\mu^2-1}}^{\sqrt{\mu^2-1}} \sqrt{a^2 + a^2 y^2}\,dy$$

$$= \frac{2\pi a^2\mu}{\sqrt{\mu^2 - 1}} \int_{-\sqrt{\mu^2-1}}^{\sqrt{\mu^2-1}} \sqrt{1 + y^2}\,dy$$

$$= \frac{\pi ab}{\sqrt{\mu^2 - 1}} \left[y\sqrt{1 + y^2} + \log\left(y + \sqrt{1 + y^2}\right) \Big|_{-\sqrt{\mu^1-1}}^{\sqrt{\mu^2-1}} \right]$$

by Problem 19-4(ix)

$$= \frac{\pi ab}{\sqrt{\mu^2 - 1}} \left[2\mu\sqrt{\mu^2 - 1} + \log(\sqrt{\mu^2 - 1} + \mu) - \log(-\sqrt{\mu^2 - 1} + \mu) \right]$$

$$= 2\pi b^2 + \frac{\pi ab}{\sqrt{\mu^2 - 1}} \log\left(\frac{\sqrt{\mu^2 - 1} + \mu}{-\sqrt{\mu^2 - 1} + \mu} \right)$$

$$= 2\pi b^2 + \frac{\pi ab}{\sqrt{\mu^2 - 1}} \log\left(\frac{[\sqrt{\mu^2 - 1} + \mu]^2}{[-\sqrt{\mu^2 - 1} + \mu] \cdot [\sqrt{\mu^2 - 1} + \mu]} \right)$$

$$= 2\pi b^2 + \frac{2\pi ab}{\sqrt{\mu^2 - 1}} \log(\sqrt{\mu^2 - 1} + \mu).$$

Note that

$$\lim_{\mu \to 1} \frac{\log(\sqrt{\mu^2 - 1} + \mu)}{\sqrt{\mu^2 - 1}} = \lim_{\mu \to 1} \frac{\dfrac{\mu}{\sqrt{\mu^2 - 1}} + 1}{\dfrac{\mu}{\sqrt{\mu^2 - 1}}} \qquad \text{by l'Hôpital's Rule}$$

$$= \lim_{\mu \to 1} \frac{\mu + \sqrt{\mu^2 - 1}}{\mu}$$

$$= 1.$$

So as $a \to b$, and thus $\mu \to 1$, the above area approaches $4\pi b^2$, the area of the sphere of radius b (Problem 11).

Case 2: $a < b$, so $\mu < 1$. The substitution

$$y = \frac{\sqrt{1 - \mu^2}}{a} x$$

gives

$$A = \frac{2\pi a \mu}{\sqrt{1 - \mu^2}} \int_{-\sqrt{1-\mu^2}}^{\sqrt{1-\mu^2}} \sqrt{a^2 - a^2 y^2} \, dy$$

$$= \frac{2\pi a^2 \mu}{\sqrt{1 - \mu^2}} \int_{-\sqrt{1-\mu^2}}^{\sqrt{1-\mu^2}} \sqrt{1 - y^2} \, dy$$

$$= \frac{\pi ab}{\sqrt{1 - \mu^2}} \left[\arcsin y + y\sqrt{1 - y^2} \Big|_{\sqrt{1-\mu^2}}^{\sqrt{1-\mu^2}} \right] \qquad \text{by Problem 19-4(viii)}$$

$$= \frac{\pi ab}{\sqrt{1 - \mu^2}} \left[2 \arcsin \sqrt{1 - \mu^2} + 2\mu\sqrt{1 - \mu^2} \right]$$

$$= 2\pi b^2 + \frac{2\pi ab}{\sqrt{1 - \mu^2}} \arcsin \sqrt{1 - \mu^2}.$$

Again, we have

$$\lim_{\mu \to 1} \frac{\arcsin \sqrt{1 - \mu^2}}{\sqrt{1 - \mu^2}} = 1,$$

either by l'Hôpital's Rule, or using $\lim_{h \to 0} (\sin h)/h = 1$.

(b) The outer portion of the torus is obtained by revolving the graph of

$$f(y) = a + \sqrt{b^2 - y^2} \qquad -b \le y \le b$$

around the vertical axis, and the inner portion is obtained by revolving the graph of

$$f(y) = a - \sqrt{b^2 - y^2} \qquad -b \le y \le b.$$

In both cases

$$f'(y) = \frac{-y}{\sqrt{b^2 - y^2}}$$

$$1 + f'(y)^2 = \frac{b^2}{\sqrt{b^2 - y^2}},$$

so the area is

$$2\pi \int_{-b}^{b} \left(a + \sqrt{b^2 - y^2} \right) \frac{b}{\sqrt{b^2 - y^2}} \, dy + 2\pi \int_{-b}^{b} \left(a - \sqrt{b^2 - y^2} \right) \frac{b}{\sqrt{b^2 - y^2}} \, dy$$

$$= 4\pi ab \int_{-b}^{b} \frac{dy}{\sqrt{b^2 - y^2}}.$$

Letting $y = bu$, $dy = b\,du$, this becomes

$$4\pi ab \int_{-1}^{1} \frac{b\,du}{\sqrt{b^2 - b^2 u^2}} = 4\pi ab \int_{-1}^{1} \frac{du}{\sqrt{1 - u^2}} = 4\pi ab \arcsin u \Big|_{-1}^{1}$$

$$= 4\pi^2 ab.$$

(Notice [Compare Problem 4] that this is the product of $2\pi b$, the radius of the re-volved circle, and $2\pi a$, the distance around which the center of this circle is revolved. This is a special case of another version of "Pappus' rule".)

13. (a) The volume is

$$\pi \int_{1}^{\infty} \frac{1}{x^2}\, dx = \pi \cdot \left[-\frac{1}{x} \Big|_{1}^{\infty} \right] = \pi.$$

(b) The surface area is

$$2\pi \int_{a}^{\infty} \frac{1}{x} \cdot \sqrt{1 + \frac{1}{x^2}}\, dx = 2\pi \int_{1}^{\infty} \frac{\sqrt{1 + x^2}}{x^2}\, dx.$$

Since

$$\frac{\sqrt{1 + x^2}}{x^2} > \frac{x}{x^2} = \frac{1}{x}$$

and $\int_{1}^{\infty} dx/x = \infty$, this surface area is infinite.

(c) The paint that covers the distant portions of the trumpet will get thinner and thinner—it's easy to paint an infinite area with a finite amount of paint if you're allowed to spread the paint as thin as you like!

1. (ii) $P_{3,0}(x) = 1 + x + x^2/2.$

(iv)

$$P_{2n,\pi}(x) = -1 + \frac{(x-\pi)^2}{2!} - \cdots + \frac{(-1)^{n+1}(x-\pi)^{2n}}{(2n)!}.$$

(vi)

$$P_{n,2}(x) = \log 2 + \frac{(x-2)}{2} - \frac{(x-2)^2}{2^2 \cdot 2} + \cdots + \frac{(-1)^{n+1}(x-2)^n}{2^n \cdot n}.$$

(viii)

$$P_{4,1}(x) = 3 + 9(x-1) + \frac{26(x-1)^2}{2!} + \frac{66(x-1)^3}{3!} + \frac{120(x-1)^4}{4!}.$$

(x) $P_{n,0}(x) = 1 - x + x^2 - x^3 + \cdots + (-1)^n x^n.$

2. (ii) $160 + 50(x-3) - 10(x-3)^2 + (x-3)^4.$
(iv) $9a + 3b + c + (6a+b)(x-3) + a(x-3)^2.$

3. (ii)

$$\sum_{i=0}^{9} \frac{(-1)^i 2^{2i+1}}{(2i+1)!} \quad \left(\text{since } \frac{2^{2n+2}}{(2n+2)!} \leq 10^{-12} \text{ for } 2n+2 \geq 20, \text{ or } n \geq 9. \right).$$

(iv)

$$\sum_{i=0}^{7} \frac{1}{i!} \quad \left(\text{since } \frac{3}{(n+1)!} \leq 10^{-4} \text{ for } n+1 \geq 8, \text{ or } n \geq 7 \right).$$

4. (i) To obtain

$$\frac{1}{(2n+2)!} < 10^{-(10^{10})} \qquad \text{or} \qquad (2n+2)! > 10^{10^{10}}$$

it certainly suffices to choose $2n+2 = 10^{10^{10}}$; we can also choose $2n+2 = 10^{10}$, since $(10^{10})!$ is clearly $> 10^{10^{10}}$. So one possible sum is

$$\sum_{i=0}^{\frac{10^{10}}{2}} \frac{(-1)^i}{(2i+1)!}.$$

(ii)

$$\sum_{i=0}^{1,000} \frac{1}{i!} \qquad \text{(since surely } (1001)! > 3 \cdot 10^{1,000}\text{)}.$$

(iii) We need to find an n with

$$\frac{10^{2n+2}}{(2n+2)!} < 10^{-20}.$$

Now

$$\frac{10^{100+k}}{(100+k)!} = \frac{10^{100}}{100!} \cdot \frac{10}{101} \cdot \frac{10}{102} \cdots \frac{10}{100+k} < \frac{10^{100}}{100!} \cdot \frac{1}{10^k},$$

so

$$\frac{10^{100+k}}{(100+k)!} < 10^{-20}$$

when

$$\frac{10^{100}}{100!} \cdot \frac{1}{10^k} < 10^{-20} \quad \text{or} \quad \frac{10^{120}}{100!} < 10^k.$$

This certainly happens for $k = 120$, so we can take $2n + 2 = 220$ or $n = 109$, giving the sum

$$\sum_{i=0}^{109} \frac{(-1)^i 10^{2i+1}}{(2i+1)!}.$$

(iv)

$$\sum_{i=0}^{234} \frac{10^i}{i!} \qquad \left(\text{since } \frac{3^{10} 10^{235}}{(235)!} < \frac{10^5 \cdot 10^{100}}{(100)!} \cdot \frac{1}{10^{135}} < 10^{-30} \right).$$

(v)

$$\sum_{i=0}^{\frac{10^{10}}{2}} (-1)^i \frac{\left(\frac{1}{10}\right)^{2i+1}}{2i+1} \qquad \left(\text{since } \frac{\left(\frac{1}{10}\right)^{2n+3}}{2n+3} < 10^{-(10^{10})} \text{ for } 2n+3 = 10^{10} \right).$$

5. (a) Let

$$\cos x = 1 - \frac{x^2}{2} + R(x), \qquad |R(x)| \leq \frac{|x|^3}{6}.$$

Then if x satisfies $\cos x = x^2$ we have

$$(*) \qquad\qquad x^2 = 1 - \frac{x^2}{2} + R(x).$$

Ignoring the term $R(x)$ gives the equation

$$3x^2 = 2,$$

with solutions $\pm\sqrt{2/3} = \pm\sqrt{6}/3$. To find bounds on the error, recall first, from Problem 11-38, that $|x| \leq 1$. So $(*)$ gives

$$|3x^2 - 2| = 2|R(x)| \leq \frac{1}{3},$$

or

$$-\frac{1}{3} \le 3x^2 - 2 \le \frac{1}{3}$$
$$\frac{5}{3} \le 3x^2 \le \frac{7}{3}$$
$$\frac{\sqrt{5}}{3} \le |x| \le \frac{\sqrt{7}}{3}.$$

We can get a better approximation if we write

$$\cos x = 1 - \frac{x^2}{2} + \frac{x^4}{24} + R(x), \qquad |R(x)| \le \frac{|x|^5}{120} \le \frac{1}{120}.$$

If x satisfies $\cos x = x^2$ we have

(**) $$x^2 = 1 - \frac{x^2}{2} + \frac{x^4}{24} + R(x);$$

ignoring $R(x)$ gives the equation

$$x^4 - 36x^2 + 24 = 0$$

with the solution

$$x^2 = 18 - 10\sqrt{3},$$

so x is approximately

(***) $$\pm\sqrt{18 - 10\sqrt{3}} = \pm.82431$$

(compared to $\sqrt{2/3} = .81649$).

Similarly, to find bounds on the error, we have

$$-\frac{1}{120} \le \frac{x^4 - 36x^2 + 24}{24} \le \frac{1}{120}$$

so x^2 lies between the solutions of

$$y^2 - 36y + (24 - \tfrac{1}{5}) = 0$$

and

$$y^2 - 36y + (24 + \tfrac{1}{5}) = 0,$$

i.e., between

$$\frac{36 - \sqrt{1200 + \frac{4}{5}}}{2} \quad \text{and} \quad \frac{36 - \sqrt{1200 - \frac{4}{5}}}{2},$$

thus between

$$18 - \sqrt{300 + \tfrac{1}{5}} \quad \text{and} \quad 18 - \sqrt{300 + \tfrac{1}{5}};$$

so

$$\sqrt{18 - \sqrt{300 + \tfrac{1}{5}}} \le |x| \le \sqrt{18 - \sqrt{300 - \tfrac{1}{5}}}$$

or
$$.82080 \le |x| \le .82780,$$
so the error in (∗∗∗) is at most .00351.

(b) Writing
$$\sin x = x + \bar{R}(x),$$
the equation $2x^2 = x \sin x + \cos^2 x$ becomes
$$2x^2 = x^2 + x\bar{R}(x) + 1 - x^2 + \frac{x^4}{4} + R(x)^2 + 2R(x)\left(1 - \frac{x^2}{2}\right).$$

Ignoring terms involving R and $\bar{R}$ gives the equation
$$x^4 - 8x^2 + 4 = 0$$
or
$$x^2 = \frac{8 \pm \sqrt{64 - 16}}{2} = 4 \pm 2\sqrt{3}.$$

Since $|x| \le 1$, we must have
$$x^2 = 4 - 2\sqrt{3}$$
$$x = \pm\sqrt{4 - 2\sqrt{3}}.$$

6. (a)
$$\arctan \tfrac{1}{2} + \arctan \tfrac{1}{3} = \arctan\left(\frac{\tfrac{1}{2} + \tfrac{1}{3}}{1 - \tfrac{1}{6}}\right) = \arctan 1 = \frac{\pi}{4}.$$

Since
$$\arctan \tfrac{1}{5} + \arctan \tfrac{1}{5} = \arctan\left(\frac{\tfrac{1}{5} + \tfrac{1}{5}}{1 - \tfrac{1}{25}}\right) = \arctan \tfrac{5}{12},$$

we have
$$4\arctan \tfrac{1}{5} = \arctan\left(\frac{\tfrac{5}{12} + \tfrac{5}{12}}{1 - \tfrac{25}{144}}\right) = \arctan \tfrac{120}{119},$$

so
$$4\arctan \tfrac{1}{5} - \arctan \tfrac{1}{239} = \arctan\left(\frac{\tfrac{120}{119} - \tfrac{1}{239}}{1 + \tfrac{120}{119} \cdot \tfrac{1}{239}}\right) = \arctan 1 = \frac{\pi}{4}.$$

(b) To compute π with an error $< 10^{-6}$, we must compute $\pi/4$ with an error $< 10^{-6}/4$, so it suffices to compute $\arctan 1/5$ and $\arctan 1/239$ with an error $< 10^{-6}/20 = 10^{-7}/2$. Now
$$\arctan x = x - \frac{x^3}{3} + \cdots + \frac{(-1)^n x^{2n+1}}{2n + 1} + R, \qquad |R| \le \frac{|x|^{2n+3}}{2n + 3}.$$

So for $x = 1/5$ and $x = 1/239$ we need

$$\frac{1}{(2n+3)5^{2n+3}} < \frac{1}{2 \cdot 10^7}$$

$$\frac{1}{(2n+3)(239)^{2n+3}} < \frac{1}{2 \cdot 10^7},$$

respectively. We can take $n = 4$ and $n = 0$, respectively. So π is

$$16 \left(\frac{1}{5} - \frac{1}{3 \cdot 5^3} + \frac{1}{5 \cdot 5^5} - \frac{1}{7 \cdot 5^7} + \frac{1}{9 \cdot 5^9} \right) - 4 \left(\frac{1}{239} \right)$$

with an error $< 10^{-6}$. To find the first 5 decimals of π we must convert each term in parentheses into a decimal. If we compute each one to 7 correct decimals, then we will introduce an extra error of at most 10^{-7}. Since we actually have

$$\frac{1}{(2 \cdot 4 + 3)5^{2 \cdot 4 + 3}} < \frac{1}{3 \cdot 10^7},$$

$$\frac{1}{3 \cdot (239)^3} < \frac{1}{3 \cdot 10^7},$$

this extra error will be no problem. The calculations are as follows:

$$\frac{1}{5} = .20000000$$

$$\frac{1}{5 \cdot 5^5} = .00006400 \qquad\qquad\qquad \frac{1}{3 \cdot 5^3} = .00266666$$

$$\frac{1}{9 \cdot 5^9} = \underline{.00000005} \qquad\qquad\qquad \frac{1}{7 \cdot 5^7} = \underline{.00000182}$$

$$\qquad\quad .20006405 \qquad\qquad\qquad\qquad\quad .00266848$$

$$-.00266848 \longleftarrow \underline{\qquad\qquad\qquad\qquad\qquad}$$

$$\qquad\quad .19739557 \qquad\qquad\qquad \frac{1}{239} = .0041841$$

$$\underline{\times 16} \qquad\qquad\qquad\qquad\qquad \underline{\times 4}$$

$$\quad 3.1583280 \qquad\qquad\qquad\qquad\quad .0167364$$

$$- \; .0167364 \longleftarrow \underline{\qquad\qquad\qquad\qquad}$$

$$\quad 3.1415916$$

The error in this result is $< 10^{-6}$; consequently we can be sure that 14159 are the first five decimals of π (because of the fortunate circumstance that the next digit in our answer is not 9!). The first ten decimals of π are

$$3.1415926535$$

7. Clearly

$$f^{(k)}(x) = \alpha(\alpha - 1) \cdots (\alpha - k + 1)(1 + x)^{\alpha - k},$$

so

$$P_{n,0}(x) = \sum_{k=0}^{n} \frac{f^{(k)}(0)}{k!} x^k = \sum_{k=0}^{n} \frac{\alpha(\alpha-1)\cdots(a-k+1)}{k!} x^k$$

$$= \sum_{k=0}^{n} \binom{\alpha}{k} x^k.$$

The Cauchy form of the remainder is

$$R_{n,0}(x) = \frac{\alpha(\alpha-1)\cdots(\alpha-n)}{n!}(1+t)^{\alpha-n-1}(x-t)^n(x-0),$$

and the Lagrange from is

$$R_{n,0}(x) = \frac{\alpha(\alpha-1)\cdots(\alpha-n)}{(n+1)}(1+t)^{\alpha-n-1}(x-0)^{n+1}.$$

8. (ii) $c_i = \sum_{j=0}^{i} a_j b_{i-j}.$

(iv) $c_0 = 0$; $c_i = a_{i-1}/i$ for $i > 0$.

9. (a) Since

$$\lim_{x \to 0} \frac{R(x)}{x^{2n+1}} = 0,$$

we have

$$0 = \lim_{x \to 0} \frac{R(x^2)}{(x^2)^{2n+1}} = \lim_{x \to 0} \frac{R(x^2)}{x^{4n+2}}.$$

Now

$$\sin(x^2) = P(x^2) + R(x^2);$$

since $Q(x) = P(x^2)$ is a polynomial of degree $4n+2$, it follows from the corollary to Theorem 3 that Q is the Taylor polynomial of degree $4n+2$ for f at 0.

(b)

$$f^{(k)}(0) = \begin{cases} 0, & k \neq 4l+2 \\ \dfrac{(-1)^l(4l+2)!}{(2l+1)!}, & k = 4l+2. \end{cases}$$

(c)

$$f^{(k)}(0) = \begin{cases} 0, & k \neq nl \\ \dfrac{g^{(l)}(0)(nl)!}{l!}, & k = nl. \end{cases}$$

10.

$$\left| \int_0^x \frac{e^t}{n!}(x-t)^n \, dt \right| = \int_x^0 \frac{e^t}{n!} |x-t|^n \, dt$$

$$\leq \int_x^0 \frac{|x-t|^n}{n!} \, dt \qquad \text{since } e^x \leq 1 \text{ for } x \leq 0$$

$$= \frac{|x|^{n+1}}{(n+1)!}.$$

11. For $-1 < x \leq t \leq 0$ we have

$$0 < 1 + x \leq 1 + t \leq 1,$$

$$0 \leq \frac{1}{1+t} \leq \frac{1}{1+x}.$$

So

$$\left| \int_0^x \frac{t^n}{1+t} \, dt \right| \leq \int_x^0 \frac{|t|^n}{1+x} \, dt \leq \frac{|x|^{n+1}}{(1+x)(n+1)}.$$

12. (a) By hypothesis,

$$-M(x-a)^n \leq g'(x) \leq M(x-a)^n \qquad \text{for } x \geq a.$$

It follows from the Mean Value Theorem that

$$\frac{-M(x-a)^{n+1}}{n+1} \leq g(x) - g(a) \leq \frac{M(x-a)^{n+1}}{n+1},$$

i.e., that $|g(x) - g(a)| \leq M(x-a)^{n+1}/(n+1)$. The case $x \leq a$ is treated similarly.

(b) For every $\varepsilon > 0$ there is a $\delta > 0$ such that $|g'(x)/(x-a)^n| \leq \varepsilon$ for $|x-a| < \delta$. This means that $|g'(x)| \leq \varepsilon |x-a|^n$ for $|x-a| < \delta$. Part (a) implies that $|g(x) - g(a)| \leq \varepsilon |x-a|^{n+1}/(n+1)$ for $|x-a| < \delta$. Since this is true for every $\delta > 0$, it follows that

$$\lim_{x \to a} \frac{g(x) - g(a)}{(x-a)^{n+1}} = 0.$$

(c) Since

$$g(x) = f(x) - \sum_{i=0}^{n} \frac{f^{(i)}(a)}{i!}(x-a)^i,$$

we have

$$g'(x) = f'(x) - \sum_{i=0}^{n} \frac{f^{(i)}(a)}{(i-1)!}(x-a)^{i-1} \quad .$$

$$= f'(x) - \sum_{j=0}^{n-1} \frac{f^{(j+1)}(a)}{j!}(x-a)^{j}$$

$$= f'(x) - \sum_{j=0}^{n-1} \frac{(f')^{(j)}(a)}{j!}(x-a)^{j}$$

$$= f'(x) - P_{n-1,a,f'}(x).$$

(d) Theorem 1 is true for $n = 1$, by the definition of f'. Now assume that Theorem 1 is true for $n - 1$, and all functions f for which $f'(a), \ldots, f^{(n-1)}(a)$ exist. If g is a function for which $g'(a), \ldots, g^{(n)}(a)$ exist, then $f = g'$ is a function for which $f'(a), \ldots, f^{(n-1)}(a)$ exist. Consequently,

$$\lim_{x \to a} \frac{g'(x) - P_{n-1,a,g'}(x)}{(x-a)^{n-1}} = 0.$$

Since $(g - P_{n,a,g})' = g' - P_{n-1,a,g'}$, it follows from part (b) that

$$\lim_{x \to a} \frac{g(x) - P_{n,a,g}(x)}{(x-a)^{n}} = 0.$$

13. Suppose $|f^{(n+1)}|$ is bounded, by some M, on some interval around a. Then for x in this interval we have

$$|R_{n,a}(x)| = \frac{|f^{(n+1)}(t)|}{n!}|x-a|^{n+1},$$

so

$$\frac{|R_{n,a}(x)|}{|x-a|^{n}} \leq M|x-a|,$$

so

$$\lim_{x \to a} \frac{R_{n,a}(x)}{(x-a)^{n}} = 0.$$

A similar proof works for the integral from of the remainder, and for the Cauchy form, if $|f^{(n+1)}|$ is assumed bounded.

14. Problem 13-23 implies that

$$R_{n,a}(x) = \int_{a}^{x} \frac{f^{(n+1)}(t)}{n!}(x-t)^{n} \, dt$$

$$= \frac{f^{(n+1)}(t)}{n!}(x-t)^{n}(x-a)$$

for some t in (a, x). Similarly, choosing $f^{(n+1)}/n!$ for f and $g(t) = (x - t)^n$, we obtain the Lagrange form. (In both cases, however, we begin with the extra assumption that $f^{(n+1)}$ is integrable.)

15. (a) Taylor's Theorem, with $n = 1$, gives

$$f(x + h) = f(x) + f'(x)h + \frac{f''(t)}{2}(x + h - t)h, \qquad t \text{ in } (x, x + h).$$

So

$$|f'(x)| \leq \frac{1}{h}|f(x + h) - f(x)| + \frac{1}{h}\frac{|f''(t)|}{2}h^2$$

$$\leq \frac{2}{h}M_0 + \frac{h}{2}M_2.$$

(b) The best inequality from (a) is obtained by choosing that $h > 0$ which minimizes $g(h) = 2M_0/h + hM_2/2$. Thus, we choose h to satisfy

$$0 = g'(h) = \frac{-2M_0}{h^2} + \frac{M_2}{2}$$

or

$$h = 2\sqrt{M_0/M_2}$$
$$g(h) = 2\sqrt{M_0 M_2}.$$

(c) The previous results can clearly be applied to any interval (a, ∞); for all $x > a$, we have

$$|f'(x)| \leq 2\sqrt{M_{0,a} M_2},$$

where $|f''(x)| \leq M_2$ for all $x > 0$, and $|f(x)| \leq M_{0,a}$ for all $x > a$. Since $f(x)$ approaches 0 as $x \to \infty$, $M_{0,a}$ can be picked arbitrarily small by choosing a large enough. Thus $f'(x) \to 0$.

(d) Problem 11-31 shows that $\lim_{x \to \infty} f''(x) = 0$, while part (c) shows that $\lim_{x \to \infty} f'(x) = 0$.

16. (a) This follows from

$$f(a + h) = f(a) + f'(a)h + \frac{f''(a)}{2}h^2 + R_{2,a}(h),$$

$$f(a - h) = f(a) - f'(a)h + \frac{f''(a)}{2}h^2 + R_{2,a}(-h),$$

since

$$\lim_{h \to 0} \frac{R_{2,a}(h)}{h^2} = \lim_{h \to 0} \frac{R_{2,a}(-h)}{h^2} = 0.$$

(b)

$$\lim_{h \to 0^+} \frac{f(0+h) + f(0-h) - 2f(0)}{h^2} = \lim_{h \to 0} \frac{h^2 - h^2 - 0}{h^2} = 0$$

and similarly for $\lim_{h \to 0^-}$.

(c) For sufficiently small h we have $f(a+h) \le f(a)$ and $f(a-h) \le f(a)$ so

$$\frac{f(a+h) + f(a-h) - 2f(a)}{h^2} \le 0.$$

(d) Using the Taylor polynomial of degree 3, we have

$$f(a+h) = f(a) + f'(a)h + \frac{f''(a)}{2}h^2 + \frac{f'''(a)h^3}{6} + R_{3,a}(h)$$

$$f(a-h) = f(a) - f'(a)h + \frac{f''(a)}{2}h^2 - \frac{f'''(a)h^3}{6} + R_{3,a}(-h).$$

Now subtract the second equation from the first.

17. If $f'' > 0$, then

$$f(x) = f(a) + f'(a)(x-a) + \int_a^x f''(t)(x-t)\,dt$$

$$> f(a) + f'(a)(x-a) \qquad \text{for } x \ne a,$$

which says that the graph of f at x lies above the tangent line through a.

18. The proof is almost exactly the same as the proof given in the text for the equation $f'' + f = 0$.

19. (a)

$$f^{(n+1)} = \left(f^{(n)}\right)' = \sum_{j=0}^{n-1} a_j f^{(j+1)} = \left(\sum_{j=0}^{n-2} a_j f^{(j+1)}\right) + a_{n-1} f^{(n)}$$

$$= \sum_{j=0}^{n-1} a_{j-1} f^{(j)} + a_{n-1} \sum_{j=0}^{n-1} a_j f^{(j)}$$

$$= \sum_{j=0}^{n-1} (a_{j-1} + a_{n-1} a_j) f^{(j)}.$$

(b) Letting $a_{-1} = a_{-2} = 0$, we have

$$f^{(n+2)} = \sum_{j=0}^{n-1} (a_{j-1} + a_{n-1}a_j) f^{(j+1)}$$

$$= \sum_{j=0}^{n-2} (a_{j-1} + a_{n-1}a_j) f^{(j+1)} + (a_{n-2} + a_{n-1}{}^2) \sum_{j=0}^{n-1} a_j f^{(j)}$$

$$= \sum_{j=0}^{n-1} (a_{j-2} + a_{n-1}a_{j-1} + a_{n-2}a_j + a_{n-1}{}^2 a_j) f^{(j)}.$$

(c) From the equation

$$f^{(n+1)} = \sum_{j=0}^{n-1} b_j{}^1 f^{(j)}$$

we obtain

$$f^{(n+2)} = \sum_{j=0}^{n-2} b_j{}^1 f^{(j+1)} + b_{n-1}{}^1 f^{(n)}$$

$$= \sum_{j=0}^{n-1} b_{j-1}{}^1 f^{(j)} + b_{n-1}{}^1 \sum_{j=0}^{n-1} a_j f^{(j)}$$

$$= \sum_{j=0}^{n-1} b_j{}^2 f^{(j)},$$

where

$$b_j{}^2 = b_{j-1}{}^1 + b_{n-1}{}^1 a_j,$$

$$|b_j{}^2| \le |b_{j-1}{}^1| + |b_{n-1}{}^1| \cdot |a_j| \le 2N^2 + 2N^3 \le 4N^3.$$

The general formula is proved similarly, by induction on k.

(d) Let $M = M_1 + \cdots + M_{n-1}$, where

$$M_i = \sup\{|f^{(i)}(t)| : 0 \le t \le x\}.$$

(e) Clearly $f^{(k)}(0) = 0$ for all k. Then by Taylor's Theorem,

$$|f(x)| = \left| \int_0^x \frac{f^{(n+k+1)}(t)}{(n+k)!} (x-t)^{n+k} \, dt \right|$$

$$\le \frac{M \cdot 2^{k+1} N^{k+2} |x|^{n+k+1}}{(n+k+1)!}$$

$$\le \frac{M \cdot |2Nx|^{n+k+1}}{(n+k+1)!}.$$

Since $|2Nx|^{n+k+1}/(n+k+1)!$ can be made as small as desired by choosing k (and hence $n+k+1$) sufficiently large, it follows that $f(x) = 0$.

(f) The difference $f = f_1 - f_2$ satisfies the same differential equation, and $f^{(j)}(0) = 0$ for $0 \leq j \leq n - 1$. So $f = 0$.

[In the case $n = 2$, the equations

$$f(0) = c_1 + c_2$$
$$f'(0) = \alpha_1 c_1 + \alpha_2 c_2$$

can always be solved if $\alpha_1 \neq \alpha_2$:

$$c_1 = \frac{\alpha_2 f(0) - f'(0)}{\alpha_2 - \alpha_1}, \qquad c_2 = \frac{\alpha_1 f(0) - f'(0)}{\alpha_1 - \alpha_2}.$$

The case $n = 3$ involves more complicated answers, but is equally straightforward. The general case, for those who know about determinants, depends upon the fact that the "Vandermonde determinant"

$$\det \begin{pmatrix} 1 & 1 & \cdots & 1 \\ \alpha_1 & \alpha_2 & \cdots & \alpha_n \\ \vdots & \vdots & & \vdots \\ \alpha_1^{n-1} & \alpha_2^{n-1} & \cdots & \alpha_n^{n-1} \end{pmatrix}$$

is non-zero if the α_i are distinct—in fact, it has the value $\prod_{i>j} \alpha_i - \alpha_j$.]

20. (a) The second derivative of $h(x) = (x - a)(b - x)$ is $h''(x) = -2$. Now applying Problem 16(c) to g with a maximum point y in (a, b), we have

$$0 \geq \text{Schwarz second derviative of } g \text{ at } y$$
$$= (\text{Schwarz second derivative of } f \text{ at } y) + 2\varepsilon$$
$$= 0 + 2\varepsilon,$$

a contradiction.

(b) We want to show that

$$f(x) = f(a) + \frac{f(b) - f(a)}{b - a}(x - a),$$

so we consider the function

$$g(x) = f(x) - \frac{f(b) - f(a)}{b - a}(x - a).$$

Then $g(a) = f(a)$ and $g(b) = f(a)$, and the Schwarz second derivative of g is 0 at all points of (a, b). So part (a), g is constant.

21. (a) Clearly $f = 0$ up to order 2 at 0. The second derivative $f''(0)$ does not exist because

$$f'(x) = \begin{cases} 0, & x = 0 \\ 4x^3 \sin \dfrac{1}{x}^2 - x \cos \dfrac{1}{x}^2, & x \neq 0 \end{cases}$$

and

$$\lim_{h \to 0} \frac{4h^3 \sin \dfrac{1}{h^2} - h \cos \dfrac{1}{h^2}}{h}$$

does not exist.

(b) Choosing $x = a + h$ and then $x = a - h$ in ($*$) we have

$$\lim_{h \to 0} \frac{1}{h^2}[f(a+h) - f(a) + hf'(a)] = 0 = \lim_{h \to 0} \frac{1}{h^2}[f(a-h) - f(a) - hf'(a)].$$

It follows that

$$0 = \lim_{h \to 0} \frac{f(a+h) + f(a-h) - 2f(a)}{h^2},$$

i.e., the Schwarz second derivative of f is 0. Problem 20 implies that f is linear, so $f''(a) = 0 = m(a)$ for all a.

(c) Let g be a function with $g'' = m$. Then the function $f - g$ satisfies ($*$) with $m = 0$. So $f - g$ is linear, by part (b). So $f'' = g'' = m$.

CHAPTER 21

1. (a) If α is a solution of the equation

(1)
$$a_n x^n + a_{n-1} x^{n-1} + \cdots + a_0 = 0,$$

then $\sqrt{\alpha}$ is a solution of the equation

$$a_n x^{2n} + a_{n-1} x^{2n-2} + \cdots + a_0 = 0.$$

(b) If α satisfies (1), then $\alpha + r$ satisfies

$$a_n (x - r)^n + a_{n-1}(x - r)^{n-1} + \cdots + a_0 = 0;$$

this equation, with rational coefficients, has the same solutions as the equation with integer coefficients that is obtained by multiplying through by a common denominator of the various coefficients.

Similarly, αr satisfies

$$a_n \left(\frac{x}{r}\right)^n + a_{n-1} \left(\frac{x}{r}\right)^{n-1} + \cdots + a_0 = 0.$$

2. Since

$$\left(\sqrt{2} + \sqrt{3}\right)^2 = 5 + 2\sqrt{6},$$
$$\left(\sqrt{2} + \sqrt{3}\right)^4 = \left(5 + 2\sqrt{6}\right)^2 = 49 + 20\sqrt{6},$$

it is clear that $\sqrt{2} + \sqrt{3}$ satisfies $x^4 - 10x^2 + 1 = 0$.

Since

$$\left[\sqrt{2}(1 + \sqrt{3})\right]^2 = 2(4 + 2\sqrt{3}) = 8 + 4\sqrt{3},$$
$$\left[\sqrt{2}(1 + \sqrt{3}\,a)\right]^4 = \left(8 + 4\sqrt{3}\right)^2 = 112 + 4\sqrt{3},$$

clearly $\sqrt{2}(1 + \sqrt{3})$ satisfies $x^4 - 16x^2 + 16 = 0$.

3. (a) If $f(p/q) = 0$, then

$$a_n x^n + a_{n-1} x^{n-1} + \cdots + a_0 = \left(x - \frac{p}{q}\right)(b_{n-1} x^{n-1} + \cdots + b_0)$$

for some $b_0, \ldots, b_{n-1}$, which will be rational numbers. Since $\alpha - p/q \neq 0$, we have

$$b_{n-1} \alpha^{n-1} + \cdots + b_0 = 0,$$

contradicting the assumption about the minimal degree of the original polynomial.

(b) Clearly $f(p/q)$ can be written as a rational number of the form r/q^n. Since $f(p/q) \neq 0$, we have $|r| \geq 1$, so $|f(p/q)| \geq 1/q^n$.

(c) If $|\alpha - p/q| < 1$, then

$$f(p/q) = \frac{f(p/q) - f(\alpha)}{p/q - \alpha} = f'(x) \qquad \text{for } |x - \alpha| < 1$$
$$< M,$$

so

$$|\alpha - p/q| > |f(p/q)|/M > 1/(Mq^n).$$

4. If α satisfied a polynomial equation of degree n, then there would be some number c with $|\alpha - p/q| > c/q^n$ for all rational p/q. Now

$$\frac{1}{10^{1!}} + \frac{1}{10^{2!}} + \cdots + \frac{1}{10^{k!}}$$

can be written as

$$\frac{p}{10^{k!}}$$

for some integer p, and

$$0 < \alpha - \frac{p}{10^{k!}} < \frac{2}{10^{(k+1)!}}.$$

Thus for every k we must have

$$\frac{c}{(10^{k!})^n} < \frac{2}{10^{(k+1)!}}$$

or

$$\frac{10^{(k+1)!}}{(10^{k!})^n} < \frac{2}{c},$$

or

$$\frac{(10^{k!})^{k+1}}{(10^{k!})^n} < \frac{2}{c},$$

or

$$(10^{k!})^{k+1-n} < \frac{2}{c},$$

which is clearly false for large enough n.

5. (a) If the elements of A and B can be arranged in the respective sequences a_1, $a_2, a_3, a_4, \ldots$ and $b_1, b_2, b_3, b_4, \ldots$, then the elements of $A \cup B$ can be arranged in the sequence

$$a_1, b_1, a_2, b_2, a_3, b_3, \ldots$$

(except that repetitions must be thrown away, if A and B have any elements in common.)

(b) Arrange the positive rational numbers in a list by following the arrows (deleting repetitions).

(c) Follow the arrows in the picture

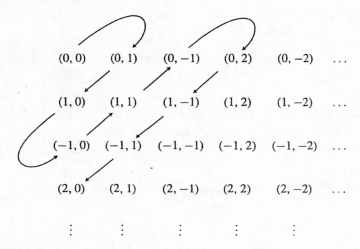

(d) Let the elements in A_i be arranged in a list $a_{i1}, a_{i2}, a_{i3}, \ldots$. Then the elements in $A_1 \cup A_2 \cup A_3 \cup \cdots$ can be arranged in the array

$$
\begin{array}{llll}
a_{11} & a_{12} & a_{13} & a_{14} \;\cdots \\
a_{21} & a_{22} & a_{23} & a_{24} \;\cdots \\
a_{31} & a_{32} & a_{33} & a_{34} \;\cdots \\
\;\;\vdots & \;\;\vdots & \;\;\vdots & \;\;\vdots
\end{array}
$$

Now use the same trick as in parts (b) and (c), deleting any repetitions.

(e) Apply part (d) with A_i the set of all triples (m, n, i). (A_i is countable, by part (c).)

(f) If the set of all n-tuples is countable, then the set of all $(n+1)$-tuples is seen to be countable by applying part (d) with A_i the set of all $(n+1)$-tuples $(a_1, a_2, \ldots, a_n, i)$.

(g) Since every such polynomial function $f(x) = a_n x^n + \cdots + a_0$ of degree n can be described by an $(n+1)$-tuple of integers $(a_0, \ldots, a_n)$, these polynomial functions can be arranged in a list $p_1, p_2, p_3, \ldots$. For each p_i, let $\alpha_{i,1}, \ldots, \alpha_{i,n}$ be its roots (if there are fewer than n roots, choose 0 for the remaining $\alpha_{i,j}$). Then

$$
\alpha_{1,1}, \ldots, \alpha_{1,n}, \alpha_{2,1}, \ldots, \alpha_{2,n}, \alpha_{3,1}, \ldots, \alpha_{3,n}, \ldots
$$

is a list of all the desired roots. Now delete repetitions.

(h) Apply part (d) with A_i the set of all roots of polynomial functions, with integer coefficients, of degree i.

6. If this number were in the list, it would be α_n for some n. But it cannot be α_n, since it differs from α_n in the nth decimal place. (This tricky construction, and others modeled on it, is known as the "Cantor diagonal method".)

7. (a) Suppose $0 < a_1 < \cdots < a_n < 1$, and $\lim\limits_{a \to a_i^+} f(x) - \lim\limits_{x \to a_i^-} f(x) > \varepsilon$. Choose

$$0 < a_1' < a_1 < a_1'' < a_2' < a_2 < a_2'' < \cdots < a_n' < a_n < a_n'' < 1.$$

Then

$$f(a_i'') - f(a_i') > \varepsilon,$$

so

$$f(1) - f(0) > \sum_{i=1}^{n} f(a_i'') - f(a_i') > n\varepsilon,$$

so $n < [f(1) - f(0)]/\varepsilon$.

(b) Let A_n be the set of all a in $[0, 1]$ with $\lim\limits_{x \to a^+} f(x) - \lim\limits_{x \to a^-} f(x) > 1/n$. Then A_n is finite, so by Problem 5(d), $A_1 \cup A_2 \cup A_3 \cup \cdots$ is countable.

8. (a) There are only countably many such intervals, since each interval is determined by a pair of rational numbers, and there are countably many rational numbers. Since each value $f(x)$ can be described in terms of these intervals (as the maximum value on such an interval), there are only countably many values of $f(x)$.

(b) If f happens to be continuous, then f cannot take on two values, for if it did, it would also take on all values in between, and hence an uncountable set of values.

(c) A minor variation of the proof in part (a) shows that if every point is either a local maximum point or a local minimum point for f, then f takes on only a countable set of values. So, again, if f is continuous, then f must be a constant.

CHAPTER 22

1. (ii)

$$\lim_{n\to\infty} \frac{n+3}{n^3+4} = \lim_{n\to\infty} \frac{\dfrac{1}{n^2} + \dfrac{3}{n^3}}{1 + \dfrac{4}{n^3}} = \frac{\lim\limits_{n\to\infty} \dfrac{1}{n^2} + \dfrac{3}{n^3}}{\lim\limits_{n\to\infty} 1 + \dfrac{4}{n^3}} = \frac{0}{1} = 0.$$

(iv) If n is even, then

$$\frac{n!}{n^n} = \frac{n(n-1)\cdots(n/2)!}{n^{n/2}\cdot n^{n/2}} \le \frac{(n/2)!}{n^{n/2}} \le \left(\frac{1}{2}\right)^{n/2};$$

similarly, if n is odd, then

$$\frac{n!}{n^n} = \frac{n(n-1)\cdots[(n-1)/2]!}{n^{(n+1)/2}\cdot n^{(n-1)/2}} \le \frac{[(n-1)/2]!}{n^{(n-1)/2}} \le \left(\frac{1}{2}\right)^{(n-1)/2}$$

(vi) $\lim\limits_{n\to\infty} (\log n)/n = 0$ (since $\lim\limits_{x\to\infty} (\log x)/x = 0$, by Problem 18-12(b)). So

$$\lim_{n\to\infty} \sqrt[n]{n} = \lim_{n\to\infty} e^{(\log n)/n} = e^0 \text{ (by Theorem 1) } = 1.$$

(viii) Suppose $a \ge b \ge 0$. Then $\sqrt[n]{a^n} \le \sqrt[n]{a^n + b^n} \le \sqrt[n]{a^n + a^n}$, i.e., $a \le \sqrt[n]{a^n + b^n} \le \sqrt[n]{2}\,a$, and $\lim\limits_{n\to\infty} \sqrt[n]{2} = 1$, by part (v).

(x) According to Problem 2-7,

$$\lim_{n\to\infty} \frac{\sum\limits_{k=1}^{n} k^p}{n^{p+1}} = \lim_{n\to\infty} \frac{\dfrac{n^{p+1}}{p+1} + An^p + Bn^{p-1} + \cdots}{n^{p+1}}$$

$$= \lim_{n\to\infty} \frac{1}{p+1} + \frac{A}{n} + \frac{B}{n^2} + \cdots = \frac{1}{p+1}.$$

2. (i)

$$\lim_{n\to\infty} \frac{n}{n+1} - \frac{n+1}{n} = \lim_{n\to\infty} \frac{n}{n+1} - \lim_{n\to\infty} \frac{n+1}{n} = 1 - 1 = 0.$$

338

(ii)

$$\lim_{n\to\infty} n - \sqrt{n+a}\sqrt{n+b} = \lim_{n\to\infty} \frac{\left(n - \sqrt{n+a}\sqrt{n+b}\right)\left(n + \sqrt{n+a}\sqrt{n+b}\right)}{n + \sqrt{n+a}\sqrt{n+b}}$$

$$= \lim_{n\to\infty} \frac{n^2 - (n+a)(n+b)}{n + \sqrt{n+a}\sqrt{n+b}}$$

$$= \lim_{n\to\infty} \frac{-(a+b)n}{n + \sqrt{n+a}\sqrt{n+b}} - \frac{ab}{n + \sqrt{n+a}\sqrt{n+b}}$$

$$= \lim_{n\to\infty} -(a+b) \cdot \frac{1}{1 + \dfrac{\sqrt{n+a}}{\sqrt{n}} \dfrac{\sqrt{n+b}}{\sqrt{n}}}$$

$$= \lim_{n\to\infty} -(a+b) \cdot \frac{1}{1 + \sqrt{1 + \dfrac{a}{n}}\sqrt{1 + \dfrac{b}{n}}}$$

$$= -\frac{(a+b)}{2}.$$

(iii)

$$\lim_{n\to\infty} \frac{2^n + (-1)^n}{2^{n+1} + (-1)^{n+1}} = \lim_{n\to\infty} \frac{1 + \dfrac{(-1)^n}{2^n}}{2 + \dfrac{(-1)^{n+1}}{2^n}} = \frac{1}{2}.$$

(iv)

$$\lim_{n\to\infty} \frac{(-1)^n \sqrt{n}\,\sin(n^n)}{n+1} = \lim_{n\to\infty} \frac{(-1)^n \sin(n^n)}{\sqrt{n}} \cdot \frac{n}{n+1} = 0 \cdot 1 = 0.$$

(v) If $a = b$, the limit is 0. If $a = -b$, the quotient is undefined for odd n, so the limit is meaningless. If $|a| > |b|$, then

$$\lim_{n\to\infty} \frac{a^n - b^n}{a^n + b^n} = \lim_{n\to\infty} \frac{1 - \left(\dfrac{b}{a}\right)^n}{1 + \left(\dfrac{b}{a}\right)^n} = \frac{1}{1} = 1.$$

Similarly, if $|a| < |b|$, then the limit is -1.

(vi) Suppose first that $c > 0$. We have

$$\lim_{x\to\infty} xc^x = \lim_{x\to\infty} e^{\log x} e^{x\log c} = \lim_{x\to\infty} e^{\log x + x\log c}.$$

Now

$$\lim_{x \to \infty} \log x + x \log c = \lim_{x \to \infty} x \left(\frac{\log x}{x} + \log c \right),$$
$$= \lim_{x \to \infty} x \log c = -\infty,$$

so $\lim_{x \to \infty} xc^x = 0$. In particular, $\lim_{n \to \infty} nc^n = 0$. The result clearly follows also for $c < 0$.

(vii) Since

$$\frac{2^{n^2}}{n!} = \frac{(2^n)^n}{n!} = \frac{2^n}{n} \cdot \frac{2^n}{n-1} \cdot \frac{2^n}{n-2} \cdots \frac{2^n}{1}$$

and each factor is ≥ 1, and in fact $\to \infty$, the whole quotient $\to \infty$.

3. (a) The sequence $\{a_n\}$ must be eventually constant, that is, there is some N such that all a_n are equal for $n > N$.

(b) All convergent subsequences are of the form

$$a_1, \ldots, a_n, 1, 1, 1, 1, 1, \ldots$$

or

$$a_1, \ldots, a_n, -1, -1, -1, -1, \ldots,$$

where $a_1, \ldots, a_n$ is some finite sequence of 1's and -1's.

(c) All convergent subsequences are of the form

$$a_1, \ldots, a_n, m, m, m, \ldots,$$

where $a_1, \ldots, a_n$ and m are natural numbers.

(d) All α in $[0, 1]$.

4. (a) Let $\{a_n\}$ be a Cauchy sequence, and suppose that $\lim_{j \to \infty} a_{n_j} = l$. For any $\varepsilon > 0$, choose J so that $|l - a_{n_j}| < \varepsilon/2$ for $j > J$. Then choose N so that $|a_n - a_m| < \varepsilon/2$ for $n, m > N$. Let $N_0 = \max(N, n_J)$. If $n > N_0$, then $|a_n - a_{n_{J+1}}| < \varepsilon/2$ and $|a_{n_{J+1}} - l| < \varepsilon/2$. Consequently, $|a_n - l| < \varepsilon/2$.

(b) Suppose that $\lim_{n \to \infty} a_n = l$, and let $\{a_{n_j}\}$ be a subsequence of $\{a_n\}$. If $\varepsilon > 0$, then there is some N such that $|l - a_n| < \varepsilon$ for $n > N$. Since $n_1 < n_2 < n_3 < \cdots$, there is some J such that $n_j > N$ for $j > J$. Thus $|l - a_{n_j}| < \varepsilon$ for $j < J$. So $\lim_{j \to \infty} a_{n_j} = l$.

6. (a) Using the inequality $\sqrt{ab} < (a+b)/2$, it is easy to prove by induction that

$$a_1 < a_n < a_{n+1} < b_{n+1} < b_n < b_1.$$

Thus the sequence $\{a_n\}$ is increasing and bounded by b_1 and the sequence $\{b_n\}$ is decreasing and bounded by a_1. So both converge.

(b) If $l = \lim\limits_{n\to\infty} a_n$ and $m = \lim\limits_{n\to\infty} b_n$, then

$$l = \lim_{n\to\infty} a_{n+1} = \lim_{n\to\infty} \sqrt{a_n b_n}$$
$$= \lim_{n\to\infty} \sqrt{a_n}\sqrt{b_n}$$
$$= \sqrt{l}\,\sqrt{m},$$

so $m = l$.

7. (a) If

$$a_k = \frac{m}{n},$$

then

$$a_{k+1} = \frac{m + 2n}{m + n} = \frac{(m+n) + n}{m+n} = 1 + \frac{n}{m+n} = 1 + \frac{1}{\dfrac{m}{n} + 1} = 1 + \frac{1}{1 + a_k}.$$

(b) It is easy to check, first, that

$$\text{if } a^2 < 2, \text{ then } \left(1 + \frac{1}{1+a}\right)^2 = \left(\frac{a+2}{a+1}\right)^2 > 2,$$

and that the same result holds with the two inequalities reversed. Since $a_1{}^2 < 2$, we thus have $a_k < \sqrt{2}$ for k odd and $a_k > \sqrt{2}$ for k even. Moreover,

$$a_{k+2} = \frac{3a_k + 4}{2a_k + 3} \quad \begin{cases} > a_k & \text{for } a_k{}^2 < 2 \\ < a_k & \text{for } a_k{}^2 > 2. \end{cases}$$

This shows that $\{a_{2n+1}\}$ is increasing and $< \sqrt{2}$ and $\{a_{2n}\}$ is decreasing and $> \sqrt{2}$, so they have limits l and m, respectively.

To show that $l = m$, we note that for both even and odd n we have

$$a_{n+1} - a_n = \frac{a_n + 2}{a_n + 1} - a_n = \frac{2 - a_n{}^2}{1 + a_n}.$$

Hence

$$l - m = \lim_{\substack{n\to\infty \\ n \text{ even}}} \frac{2 - a_n{}^2}{1 + a_n} = \frac{2 - m^2}{1 + m}$$

and also

$$m - l = \lim_{\substack{n\to\infty \\ n \text{ odd}}} \frac{2 - a_n{}^2}{1 + a_n} = \frac{2 - l^2}{1 + l}.$$

This gives

$$l + lm - m = 2$$
$$m + lm - l = 2;$$

hence $l = m = \sqrt{2}$.

(c) This is a straightforward generalization of part (b). For convenience, let

$$c = a^2 + b.$$

We consider

$$\alpha_1 = a, \qquad \alpha_{k+1} = a + \frac{b}{a + \alpha_k} = \frac{a\alpha_k + c}{a + \alpha_k}.$$

Note that $\alpha_1{}^2 < c$. Also, the inequality

$$\left(\frac{a\alpha + c}{a + \alpha} \right)^2 > c$$

is equivalent to

$$(a^2 - c)\alpha^2 > c(a^2 - c)$$

or simply

$$\alpha^2 < c.$$

So we see that

$$\text{if } \alpha^2 \lesseqgtr c \text{ then } \left(\frac{a\alpha + c}{a + \alpha} \right)^2 \gtreqless c.$$

So $\{a_{2n+1}\}$ is increasing and $< \sqrt{a^2 + b}$ and $\{a_{2n}\}$ is decreasing and $> \sqrt{a^2 + b}$, so they have limits l and m. Moreover,

$$\alpha_{n+1} - \alpha_n = \frac{a\alpha_n + c}{a + \alpha_n} - \alpha_n = \frac{c - \alpha_n{}^2}{a + \alpha_n}$$

for both even and odd n. As before, this gives

$$l + lm - m = c$$
$$m + lm - l = c;$$

hence $l = m = \sqrt{c}$.

9. (ii) $\int_0^2 e^x \, dx = e^2 - 1$.

(iv) 0, since

$$\frac{1}{n^2} + \frac{1}{(n+1)^2} + \cdots + \frac{1}{(2n)^2} \leq n \cdot \frac{1}{n^2} \leq \frac{1}{n}.$$

(vi)

$$\int_0^1 \frac{1}{1 + x^2} \, dx = \pi/4.$$

10. (a) If $a = 1 + h$, then $a^n = (1 + h)^n \geq 1 + nh$. Since $h > 0$, clearly $\lim_{n \to \infty} nh = \infty$.

(b) $\lim_{n \to \infty} a^n = 0$, because $\lim_{n \to \infty} 1/a^n = \infty$, by part (a).

(c) If $\sqrt[n]{a} = 1 + h$, then $a = (1 + h)^n \geq 1 + nh$, so $h \leq (a - 1)/n$. Thus $1 \leq \sqrt[n]{a} \leq 1 + (a - 1)/n$, so $\lim_{n \to \infty} \sqrt[n]{a} = 1$.

(d) $\lim_{n \to \infty} \sqrt[n]{a} = 1 \bigg/ \left(\lim_{n \to \infty} \sqrt[n]{1/a} \right) = 1$, by part (c).

(e) If $\sqrt[n]{n} = 1 + h$, then

$$n = (1 + h)^n \geq \frac{n(n - 1)}{2} h^2,$$

so

$$h \leq \sqrt{\frac{2}{n - 1}},$$

so

$$1 \leq \sqrt[n]{n} \leq 1 + \sqrt{\frac{2}{n - 1}},$$

so $\lim_{n \to \infty} \sqrt[n]{n} = 1$.

11. (a) Suppose that $\lim_{n \to \infty} a_n = l$. Choose N so that $|a_n - l| < 1$ for $n > N$. Then $|a_n| < \max(|l| + 1, |a_1|, \ldots, |a_N|)$ for all n.

(b) Choose N so that $|a_n - 0| < a_1$ for $n > N$. Then the maximum of $a_1, a_2, \ldots, a_N$ is the maximum of a_n for all n.

12. (a) This relation is equivalent to

$$\frac{1}{n + 1} < \int_n^{n+1} \frac{1}{x} \, dx < \frac{1}{n},$$

which is true because

$$\frac{1}{n + 1} < \frac{1}{x} < \frac{1}{n} \qquad \text{for } x \text{ in } (n, n + 1).$$

(b) Since

$$a_n - a_{n+1} = \log(n + 1) - \log n - \frac{1}{n + 1}$$
$$> 0 \qquad \text{by part (a)},$$

the sequence $\{a_n\}$ is decreasing.

To prove that $a_n \geq 0$, add the inequalities

$$\log(j + 1) - \log j < \frac{1}{j}$$

for $j = 1, \ldots, n - 1$, to obtain

$$\log n < 1 + \cdots + \frac{1}{n-1}.$$

13. (a) Since f is increasing

$$f(i) < \int_i^{i+1} f(x)\, dx < f(i+1);$$

add these inequalities for $i = 1, \ldots, n - 1$.

(b) From part (a) we have

$$\log 1 + \cdots + \log(n-1) < \int_1^n \log x \, dx < \log 2 + \cdots + \log n$$
$$\log(n-1)! < n \log n - n_1 < \log n!$$
$$(n-1)! < \frac{n^n}{e^{n-1}} < n!.$$

So

$$\frac{n^n}{e^{n-1}} < n! < \frac{(n+1)^{n+1}}{e^n} < \frac{n^{n+1}}{e^{n-1}}.$$

14. (a) The tangent line has slope $f'(x_0)$ so it is the graph of

$$g(x) = f'(x_0)(x - x_0) + f(x_0). \quad .$$

So x_1 is the solution of

$$0 = g(x_1) = f'(x_0)(x_1 - x_0) + f(x_0)$$
$$x_1 - x_0 = -\frac{f(x_0)}{f'(x_0)}$$
$$x_1 = x_0 - \frac{f(x_0)}{f'(x_0)}.$$

(b) From $f'(x_0) > 0$ we immediately have $x_0 > x_1$. We also know that f is convex by Theorem 2 of the Appendix to Chapter 11. So by Theorem 1 of that Appendix, the graph lies above the tangent line through $(x_0, f(x_0))$. So clearly $x_1 > c$. Obviously $f'(x_1) > 0$, for otherwise the convex function f already has its (unique) minimum between x_1 and x_0, so doesn't even have a zero. This shows that $x_1 > x_2$, etc.

(c) We have

$$\delta_{k+1} = x_{k+1} - c = x_k - c + (x_{k+1} - x_k)$$

$$= \delta_k - \frac{f(x_k)}{f'(x_k)}$$

$$= \frac{f(x_k)}{f'(\xi_k)} - \frac{f(x_k)}{f'(x_k)}.$$

Hence

$$\delta_{k+1} = \frac{f(x_k)}{f'(\xi_k)f'(x_k)} \cdot [f'(x_k) - f'(\xi_k)]$$

$$= \frac{f(x_k)}{f'(\xi_k)f'(x_k)} \cdot f''(\eta_k)(x_k - \xi_k)$$

$$= \delta_k \frac{f''(\eta_k)}{f'(x_k)} \cdot \delta_k$$

$$= \frac{f''(\eta_k)}{f'(x_k)} \cdot \delta_k{}^2.$$

(d) If

$$\delta_0 = x_0 - c < \frac{m}{M},$$

then

$$\frac{M}{m}\delta_0 = \alpha < 1, \qquad \text{for some } \alpha.$$

Then from $(*)$ we have first

$$\delta_1 \leq \frac{M}{m}\delta_0{}^2$$

$$\frac{M}{m}\delta_1 \leq \left(\frac{M}{m}\delta_0\right)^2 = \alpha^2.$$

Then

$$\delta_2 \leq \frac{M}{m}\delta_1{}^2$$

$$\frac{M}{m}\delta_2 \leq \left(\frac{M}{m}\delta_1\right)^2 \leq \alpha^4,$$

etc. Since $\alpha < 1$ the powers $\alpha^k \to 0$, so $\delta_n \to 0$, i.e., Newton's method works.

(e)

$$x_{n+1} = x_n - \frac{x_n{}^2 - A}{2x_n}.$$

15. To do this Problem you will need a calculator, of course.

(i) $x_1 = 1$, $x_2 = .7081$, $x_3 = .6055$, $x_4 = .5988$ [$f(x_4) = 7.97 \times 10^{-5}$].

(ii) $x_1 = 1$, $x_2 = .8382$, $x_3 = .8258$, $x_4 = .8243$; [$f(x_4) = 3.9 \times 10^{-4}$].

(iii) $x_1 = 1$, $x_2 = .75$, $x_3 = .68604$, $x_4 = .68234$; [$f(x_4) = 2.9 \times 10^{-5}$].

(iv) $x_1 = 1$, $x_2 = .6667$, $x_3 = .6527$, $x_4 = .6527$; [$f(x_4) = 9.13 \times 10^{-6}$].

16. If $\varepsilon > 0$, pick N so that $|a_n - l| < \varepsilon$ for $n \geq N$. Then

$$|a_N + a_{N+1} + \cdots + a_{N+M} - Ml| < \varepsilon M,$$

so

$$\left| \frac{1}{N+M}[a_N + a_{N+1} + \cdots + a_{N+M}] - \frac{Ml}{N+M} \right| < \frac{\varepsilon M}{N+M} < \varepsilon.$$

Choose M so that

$$\left| \frac{Ml}{N+M} - l \right| < \varepsilon \qquad \text{and} \qquad \left| \frac{1}{N+M}[a_1 + \cdots + a_N] \right| < \varepsilon.$$

Then

$$\left| \frac{1}{N+M}[a_1 + \cdots + a_{N+M}] - l \right| < 3\varepsilon.$$

17. Let $a_n = f(n+1) - f(n)$. Then $\lim\limits_{n \to \infty} a_n = 0$. So by Problem 16

$$0 = \lim_{n \to \infty} \frac{a_1 + \cdots + a_n}{n} = \lim_{n \to \infty} \frac{f(n) - f(1)}{n} = \lim_{n \to \infty} \frac{f(n)}{n}.$$

Since f is continuous, it follows easily that $\lim\limits_{n \to \infty} f(x)/x = 0$.

18. If $\varepsilon > 0$, pick N so that $|a_{n+1}/a_n - l| < \varepsilon$ for $n \geq N$. Then

$$l - \varepsilon < \frac{a_{n+1}}{a_n} < l + \varepsilon \qquad \text{for } n \geq N,$$

so

$$(l - \varepsilon)^m < \frac{a_{n+m}}{a_{n+m-1}} \cdot \frac{a_{n+m-1}}{a_{n+m-2}} \cdots \frac{a_{n+1}}{a_n} < (l + \varepsilon)^m,$$

so

$$\left| \sqrt[m]{\frac{a_{n+m}}{a_n}} - l \right| < \varepsilon.$$

Now

$$\sqrt[n+m]{a_{n+m}} = \sqrt[n+m]{\frac{a_{n+m}}{a_n}} \cdot \sqrt[n+m]{a_n}$$

$$= \left(\sqrt[m]{\frac{a_{n+m}}{a+n}} \right)^{m/n+m} \cdot \sqrt[n+m]{a_n}.$$

Since $\lim\limits_{m \to \infty} \sqrt[n+m]{a_n} = 1$, it follows that $\sqrt[n+m]{a_{n+m}}$ can be made within 2ε of l by choosing m sufficiently large.

19. (a) Suppose $\lim\limits_{n \to \infty} a_n = l > 1$. Since $l - 1 > 0$, there would be some n with $|l - a_n| < l - 1$, and hence $a_n > 1$, a contradiction. Similarly, we cannot have $l < 0$.

(b) $a_n = 1/n$.

20. Let us denote

$$\underbrace{f(f(f \ldots f(x) \ldots))}_{k \text{ times}}$$

by $f^k(x)$. Then by Theorem 1,

$$f(l) = f\left(\lim_{k \to \infty} f^k(x)\right) = \lim_{k \to \infty} f\left(f^k(x)\right)$$
$$= \lim_{k \to \infty} f^{k+1}(x) = l.$$

21. (a) Suppose $f(x) > x$. Since f is increasing, $f(f(x)) > f(x)$. Consequently, $f(f(f(x))) > f((f(x)))$, etc. Thus the sequence $x, f(x), f(f(x)), \ldots$ is increasing, and bounded by 1, so it has a limit. The proof when $x < f(x)$ is similar.

(b) There is some m with $g(m) = m$ (by Problem 7-11). According to part (a), the sequence $f^k(m)$ has a limit l, which is a fixed point for f (using the notation introduced in the solution to Problem 20). Moreover,

$$f^k(m) = f^k(g(m)) = g(f^k(m)),$$

since $f \circ g = g \circ f$. Hence, by Theorem 1,

$$l = \lim_{k \to \infty} f^k(m) = \lim_{k \to \infty} g\left(f^k(m)\right) = g\left(\lim_{k \to \infty} f^k(m)\right) = g(l).$$

22. (a)

$$c^m + c^{m+1} + \cdots + c^n = c^m(1 + c + \cdots + c^{n-m})$$
$$= \frac{c^m(1 - c^{n-m+1})}{1 - c}$$
$$= \frac{c^m - c^{n+1}}{1 - c}.$$

(b) Since $|c| < 1$, we have $\lim\limits_{n \to \infty} c^m = \lim\limits_{n \to \infty} c^{n+1} = 0$.

(c)

$$|x_n - x_m| = |(x_n - x_{n+1}) + (x_{n+1} - x_{n+2}) + \cdots + (x_{m-1} - x_m)|$$
$$\leq |x_n - x_{n+1}| + |x_{n+1} - x_{n-2}| + \cdots + |x_{m-1} - x_m|$$
$$\leq c^n + \cdots + c^{m-1},$$

so $\lim\limits_{m,n \to \infty} |x_n - x_m| = 0$, by part (b).

23. (a) If $c = 0$, then f is constant, so continuous. If $c \neq 0$ and $\varepsilon > 0$, then $|f(x) - f(a)| < \varepsilon$ for $|x - a| < \varepsilon/c$.

(b) If $f(l) = l$ and $f(m) = m$, then

$$|l - m| = |f(l) - f(m)| \leq c|l - m|,$$

so $l = m$, since $c < 1$.

(c) If x is any point in $\mathbf{R}$ and

$$x_n = f^n(x) = \underbrace{f(f(f \ldots f(x) \ldots))}_{n \text{ times}},$$

then

$$|x_n - x_{n+1}| = |f(x_{n-1}) - f(x_n)| \leq c|x_{n-1} - x_n| \leq c^2|x_{n-1} - x_{n-2}|$$
$$\leq \cdots \leq c^{n-1}|x_2 - x_1|.$$

Consequently, Problem 22(c) implies that $\{x_n\}$ is a Cauchy sequence, so converges. It converges to a fixed point, by Problem 20.

24. (a) If $f(x) = x$ and $f(y) = y$, then by the Mean Value Theorem, for some ξ between x and y we have

$$f'(\xi) = \frac{f(y) - f(x)}{y - x} = 1,$$

a contradiction.

(b) The Mean Value Theorem gives

$$|f(x) - f(y)| = |f'(\xi)| \cdot |x - y| \qquad \text{for some } \xi \text{ between } x \text{ and } y$$
$$\leq c|x - y|,$$

so the result follows from the previous Problem.

(c) Let $f(x) = x + 1$.

25. Since f is continuous at b we have

$$f(b) = \lim_{n \to \infty} f(b_n) = \lim_{n \to \infty} b_{n+1} = \lim_{n \to \infty} b_n = b.$$

Choose n so that $b_n, b_{n+1}, \ldots$ are all in the interval around b on which $|f'| > 1$. Then

$$\frac{|f(b_n) - f(b)|}{|b_n - b|} > 1 \qquad \text{or} \qquad |b_{n+1} - b| > |b_n - b|.$$

Similarly,

$$|b_{n+2} - b| > |b_{n+1} - b| > |b_n - b|,$$
$$|b_{n+3} - b| > |b_n - b|, \ldots .$$

This contradicts the fact that $\lim\limits_{m\to\infty} b_m = b$.

26. (a)

$$a^b = a^{\left(\lim\limits_{n\to\infty} b_n\right)} = \lim\limits_{n\to\infty} a^{b_n}$$
$$= \lim\limits_{n\to\infty} b_{n+1} = b.$$

(b) If $f(x) = x^{1/x} = e^{(\log x)/x}$, then

$$f'(x) = x^{1/x}\left[\frac{1}{x^2} - \frac{\log x}{x^2}\right] \quad \begin{cases} < 0 & \text{for } x > e \\ > 0 & \text{for } x < e. \end{cases}$$

Since $(\log x)/x \to 0$ as $x \to \infty$ and $-\infty$ as $x \to 0^+$ (Problem 18-34), $f(x) \to 1$ as $x \to \infty$ and $f(x) \to 0$ as $x \to 0^+$.

In particular, $0 < f(x) \le e^{1/e}$ for all $x > 0$. So $0 < a \le e^{1/e}$.

(c) Since $1 \le a$ we have

$$a \le a^a, \qquad \text{i.e.,} \qquad b_1 \le b_2.$$

And if $b_n \le b_{n+1}$, then

$$b_{n+1} = a^{b_n} \le a^{b_{n+1}} = b_{n+2},$$

so by induction $\{b_n\}$ is increasing. Moreover, if $b_n \le e$, then

$$b_{n+1} = a^{b_n} \le a^e \le (e^{1/e})^e = e.$$

(d) Choosing $f(x) = a^x$ in Problem 25, we see that if b exists then

$$|f'(b)| \le 1,$$

where

$$f'(b) = a^b(\log a) = b\log a = \log(a^b) = \log b,$$

so

$$-1 \le \log b \le 1,$$

or

$$e^{-1} \le b \le e.$$

Since $a = b^{1/b}$ we have

$$(e^{-1})^{1/e^{-1}} \le a \le e^{1/e}$$

or

$$e^{-e} \le a \le e^{1/e}.$$

(e) Since

$$f'(x) = \frac{\log x (\log a) a^x - a^x/x}{(\log x)^2}$$

$$= \frac{a^x}{(\log x)^2} \left[\log x \log a - \frac{1}{x} \right]$$

we need to show that

$$\log x \log a \le \frac{1}{x} \qquad \text{on } (0, 1)$$

or

$$\log a \ge \frac{1}{x \log x}.$$

Now $x \log x \to -\infty$ as $x \to 0^+$ or $x \to 1^-$, so the maximum of $1/x \log x$ occurs for

$$0 = \left[x \cdot \frac{1}{x} + \log x \right]$$

$$x = e^{-1}.$$

Thus we need to show

$$\log a \ge \frac{1}{e^{-1} \log(e^{-1})} = -e,$$

which is true since $a \ge e^{-e}$.

(f) We know from part (b) that the graph of $g(x) = x^{1/x}$ increases from 0 to 1 on $[0, 1]$ (and then increases further on $[1, e]$ and then decreases to 1 on $[e, \infty)$). So there is a unique b with $a = b^{1/b}$ (we don't even need $e^{-e} \le a$ for this, just $a \le 1$). For $0 < x < 1$ we have $1/x > 1$ so $x^{1/x} < x$ (the signs reverse since $\log x < 0$), so

$$a = b^{1/b} < b.$$

For $x < b$ we have

$$a^x > a^b = b$$

so

$$a^{a^x} < a^b = b$$

(the inequalities reverse since $a < 1$). Moreover, part (e) shows that for $0 < x < b$ we have

$$\frac{a^x}{\log x} > \frac{a^b}{\log b}$$

so

$$a^x \log b > a^b \log x \qquad \text{(since } \log b, \log x < 0)$$
$$a^x(\log(a^b)) > b \log x$$
$$a^x b \log a > b \log x$$
$$a^x \log a > \log x$$
$$\log(a^{a^x}) > \log x$$
$$a^{a^x} > x.$$

In particular,

$$0 < a < b \qquad \text{so} \qquad x < a^{a^a} < b$$
$$0 < a^{a^a} < b \qquad \text{so} \qquad a^a < a^{a^{a^a}} < b, \quad \text{etc.}$$

So the sequence b_{2n+1} is increasing and bounded by b, so has a limit l. Clearly

$$l = \lim_{n\to\infty} b_{2n+1} = \lim_{n\to\infty} b_{2n+3}$$

$$= \lim_{n\to\infty} a^{a^{b_{2n+1}}} = a^{a^{\left(\lim_{n\to\infty} b_{2n+1}\right)}} = a^{a^l}.$$

(g) Since $a^b = b$ we have $a^{a^b} = a^b = b$. Since we also have $a^{a^l} = l$ we must have $l = b$ by part (e).

(h)

$$\lim_{n\to\infty} b_{2n+2} = \lim_{n\to\infty} a^{b_{2n+1}} = a^{\lim_{n\to\infty} b_{2n+1}} = a^b = b.$$

27. (a) Clearly $\{y_n\}$ is decreasing, and bounded below (by a lower bound for $\{x_n\}$).

(b) (i) 0.

(ii) 0.

(iii) 1.

(iv) 1.

(c) $\lim_{n\to\infty} x^n = \lim_{n\to\infty} z_n$, where $z_n = \inf\{x_n, x_{n+1}, x_{n+2}, \dots\}$. Since $z_n \le y_n$ for each n, it is clear that $\underline{\lim}_{n\to\infty} x_n \le \overline{\lim}_{n\to\infty} x_n$.

(d) Suppose first that $\lim_{n\to\infty} x_n = l$. If $\varepsilon > 0$ there is some N with $|x_n - l| < \varepsilon$ for $n \ge N$. So $x_N < l + \varepsilon$, $x_{N+1} < l + \varepsilon$, $\dots$, so $y_N \le l + \varepsilon$. Similarly $z_N \ge l - \varepsilon$. Since this is true for all $\varepsilon > 0$, we have $\underline{\lim}_{n\to\infty} x_n = l = \overline{\lim}_{n\to\infty} x_n$.

Conversely, suppose that $\varliminf\limits_{n\to\infty} x_n = \varlimsup\limits_{n\to\infty} x_n = l$. Then for any $\varepsilon > 0$ there is some N with $l - \varepsilon < z_N < y_N < l + \varepsilon$. This implies that $l - \varepsilon < x_n < l + \varepsilon$ for every $n \geq N$.

(e) Let $l = \varlimsup\limits_{n\to\infty} x_n = \lim\limits_{n\to\infty} y_n$. If $a < l$, then $a < y_n$ for all n. Consequently, $a < x_n$ for infinitely many x_n, so a is not an almost upper bound for A. On the other hand, if $a > l$, then $a > y_n$ for all but finitely many n. Consequently, $a > x_n$ for all but finitely many n, so a is an almost upper bound. Thus, l is the greatest lower bound of all almost upper bounds of A.

28. (a) First choose $\delta > 0$ so that $|f(x) - f(y)| < 1$ for rational x and y with $|x - y| < \delta$. Since $x_n \to x$, there is some N so that $|x_n - x| < \delta/2$ for $n \geq N$. Hence $|x_n - x_m| < \delta$ for $m, n \geq N$. So $|f(x_m) - f(x_{N+1})| < 1$ for $m \geq N$. This shows that the sequence $\{f(x_n)\}$ is bounded. It follows that it has a convergent subsequence, say $f(x_{n_1})$, $f(x_{n_2})$, ... approaches the limit l. We claim finally that the original sequence $\{f(x_n)\}$ approaches l. In fact, for any $\varepsilon > 0$ we can choose K so large that $|f(x_{n_k}) - l| < \varepsilon/2$ for $k > K$ and also choose δ so that $|f(x) - f(y)| < \varepsilon/2$ for rational x and y with $|x - y| < \delta$. Finally, we choose N so large that $|x - x_n| < \delta/2$ for $n \geq N$. Then for $n \geq N$ we have, for any $k > K$ with $x_{n_k} > N$,

$$|x_n - x_{n_k}| \leq |x_n - x| + |x_{n_k} - x|$$
$$< \delta/2 + \delta/2,$$

so

$$|f(x_n) - l| \leq |f(x_n) - f(x_{n_k})| + |f(x_{n_k}) - l|$$
$$\leq \varepsilon/2 + \varepsilon/2.$$

(b) Given another sequence $\{y_n\}$ with $\lim\limits_{n\to\infty} y_n = x$, consider the sequence $x_1, y_1, x_2, y_2, \ldots$. This also approaches x, so the sequence $f(x_1), f(y_1), f(x_2), f(y_2), \ldots$ has a limit, which must be the limit of the two sequences $\{f(x_n)\}$ and $\{f(y_n)\}$.

(c) Given $\varepsilon > 0$ choose $\delta > 0$ so that for rational x and y with $|x - y| < \delta$ we have

(1) $$|f(x) - f(y)| < \varepsilon/3.$$

If z and w are any two numbers in the interval with $|z - w| < \delta$, then by the definition of $\bar{f}$ we can choose rational x and y so that

(2) $$|f(x) - \bar{f}(z)| < \varepsilon/3$$
(3) $$|f(y) - \bar{f}(w)| < \varepsilon/3.$$

Moreover, by choosing x sufficiently close to z and y sufficiently close to w, we can insure that $|x - y| < \delta$, so that (1) holds. It follows that $|\bar{f}(z) - \bar{f}(w)| < \varepsilon$.

29. (a) Since $a^y = a^{y-x} \cdot a^x$ we just need that $a^z > 1$ for rational $z > 0$, and this follows immediately from the elementary definitions.

(b) Problem 10 (c), (d) shows that for large enough n we have

$$1 - \varepsilon < a^{-1/n} < a^{1/n} < 1 + \varepsilon \qquad a > 1$$
$$1 - \varepsilon < a^{1/n} < a^{-1/n} < 1 + \varepsilon \qquad a < 1.$$

So, by (a), for rational x with $|x| < 1/n$ we have $|a^x - 1| < \varepsilon$.

(c) For rational x, y in the closed interval $[-M, M]$ we have

$$|a^x - a^y| = |a^y| \cdot |a^{x-y} - 1|$$
$$\leq \max(a^M, -a^{-M})|a^{x-y} - 1|.$$

Since $|a^{x-y} - 1|$ can be made $< \varepsilon$ by making $|x - y|$ sufficiently small, f is uniformly continuous on $[-M, M]$.

(d) If x_n and y_n are rational and $x_n \to x$ and $y_n \to y$, then $x_n + y_n \to x + y$, so

$$\bar{f}(x + y) = \lim_{n \to \infty} f(x_n + y_n) = \lim_{n \to \infty} f(x_n)f(y_n)$$
$$= \lim_{n \to \infty} f(x_n) \lim_{n \to \infty} f(y_n)$$
$$= \bar{f}(x)\bar{f}(y).$$

If $x < y$ then we can choose rational $x_n \to x$ and $y_n \to y$ with all $x_n <$ all y_m, so

$$\bar{f}(x) = \lim_{n \to \infty} f(x_n) \leq \lim_{n \to \infty} f(y_n) = \bar{f}(y).$$

To prove strict inequality choose rational x', y' with $x < x' < y' < y$. Then

$$\bar{f}(x) \leq \bar{f}(x') = f(x') < f(y') = \bar{f}(y') \leq \bar{f}(y).$$

30. (a) (i) 0.

(ii) 0 and $1/n$ for each natural number n.

(iii) -1 and 1.

(iv) No limit points.

(v) All real numbers.

(b) If there are infinitely many points a of A satisfying $|x - a| < \varepsilon$, then there is surely one such a with $a \neq x$. Conversely, if there were only finitely many such points $a_1, \ldots, a_n$ and $\varepsilon_1 > 0$ is the minimum of all those $|x - a_i|$ which are $\neq 0$, then there would be no points a in A satisfying $|x - a| < \varepsilon_1$.

(c) For any $\varepsilon > 0$, the number $\overline{\lim} A - \varepsilon$ is *not* an almost upper bound of A, so there are infinitely many numbers y in A with $y > \overline{\lim} A + \varepsilon$. Moreover, there cannot be infinitely many such numbers y with $y > \overline{\lim} A + \varepsilon$, for in this case no numbers between $\overline{\lim} A$ and $\overline{\lim} A + \varepsilon$ could be almost upper bounds of A, so $\overline{\lim} A + \varepsilon$ would be a larger lower bound for the set of all almost upper bounds. This shows that there are infinitely many numbers y in A between $\overline{\lim} A - \varepsilon$ and $\overline{\lim} A + \varepsilon$. Consequently, $\overline{\lim} A$ *is* a limit point of A. If there were another limit point $\alpha > \overline{\lim} A$, then no number less that α could be an almost upper bound so α would be a lower bound for the set of almost upper bounds, a contradiction. The proof for $\underline{\lim} A$ is similar.

(d) Choose distinct points $x_1, x_2, x_3, \ldots$ in A. The sequence $\{x_n\}$ is bounded, since A is contained in $[a, b]$. So there is a convergent subsequence $\{x_{n_j}\}$. Let $l = \lim_{j \to \infty} x_{n_j}$. For any $\varepsilon > 0$ there is some J such that $|l - x_{n_j}| < \varepsilon$ for all $j > J$. Since the x_{n_j} are distinct, this shows that there are infinitely many a in A with $|l - a| < \varepsilon$.

(e) Choose a sequence of intervals $I_1, I_2, I_3, \ldots$ with $I_1 = [a, b]$, and each I_{j+1} a half of I_j, such that each I_j contains infinitely many points of A. If x is the point in all I_j, then x is a limit point of A.

31. (a) Choose x_n with $f(x_n) > n$. There is a subsequence x_{n_j} which converges to a point x, which is in $[a, b]$. Thus for every $\varepsilon > 0$ there are infinitely many x_{n_j} with $|x - x_{n_j}| < \varepsilon$, and consequently f is unbounded on $[x - \varepsilon, x + \varepsilon]$, contradicting the fact that f is continuous at x.

(b) Given $\varepsilon > 0$, suppose there is no $\delta > 0$ such that $|f(x) - f(y)| < \varepsilon$ for all x, y with $|x - y| < \delta$. Then for each n there are points x_n, y_n such that $|x_n - y_n| < 1/n$ but $|f(x_n) - f(y_n)| \geq \varepsilon$. Choose subsequences x_{n_j} and y_{n_j} converging to points x, y in $[a, b]$. Then

$$|x - y| = \lim_{j \to \infty} |x_{n_j} - y_{n_j}| = 0,$$

so $x = y$, but

$$|f(x) - f(y)| = \lim_{j \to \infty} |f(x_{n_j}) - f(y_{n_j})|$$
$$\geq \varepsilon,$$

a contradiction.

32. (a) Let $\#(n)$ be the number of j for which j/n is in $[a, b]$. To estimate $\#(n)$, let j/n be the smallest such fraction in $[a, b]$ and k/n the largest. Then $(j-1)/n < a \leq j/n$ and $k/n \leq b < (k+1)/n$.

So

$$\frac{k}{n} - \frac{j}{n} \leq b - a < \frac{k+1}{n} - \frac{j-1}{n},$$
$$k - j \leq n(b - a) < k - j + 2.$$

Since $\#(n) = k - j + 1$, we also have

$$k - j < \#(n) < k - j + 2,$$

so

$$|\#(n) - n(b - a)| < 2.$$

Adding these inequalities for $1, \ldots, n$ we obtain

$$|\#(1) + \cdots + \#(n) - [1 + \cdots + n](b - a)| < 2n.$$

Consequently,

$$\left| \frac{\#(1) + \cdots + \#(n)}{1 + \cdots + n} - (b - a) \right| < \frac{2n}{1 + \cdots + n}.$$

Since $\lim_{n \to \infty} 2n/(1 + \cdots + n) = 0$, this shows that

$$\frac{\#(1) + \cdots + \#(n)}{1 + \cdots + n} \quad \text{approaches} \quad (b - a).$$

Of course $\#(1) + \cdots + \#(n) = N(1 + \cdots + n; a, b)$. For an arbitrary number m, let n be the largest number with $1 + \cdots + n \leq m$. Then

$$m - (1 + \cdots + n) \leq n.$$

Clearly,

$$\left| N(m; a, b) - [\#(1) + \cdots + \#(n)] \right| \leq m - (1 + \cdots + n) \leq n.$$

Consequently

$$(1) \qquad \left| \frac{N(m; a, b)}{m} - \frac{[\#(1) + \cdots + \#(n)]}{m} \right| \leq \frac{n}{m}$$

$$\leq \frac{n}{1 + \cdots + n} \to 0, \quad \text{as } n \to \infty.$$

Moreover,

$$\frac{\#(1) + \cdots + \#(n)}{m} = \frac{\#(1) + \cdots + \#(n)}{1 + \cdots + n} \cdot \frac{1 + \cdots + n}{m};$$

since

$$\frac{1 + \cdots + n}{1 + \cdots + (n + 1)} \leq \frac{1 + \cdots + n}{m} \leq \frac{1 + \cdots + n}{1 + \cdots + n} = 1,$$

it follows that $[\#(1) + \cdots + \#(n)]/m$ can be made as close to $[\#(1) + \cdots + \#(n)]/(1 + \cdots + n)$ as desired by choosing m (and hence n) sufficiently large. Since the latter expression can be made as close as desired to $b - a$ by choosing m sufficiently large, it follows from (1) that $\lim_{m \to \infty} N(m; a, b)/m = b - a$.

(b) Consider the special case where $s(x) = c$ for x in $[a, b]$ and $s(x) = 0$ for other x in $[0, 1]$. Then

$$\lim_{n \to \infty} \frac{s(a_1) + \cdots + s(a_n)}{n} = \lim_{n \to \infty} c \cdot \frac{N(n; a, b)}{n} = c(b - a) = \int_0^1 s.$$

This holds, in particular, when $a = b$. A similar proof works when $s(x) = c$ for x in (a, b). Any step function s can be written $s = s_1 + \cdots + s_m$ where s_i is one of these special kinds. Then

$$\int_0^1 s = \sum_{i=1}^m \int_0^1 s_i = \sum_{i=1}^m \lim_{n \to \infty} \frac{s_i(a_1) + \cdots + s_i(a_n)}{n}$$

$$= \lim_{n \to \infty} \frac{s(a_1) + \cdots + s(a_n)}{n}.$$

(c) Let $\varepsilon > 0$. According to Problem 13-16, there is a step function $s \leq f$ with $\int_a^b [f - s] < \varepsilon$. Thus for sufficiently large n,

$$-\varepsilon + \int_a^b f < \int_a^b s < \frac{s(a_1) + \cdots + s(a_n)}{n} + \varepsilon \leq \frac{f(a_1) + \cdots + f(a_n)}{n} + \varepsilon.$$

Similarly, since there is a step function $s \geq f$ with $\int_a^b [s - f] < \varepsilon$, we have

$$-2\varepsilon + \int_a^b f < \frac{f(a_1) + \cdots + f(a_n)}{n} < 2\varepsilon + \int_a^b f$$

for sufficiently large n.

33. (a) If there were infinitely many such points a in $[0, 1]$, then the set of all such points would have a limit point x in $[0, 1]$. For every $\delta > 0$ there would be some a with $|a - x| < \delta/2$ and $|\lim_{y \to a} f(y) - f(a)| > \varepsilon$. Consequently there would be a' with $|a' - a| < \delta/2$ (and consequently $|a' - x| < \delta$) such that $|f(a') - f(a)| > \varepsilon$. But since $\lim_{y \to x} f(x) = l$ for some l, there is some $\delta > 0$ such that $|f(y) - l| < \varepsilon/2$ for $|y - x| < \delta$. In particular, if $|a - x| < \delta$ and $|a' - x| < \delta$, then $|f(a) - f(a')| \leq |f(a) - l| + |f(a') - l| < \varepsilon$, a contradiction.

(b) By part (a), the set A_n of points a where $|\lim_{y \to a} f(y) - f(a)| > 1/n$ is finite. By Problem 21-5, the union $A_1 \cup A_2 \cup A_3 \cup \cdots$ is countable. This union is the same as the set of all points a at which f is discontinuous.

CHAPTER 23

1. (ii) Convergent, by Leibnitz's Theorem. The series is not absolutely convergent, since

$$1 + \frac{1}{3} + \frac{1}{5} + \frac{1}{7} + \cdots > \frac{1}{2} + \frac{1}{4} + \frac{1}{6} + \frac{1}{8} + \cdots = \frac{1}{2}\left(1 + \frac{1}{2} + \frac{1}{3} + \frac{1}{4} + \cdots\right).$$

(iv) Convergent, by Leibnitz's Theorem. (The function $f(x) = (\log x)/x$ is decreasing for $x \geq e$, since $f'(x) = (1 - \log x)/x^2$.) The series is not absolutely convergent (see (viii)).

(vi) Divergent, since

$$\frac{1}{\sqrt[3]{n^2 + 1}} \geq \frac{1}{2n^{2/3}}.$$

(viii) Divergent, since

$$\int_1^N \frac{\log x}{x}\, dx = \frac{(\log N)^2}{2} \to \infty \text{ as } N \to \infty,$$

and $f(x) = (\log x)/x$ is decreasing for $x \geq e$ (see (iv)).

(x) Divergent, since

$$\frac{1}{(\log n)^k} > \frac{1}{n}$$

for sufficiently large n (Problem 18-12).

(xii) (Absolutely) convergent, by (xi).

(xiv) Divergent, since

$$\sin \frac{1}{n} > \frac{1}{2n},$$

for sufficiently large n.

(xvi) Convergent, since

$$\int_2^N \frac{1}{x(\log x)^2}\, dx = -\frac{1}{\log N} + \frac{1}{\log 2} \to \frac{1}{\log 2} \text{ as } N \to \infty.$$

(xviii) Convergent, since

$$\lim_{n\to\infty} \frac{(n + 1)!/(n + 1)^{n+1}}{n!/n^n} = \lim_{n\to\infty} \frac{(n + 1)n^n}{(n + 1)^{n+1}}$$

$$= \lim_{n\to\infty} \frac{1}{\left(1 + \dfrac{1}{n}\right)^n} = \frac{1}{e},$$

by Problem 18-16.

357

(xx) Divergent, since

$$\lim_{n\to\infty} \frac{3^{n+1}(n+1)!/(n+1)^{n+1}}{3^n n!/n^n} = \lim_{n\to\infty} \frac{3(n+1)n^n}{(n+1)^{n+1}}$$

$$= \lim_{n\to\infty} \frac{3}{\left(1+\dfrac{1}{n}\right)^n} = \frac{3}{e},$$

by Problem 18-16.

2. (a) According to Problem 22-13,

$$\frac{e^n n!}{n^n} > e,$$

so the series certainly diverges.

(b) Since

$$\lim_{n\to\infty} \frac{(n+1)^{n+1}/a^{n+1}(n+1)!}{n^n/a^n n!} = \lim_{n\to\infty} \frac{(n+1)^{n+1}}{a(n+1)n^n}$$

$$= \lim_{n\to\infty} \frac{1}{a}\left(1+\frac{1}{n}\right)^n = \frac{e}{a},$$

the series converges for $a > e$ and diverges for $a < e$. By Problem 22-13,

$$\frac{n^n}{e^n n!} = \frac{n^n}{(n+1)^{n+1}} \cdot \frac{(n+1)^{n+1}}{e^n n!} > \frac{n^n}{(n+1)^{n+1}}$$

$$= \frac{1}{n+1}\left(\frac{n}{n+1}\right)^n = \frac{1}{(n+1)\left(1+\dfrac{1}{n}\right)^n} > \frac{1}{2e(n+1)}$$

for sufficiently large n, so $\sum_{n=1}^{\infty} n^n/e^n n!$ diverges.

3. (a) The function $f(y) = e^y/y^y$ is decreasing for $y \geq 1$, since

$$f'(y) = \frac{y^y e^y - e^y y^y(1+\log y)}{y^{2y}} = \frac{e^y}{y^y}(-\log y).$$

Now the series $\sum_{n=1}^{\infty} (e/n)^n$ clearly converges, since $(e/n)^n \leq e^2/n^2$ for $n \geq 2$, so the integral also converges.

(b) Since $f(x) = (\log x)^{-\log x}$ is clearly decreasing for $x \geq 1$, the series converges if $\int_0^{\infty} (\log x)^{-\log x}\, dx$ exists. The substitution $y = \log x$, $dx = e^y\, dy$, changes this integral to

$$\int_1^{\infty} \frac{e^y}{y^y}\, dy,$$

which exists, by part (a).

(c) The substitution $y = \log x$, $dx = e^y \, dy$ changes $\int_1^\infty (\log x)^{-\log(\log x)} \, dx$ to

$$\int_1^\infty \frac{e^y}{y^{\log y}} \, dy.$$

Now

$$\frac{e^y}{y^{\log y}} = \frac{e^y}{e^{(\log y)^2}} = e^{y-(\log y)^2} = e^{y(1-(\log y)^2/y)}.$$

Since $\lim_{y\to\infty} (\log y)^2/y = 0$ (Problem 18-12), it follows that $e^y/y^{\log y}$ is close to e^y for large y, so the integral certainly diverges.

4. Note that

$$\frac{1}{n^{1+1/n}} = \frac{1}{n} \cdot \frac{1}{n^{1/n}}$$

and apply Theorem 2 to $a_n = 1/(n^{1+1/n})$ and $b_n = 1/n$: We have

$$\lim_{n\to\infty} a_n/b_n = \lim_{n\to\infty} \frac{1}{n^{1/n}} = 1 \qquad \text{(Problem 22-10(e)).}$$

Since $\sum_{n=1}^\infty 1/n$ diverges the given series also diverges.

5. (b) Define $\{a_n\}$ inductively as follows:

$$a_1 = [10x],$$
$$a_n = [10^n x - (10^{n-1}a_1 + \cdots + 10a_{n-1})].$$

For each n we have

$$0 \le 10^n x - (10^{n-1}a_1 + \cdots + 10a^{n-1}) - a_n < 1,$$

so

(*) $$0 \le 10^{n+1}x - (10^n a_1 + \cdots + 10^2 a_{n-1} + 10a_n) < 10,$$

so $0 \le a_{n+1} \le 9$ for each n. Moreover, from (*) we have

$$0 \le x - (a_1 10^{-1} + a_2 10^{-2} + \cdots + 10^{-n}a_n) < 10^{-n},$$

so $x = \sum_{n=1}^\infty a_n 10^{-n}$.

(c) Let $\alpha = 10^k a_1 + \cdots + 10 a_{k-1} + a_k$. Then

$$\sum_{n=1}^{\infty} a_n 10^{-n} = \frac{\alpha}{10^k} + \frac{\alpha}{10^{2k}} + \frac{\alpha}{10^{3k}} + \cdots$$

$$= \frac{1}{10^k} \left[1 + \frac{1}{10} + \left(\frac{1}{10}\right)^2 + \cdots \right]$$

$$= \frac{\alpha}{10^k} \cdot \frac{1}{1 - \dfrac{1}{10}}$$

$$= \frac{9\alpha}{10^{k-1}}.$$

(d) The number a_n in part (b) satisfies

$$0 \le \frac{10^n p}{q} - (10^{n-1} a_1 + \cdots + 10 a_{n-1} + a_n) < 1.$$

Now $10^n p/q$ can be written as $k + r/q$ where k is an integer and $0 \le r < q - 1$. In this case $a_{n-1} = [10r/q]$. Since there are at most q different fractions r/q, there will have to be some m and n with $m > n$ and $a_{n+1} = [10r/q] = a_{m+1}$. It is easy to see that we will then have $a_{n+2} = a_{m+2}$, etc.

6. The proof of Leibnitz's Theorem shows that if N is even then

$$s_N \le \sum_{n=1}^{\infty} (-1)^{n+1} a_n \le s_{N+1},$$

so $\left| \sum_{n=1}^{\infty} (-1)^{n+1} a_n - s_N \right| \le s_{N+1} - s_N = a_{N+1} \le a_N$. (Strict inequality holds unless $s_N = s_{N+1}$, or $a_{N+1} = 0$.) The proof is similar if N is odd.

7. Suppose $r < 1$. Choose s with $r < s < 1$. There is some N such that $\sqrt[n]{a_n} \le s$ for $n \ge N$. Then $a_n \le s^n$, so

$$\sum_{n=N}^{\infty} a_n \le \sum_{n=N}^{\infty} s^n$$

converges. If $r > 1$, and $r > s > 1$, then there is some N such that $\sqrt[n]{a_n} \ge s$ for $n \ge N$. Thus $a_n \ge s^n \ge 1$, so $\sum_{n=1}^{\infty} a_n$ does not converge.

8. We have

$$c_n = \sum_{k=1}^{n} \frac{(-1)^k (-1)^{n+1-k}}{\sqrt{k}\sqrt{n+1-k}} = \pm \sum_{k=1}^{n} \frac{1}{\sqrt{k}\sqrt{n+1-k}}.$$

We can easily obtain estimates on the individual terms in this sum. In fact the minimum of $\sqrt{x}\sqrt{n+1-x}$ [occurs at the minimum of $x(n+1-x)$ which] occurs at $x = (n+1)/2$, so each

$$\frac{1}{\sqrt{k}\sqrt{n+1-k}} \geq \frac{1}{\sqrt{\dfrac{n+1}{2}}\sqrt{\dfrac{n+1}{2}}} = \frac{2}{n+1}.$$

There are n such terms, so

$$|c_n| \geq \frac{2n}{n+1} \geq 1.$$

9. The sequence $1, -1, -1, 1, -1, \ldots$ is Cesaro summable to $1/2$.

10. Say $|na_n| \leq M$. We have

$$\sigma_n = \frac{a_1 + (a_1 + a_2) + \cdots + (a_1 + \cdots + a_n)}{n} = \frac{na_1 + (n-1)a_n + \cdots + a_n}{n},$$

so

$$\frac{n}{n+1}\sigma_n = \frac{na_1}{n+1} + \frac{(n-1)a_2}{n+1} + \cdots + \frac{a_n}{n+1}$$

and

$$s_n - \frac{n}{n+1}\sigma_n$$

$$= \left(1 - \frac{n}{n+1}\right)a_1 + \left(1 - \frac{(n-1)}{n+1}\right)a_2 + \cdots + \left(1 - \frac{1}{n+1}\right)a_n$$

$$= \frac{a_1 + 2a + \cdots + na_n}{n+1}$$

(the essentially irrelevant factor $n/(n+1)$ was simply used to get this rather than an expression with $a_2 + 2a_3 + \cdots$ in the numerator). So

$$\left| s_n - \frac{n}{n+1}\sigma_n \right| \leq \frac{n}{n+1}M,$$

hence is bounded. Since $\{\sigma_n\}$ approaches a limit, $|\sigma_n|$ is bounded. Hence $|s_n|$ is bounded. Since $a_n > 0$, this means that $\sum_{n=1}^{\infty} a_n$ converges.

11. (a) Choose m so that $a_1, \ldots, a_n$ appear among $b_1, \ldots, b_m$.

(b) This follows immediately from part (a), since $\sum_{n=1}^{\infty} a_n$ is the last upper bound of all partial sums s_n.

(c) The reverse inequality $\sum_{n=1}^{\infty} b_n \leq \sum_{n=1}^{\infty} a_n$ follows from part (b), since $\{a_n\}$ is also a rearrangement of $\{b_n\}$. It follows that $\sum_{n=1}^{\infty} b_n$ exists and equals $\sum_{n=1}^{\infty} a_n$.

(d) Let $\{p_n\}$ and $\{q_n\}$ be the series formed of the positive and negative terms, respectively, of $\{a_n\}$, and let $\{p_n{}'\}$ and $\{q_n{}'\}$ be defined similarly for $\{b_n\}$. Then $\{p_n{}'\}$ is a rearrangement of $\{p_n\}$ and similarly for $\{q_n\}$. So by part (c), $\sum p_n{}' = \sum p_n$ and $\sum q_n{}' = \sum q_n$, the sums on the right existing because $\{a_n\}$ is absolutely convergent. Therefore, $\{b_n\}$ is absolutely convergent, and $\sum_{n=1}^{\infty} b_n = \sum p_n' - \sum q_n' = \sum p_n - \sum q_n = \sum_{n=1}^{\infty} a_n$.

12. (a) Let $b_j = a_{n_j}$. Then

$$|b_j| + |b_{j+1}| + \cdots + |b_k| = |a_{n_j}| + |a_{n_{j+1}}| + \cdots + |a_{n_k}|$$
$$\leq |a_{n_j}| + |a_{n_j+1}| + |a_{n_j+2}| + \cdots + |a_{n_k}|.$$

Consequently, $\lim_{j,k \to \infty} |b_j| + \cdots + |b_k| = 0$.

(b) If $\sum_{n=1}^{\infty} a_n$ does not converge absolutely, then either $\sum p_n$ or $\sum q_n$ diverges, where $\sum p_n$ is the series of positive terms, and $\sum q_n$ is the series of negative terms. Choose the appropriate one as $\sum b_n$.

(c) The series $a_1 + a_3 + a_5 + \cdots$ and $a_2 + a_4 + a_6 + \cdots$ both converge by part (a). The same is true of the series $a_1 + 0 + a_3 + 0 + a_5 + \cdots$ and $0 + a_2 + 0 + a_4 + 0 + a_6 + \cdots$, whose sum is $\sum_{n=1}^{\infty} a_n$.

13. For every N, we have

$$\left| \sum_{n=1}^{N} a_n \right| \leq \sum_{n=1}^{N} |a_n| \leq \sum_{n=1}^{\infty} |a_n|.$$

Since $\sum_{n=1}^{\infty} a_n = \lim_{N \to \infty} \sum_{n=1}^{N} a_n$ the result follows.

14. Choose $\delta > 0$ so that $|\sin x| \geq 1/2$ on $(k\pi + \pi/2 - \delta, k\pi + \pi/2 + \delta)$. Then

$$\int_{k\pi+\pi/2-\delta}^{k\pi+\pi/2+\delta} \left| \frac{\sin x}{x} \right| dx \geq \frac{\delta}{k\pi + \pi/2}.$$

Since the series

$$\sum_{k=1}^{\infty} \frac{1}{k\pi + \pi/2}$$

diverges, the same is true for the integral.

15. Let f be the function whose graph is show below.

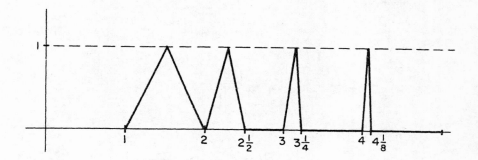

16. For the partition

$$P = \left\{ 0, \frac{2}{(2n+1)\pi}, \ldots, \frac{2}{\pi}, 1 \right\}$$

we have

$$\ell(f, P) > \frac{2}{\pi} + \frac{2}{3\pi} + \cdots + \frac{2}{(2n+1)\pi} = \frac{2}{\pi}\left(1 + \frac{1}{3} + \cdots + \frac{1}{2n+1}\right),$$

and these sums are not bounded.

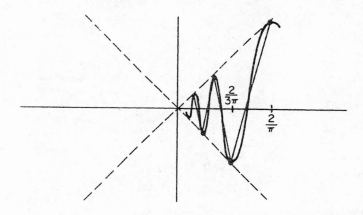

18. (a)

$$\frac{\dbinom{\alpha}{k+1} r^{k+1}}{\dbinom{\alpha}{k} r^k} = \frac{r\alpha(\alpha-1)\cdots(\alpha-k)/(k+1)!}{(\alpha-1)\cdots(\alpha-k+1)/k!} = r\frac{\alpha-k}{k+1}.$$

Clearly

$$\lim_{k\to\infty}\left|\frac{\alpha-k}{k+1}\right|=1,$$

so

$$\lim_{k\to\infty}\frac{\left|\binom{\alpha}{k+1}r^{k+1}\right|}{\left|\binom{\alpha}{k}r^{k}\right|}=|r|.$$

(b)

$$|R_{n,0}(x)|=\left|\binom{\alpha}{n+1}x^{n+1}(1+t)^{\alpha-n-1}\right|$$

$$\leq\left|\binom{\alpha}{n+1}x^{n+1}\right|\to 0,\qquad\text{by part (a)}.$$

(c) We have $0<1+x<1+t\leq 1$. If $\alpha-1\geq 0$, then $(1+t)^{\alpha-1}\leq 1$. If $\alpha-1<0$, then $(1+t)^{\alpha-1}<(1+x)^{\alpha-1}$. So $(1+t)^{\alpha-1}\leq M$, so $|x(1+t)^{\alpha-1}|\leq|x|M$. Moreover, since $-1<x$ and $t\leq 0$, we have

$$-t\geq xt,$$

$$0>x-t\geq x+xt,$$

$$0<\frac{x-t}{x}\leq 1+t,\qquad\text{since }x<0$$

$$0<1-t/x\leq 1+t,$$

$$0<\frac{1-t/x}{1+t}\leq 1,\qquad\text{since }1+t>0.$$

Thus

$$|R_{n,0}(0)|=\left|(n+1)\binom{\alpha}{n+1}x(1+t)^{\alpha-1}\left(\frac{x-t}{1+t}\right)^{n}\right|$$

$$\leq|n\alpha M|\cdot\left|\binom{\alpha-1}{n}x^{n}\right|\to 0\qquad\text{by part (a)}.$$

19. (a) According to Problem 19-35(b), if $m\leq a_1+\cdots+a_n\leq M$, then

$$b_k m\leq a_k b_k+\cdots+a_n b_n\leq b_k M.$$

Since $\lim_{k\to\infty}b_k=0$, this shows that $\lim_{k,n\to\infty}a_k b_k+\cdots+a_n b_n=0$.

(b) Let $a_n=(-1)^{n+1}$; the partial sums are bounded. So if $b_1\geq b_2\geq b_3\geq\cdots\geq 0$ and $\lim_{n\to\infty}b_n=0$, then $\sum_{n=1}^{\infty}(-1)^{n+1}b_n$ converges.

(c) Choose $b_n = 1/n$ and $a_n = \cos nx$. The partial sums of $\{a_n\}$ are bounded because, by Problem 15-33,

$$|\cos x + \cdots + \cos nx| = \left| \frac{\sin(n + 1/2)x}{2 \sin x/2} - \frac{1}{2} \right| \leq \left| \frac{\sin(n + 1/2)x}{2 \sin x/2} \right| + \frac{1}{2}$$

$$\leq \frac{1}{2|\sin x/2|} + \frac{1}{2}.$$

(d) It clearly suffices to consider the case where $\{b_n\}$ is nonincreasing. Then so is $\{b_n - b\}$, and $\lim\limits_{n \to \infty} b_n - b = 0$. Since $\sum_{n=1}^{\infty} a_n$ converges, the partial sums of $\{a_n\}$ are surely bounded, so by Dirichlet's test $\sum_{n=1}^{\infty} a_n b_n - a_n b$ converges. Since $\sum_{n=1}^{\infty} a_n b$ also converges, this implies that $\sum_{n=1}^{\infty} a_n b_n$ converges.

20. Since

$$a_1 \geq a_2 \geq \underbrace{a_3 \geq a_4} \geq \underbrace{a_5 \geq a_6 \geq a_7 \geq a_8} \geq \cdots,$$

we have

$$a_2 \leq a_1 + a_2$$
$$2a_4 \leq a_3 + a_4$$
$$4a_8 \leq a_5 + a_6 + a_7 + a_8,$$
$$\text{etc.}$$

So

$$\sum_{n=1}^{N} 2^n a_{2^n} \leq \sum_{k=1}^{2^N} a_k \leq \sum_{k=1}^{\infty} a_k.$$

21. (a) By Problem 2-21 we have

$$|a_n b_n + \cdots + a_m b_m| \leq \sqrt{a_n{}^2 + \cdots + a_m{}^2} \sqrt{b_n{}^2 + \cdots + b_m{}^2}.$$

This shows that the Cauchy condition for $\{a_n{}^2\}$ and $\{b_n{}^2\}$ implies the Cauchy condition for $\{a_n b_n\}$.

(b) Apply part (a) with $b_n = 1/n^\alpha$.

22. Choose n so that $a_n + \cdots + a_m < \varepsilon$ for $m \geq n$. Then

$$(m - n)a_m \leq a_n + \cdots + a_m < \varepsilon.$$

Since $\lim\limits_{m \to \infty} m/(m + n) = 1$ and

$$m a_m = \frac{m}{m - n} \cdot (m - n)a_n,$$

it follows that $m a_m < 2\varepsilon$ for sufficiently large m.

23. Let $\{a_n\}$ be

$$1, \; -\frac{1}{2}, \; -\frac{1}{2}, \; \frac{1}{3}, \; \frac{1}{3}, \; \frac{1}{3}, \; -\frac{1}{4}, \; -\frac{1}{4}, \; -\frac{1}{4}, \; -\frac{1}{4}, \cdots.$$

24. We can assume $a_k \to 0$, for otherwise we will not have $a_k/(1 + a_k) \to 0$, so that divergence will be automatic. We might as well assume that $a_k < 1$ for all k, so that $1 + a_k > 2a_k$. Then

$$a_k - \frac{a_k}{1 + a_k} = \frac{a_k{}^2}{1 + a_k} < \frac{a_k{}^2}{2a_k} < \frac{a_k}{2},$$

so

$$\sum_{k=1}^n \frac{a_k}{1 + a_k} < \frac{\displaystyle\sum_{k=1}^n a_k}{2}.$$

Thus, the partial sums on the left are unbounded.

The converse hold trivially, since $a_k > a_k/(1 + a_k)$ (for $a_k > -1$).

25. (a) Since $0 \neq l = \lim_{n\to\infty} p_n$ we have

$$\lim_{n\to\infty} (1 + a_n) = \lim_{n\to\infty} \frac{p_n}{p_{n-1}} = \frac{\displaystyle\lim_{n\to\infty} p_n}{\displaystyle\lim_{n\to\infty} p_{n-1}} = \frac{l}{l} = 1;$$

consequently $\lim_{n\to\infty} a_n = 0$. Note, in particular, that $1 + a_n > 0$ for sufficiently large n. In the remaining parts we assume $1 + a_n > 0$ for all n, which is really no restriction, since a finite number of terms do not affect the question of convergence.

(b) We have

$$\log p_n = \sum_{i=1}^n \log(1 + a_n).$$

If $p_n \to l \neq 0$, then

$$\log l = \lim_{n\to\infty} \log p_n = \lim_{n\to\infty} \sum_{i=1}^n \log(1 + a_i) = \sum_{i=1}^\infty \log(1 + a_i).$$

Conversely, if

$$s = \sum_{i=1}^\infty \log(1 + a_i) = \lim_{n\to\infty} \sum_{i=1}^n \log(1 + a_i) \quad \text{exists,}$$

then

$$0 \neq e^s = e^{\left(\lim\limits_{n \to \infty} \Sigma_{i=1}^n \log(1+a_i) \right)}$$

$$= \lim\limits_{n \to \infty} e^{\Sigma_{i=1}^n \log(1+a_i)}$$

$$= \lim\limits_{n \to \infty} e^{\log P_n}$$

$$= \lim\limits_{n \to \infty} P_n.$$

(c) We have

$$\frac{a_n}{1 + a_n} \leq \log(1 + a_n) \leq a_n,$$

by looking at a lower and upper sum for $\log(1 + a_n) = \int_1^{1+a_n} 1/x \, dx$. So if $\sum a_n$ converges, then $\sum \log(1 + a_n)$ converges, and hence the product converges by part (b). Conversely, if the product converges, then $\sum \log(1 + a_n)$ converges, so $\sum a_n/(1 + a_n)$ converges. It follows from Problem 24 that $\sum a_n$ also converges.

Counterexamples without the hypothesis $a_n \geq 0$ can be obtained as follows. Since

$$\log(1 + x) = x - \frac{x^2}{2} + \frac{x^3}{3} - \cdots$$

we have

$$\lim\limits_{x \to 0} \frac{x - \log(1 + x)}{x^2} = \frac{1}{2},$$

so for sufficiently small x we have

(∗) $$\tfrac{1}{4}x^2 \leq x - \log(1 + x) \leq \tfrac{3}{4}x^2.$$

Now suppose that both $\sum a_n$ and $\sum \log(1 + a_n)$ converge. Then by the Cauchy criterion we have

$$\lim\limits_{m,n \to \infty} \sum_{i=n+1}^m a_i - \log(1 + a_i) = 0.$$

It follows from (∗) that

$$\lim\limits_{m,n \to \infty} \sum_{i=n+1}^m a_i^2 = 0,$$

so that $\sum a_n^2$ must converge also. Now

$$\sum_{n=2}^{\infty} \frac{(-1)^n}{\sqrt{n}} \text{ converges,}$$

but $\sum 1/n$ diverges, so

$$\prod_{n=2}^{\infty} \left(1 + \frac{(-1)^n}{\sqrt{n}} \right) \text{ diverges.}$$

[We can actually conclude from (∗) that if $\sum a_n$ converges, then $\sum \log(1 + a_n)$ converges if and only if $\sum a_n{}^2$ converges. Also, if $\sum a_n{}^2$ converges, then $\sum \log(1 + a_n)$ converges if and only if $\sum a_n$ converges.]

We can also find an example where $\sum a_n$ diverges, but $\prod(1 + a_n)$ converges. The simple first guess, $a_n = -1/n$, doesn't quite work since

$$\prod_{n=2}^{\infty} \left(1 - \frac{1}{n}\right)$$

doesn't converge by our definition. In fact, if $\sum a_n$ is divergent with all $a_n < 0$, then $\sum \log(1 + a_n)$ is also divergent. In fact, writing (∗) as

$$\tfrac{1}{4}x^2 - x \leq -\log(1 + x)$$

we see that

$$\sum_{n=1}^{\infty} -\log(1 + a_n) \geq \sum_{n=1}^{\infty} -a_n + \tfrac{1}{4}\sum_{n=1}^{\infty} a_n{}^2 = \infty.$$

Here is a simple genuine counterexample (compare the answer to Problem 23):

$$\{a_n\} = \underbrace{1, -\tfrac{1}{2}}_{1 \text{ pair}}, \underbrace{\tfrac{1}{3}, -\tfrac{1}{4}, \tfrac{1}{3}, -\tfrac{1}{4}, \tfrac{1}{3}, -\tfrac{1}{4}}_{3 \text{ pairs}}, \underbrace{\tfrac{1}{5}, -\tfrac{1}{6}, \tfrac{1}{5}, -\tfrac{1}{6}, \tfrac{1}{5}, -\tfrac{1}{6}}_{5 \text{ pairs}}, \cdots \quad \cdots$$

This clearly diverges, since

$$
\begin{aligned}
1 - \tfrac{1}{2} &&&= \tfrac{1}{2} \\
\tfrac{1}{3} - \tfrac{1}{4} + \tfrac{1}{3} - \tfrac{1}{4} + \tfrac{1}{3} - \tfrac{1}{4} &= 1 - \tfrac{3}{4} &&= \tfrac{1}{4} \\
\tfrac{1}{5} - \tfrac{1}{6} + \cdots && = 1 - \tfrac{5}{6} &= \tfrac{1}{6}
\end{aligned}
$$

$$\cdots$$

But

$$\prod_{n=1}^{\infty}(1 + a_n)$$

$$= (1 + 1) \cdot \left(1 - \frac{1}{2}\right) \cdot \left(1 + \frac{1}{3}\right) \cdot \left(1 - \frac{1}{4}\right) \cdots \left(1 + \frac{1}{5}\right) \cdot \left(1 - \frac{1}{6}\right) \cdots$$

$$= \quad 2 \quad \cdot \quad \frac{1}{2} \quad \cdot \quad \frac{4}{3} \quad \cdot \quad \frac{3}{4} \quad \cdots \quad \frac{6}{5} \quad \cdot \quad \frac{5}{6} \quad \cdots$$

$$= 1.$$

26. **(a)** We have

$$\prod_{k=2}^{n}\left(1 - \frac{1}{k^2}\right) = \prod_{k=2}^{n}\frac{k^2 - 1}{k^2} = \prod_{k=2}^{n}\frac{(k-1)(k+1)}{k \cdot k}$$

$$= \frac{(2-1)(2+1)}{2 \cdot 2} \cdot \frac{(3-1)(3+1)}{3 \cdot 3} \cdots \frac{(n-1)(n+1)}{n \cdot n}.$$

Each factor $(k + 1)$, except for $(n + 1)$, cancels a $k + 1$ in the denominator of the next fraction, and each factor $(k - 1)$, except for $(2 - 1)$, cancels a $k - 1$ in the denominator of the previous fraction, so the product is just

$$\frac{1}{2} \cdot \frac{n+1}{n}$$

which approaches $1/2$.

(b) Note that

$$(1 + x^2)(1 + x^4) = 1 + x^2 + x^4 + x^6$$
$$(1 + x^2)(1 + x^4)(1 + x^8) = (1 + x^2 + x^4 + x^6)(1 + x^8)$$
$$= (1 + x^2 + x^4 + x^6 + x^8 + x^{10} + x^{12} + x^{14},$$

and in general

$$\prod_{k=1}^{n} (1 + x^{2^k}) = 1 + x^2 + x^4 + \cdots + x^{2^{k+1}-2}$$

so the infinite product is

$$1 + x^2 + x^4 + \cdots = \frac{1}{1 - x^2}.$$

27. **(a)** If $1/(n + 1) < p/q < 1/n$, then $np < q$ and

$$\frac{p}{q} - \frac{1}{n+1} = \frac{pn + p - q}{q(n+1)}.$$

The numerator $pn + p - q$ is $< q + p - q = p$. Of course, the numerator may be even smaller when the fraction is expressed in lowest terms. Notice, moreover, that

$$\frac{p}{q} - \frac{1}{n+1} < \frac{1}{n} - \frac{1}{n+1} = \frac{1}{n+1},$$

so that $p/q - 1/(n + 1)$ must be a fraction with denominator $> n + 1$.

(b) Part (a) proves the result for $x \le 1$. For $x > 1$, since $\sum 1/k$ diverges there is some $n \ge 1$ with

$$\frac{1}{1} + \frac{1}{2} + \cdots + \frac{1}{n} \le x \le \frac{1}{1} + \frac{1}{2} + \cdots + \frac{1}{n} + \frac{1}{n+1}.$$

If either inequality is an equality we are done. Otherwise

$$0 < x - \left(1 + \frac{1}{2} + \cdots + \frac{1}{n}\right) < \frac{1}{n+1}.$$

It follows from part (a) that $x - (1 + 1/2 + \cdots + 1/n)$ can be written as a finite sum of distinct numbers of the form $1/k$ with $k > n$, which gives the desired expression for x.

CHAPTER 24

1. (ii) $f(x) = \lim\limits_{n\to\infty} f_n(x) = 0$. The sequence $\{f_n\}$ converges uniformly to f (in fact, is eventually 0) on $[a, b]$, but does not converge uniformly to f on $\mathbf{R}$.

(iv)

$$f(x) = \lim_{n\to\infty} e^{-nx^2} = \begin{cases} 1, & x = 0 \\ 0, & x \neq 0. \end{cases}$$

$\{f_n\}$ does not converge uniformly to f.

2. (i) $f(x) = \lim\limits_{n\to\infty} f_n(x) = 0$, since $\lim\limits_{n\to\infty} x^n = 0$ for $0 \le x < 1$ and $x^n = x^{2n}$ for $x = 1$. The maximum of $f_n(x) - f(x)$ occurs when

$$nx^{n-1} - 2nx^{2n-1} = 0$$

$$x = 1/\sqrt[n]{2};$$

the maximum is $1/2 - 1/4 = 1/4$. So convergence is not uniform.

(ii) We have

$$f(x) = \lim_{n\to\infty} \frac{nx}{1+n+x} = x.$$

Since

$$f(x) - f_n(x) = x - \frac{nx}{1+n+x}$$

is close to $x - n$ for large x, convergence is not uniform.

(iii) $f(x) = \lim\limits_{n\to\infty} f_n(x) = x$. We could write $f(x) - f_n(x) = x - \sqrt{x^2 + 1/n^2}$ as a fraction (as in the hint), by multiplying and dividing by $x + \sqrt{x^2 + 1/n^2}$. Actually, it's easier to apply the Mean Value Theorem:

$$f(x) - f_n(x) = \sqrt{x^2} - \sqrt{x^2 + \frac{1}{n^2}}$$

$$= \frac{1}{2n^2\sqrt{\xi}}$$

for some ξ with

$$x^2 < \xi < x^2 + \frac{1}{n^2},$$

hence $x < \sqrt{\xi}$, or

$$1/x > 1/\sqrt{\xi}.$$

So

$$f(x) - f_n(x) < \frac{1}{n^2} \cdot \frac{1}{2x}$$

$$< \frac{1}{2n^2 a}.$$

So convergence is uniform.

(iv) $f(x) = |x|$. As in (iii) we have

$$|f(x) - f_n(x)| < \frac{1}{2n^2a} \qquad \text{on } (-\infty, a] \text{ and } [a, \infty).$$

On $[-a, a]$ we have

$$|f(x) - f_n(x)| < a + \sqrt{a^2 + \frac{1}{n^2}}$$

$$< 3a \qquad \text{for } n \text{ large.}$$

So by first taking a small, and then n sufficiently large, we can make $|f(x) - f_n(x)|$ as small as desired on all of **R**.

(v) $f(x) = 0$ and

$$f(x) - f_n(x) = \sqrt{x} - \sqrt{x + \frac{1}{n}}$$

$$= \frac{1}{2n\sqrt{\xi}} \qquad x < \xi < x + \frac{1}{n}$$

$$< \frac{1}{2nx}$$

$$< \frac{1}{2na}.$$

So convergence is uniform.

(vi) Convergence is still uniform, arguing as in (iv).

(vii) We have

$$(*) \qquad f_n(x) = n\left(\sqrt{x + \frac{1}{n}} - \sqrt{x}\right) = n \cdot \frac{1}{2n\sqrt{\xi}}, \qquad x < \xi < x + \frac{1}{n}$$

so

$$f(x) = \lim_{n\to\infty} f_n(x) = \frac{1}{2\sqrt{x}}.$$

Moreover, on $[a, \infty)$ we have

$$f(x) - f_n(x) = \frac{1}{2\sqrt{x}} - n\left(\sqrt{x + \frac{1}{n}} - \sqrt{x}\right)$$

$$= \frac{1}{2\sqrt{x}} - \frac{1}{2\sqrt{\xi}} \qquad a < x < \xi < x + \frac{1}{n}$$

$$= \frac{\sqrt{\xi} - \sqrt{x}}{2\sqrt{x}\sqrt{\xi}}$$

$$< \frac{\sqrt{\xi} - \sqrt{x}}{2a} \to 0, \qquad \text{by } (*).$$

So convergence is uniform.

(viii) For $x = 0$ we have

$$f_n(0) = n\sqrt{\frac{1}{n}} = \sqrt{n},$$

so $\lim\limits_{n\to\infty} f_n(x)$ does not exist. (Convergence cannot even be uniform on $(0, \infty)$, for each f_n is bounded, so certainly not close to $f(x) = 1/2\sqrt{x}$ near 0.)

3. (ii) $\log(-a) - \dfrac{x}{a} - \dfrac{x^2}{2a^2} - \dfrac{x^3}{3a^3} - \dfrac{x^4}{4a^4} - \cdots$.

(iv) $\displaystyle\sum_{k=0}^{\infty} \binom{-1/2}{k} x^{2k}$.

4. (ii) $1/(1 + x^3)$.

5. (i) Since

$$\cos x = \sum_{n=0}^{\infty} \frac{(-1)^n x^{2n}}{(2n)!},$$

the sum is $\cos(2\pi) = 1$.

(ii) Since

$$e^x = \sum_{n=0}^{\infty} \frac{x^n}{n!}$$

$$e^{-x} = \sum_{n=0}^{\infty} \frac{(-1)^n x^n}{n!},$$

we have

$$\sum_{n=0}^{\infty} \frac{x^{2n}}{(2n)!} = \frac{e^x + e^{-x}}{2},$$

so the sum is $(e + e^{-1})/2$.

(iii) Since

$$\log(1 + x) = \sum_{n=1}^{\infty} \frac{(-1)^{n+1} x^n}{n}$$

$$\log(1 - x) = -\sum_{n=1}^{\infty} \frac{x^n}{n},$$

we have

$$\sum_{n=0}^{\infty} \frac{x^{2n+1}}{2n + 1} = \frac{1}{2}[\log(1 + x) - \log(1 - x)],$$

so the sum is

$$\frac{1}{2}\left[\log\frac{3}{2} - \log\frac{1}{2}\right] = \frac{1}{2}\log\left(\frac{3}{2}\bigg/\frac{1}{2}\right) = \frac{1}{2}\log 3.$$

(iv) If

$$f(x) = \sum_{n=0}^{\infty} nx^n,$$

then

$$\frac{f(x)}{x} = \sum_{n=1}^{\infty} nx^{n-1} = \left(\sum_{n=1}^{\infty} x^n\right)'$$

$$= \left(\frac{x}{1-x}\right)' = \frac{1}{(1-x)^2},$$

so

$$f(x) = \frac{x}{(1-x)^2},$$

so the sum is

$$\frac{(1/2)}{(1-1/2)^2} = 2.$$

(v) If

$$f(x) = \sum_{n=0}^{\infty} \frac{x^n}{n+1},$$

then

$$xf(x) = \sum_{n=0}^{\infty} \frac{x^{n+1}}{n+1} = -\log(1-x)$$

so

$$f(x) = -\frac{\log(1-x)}{x},$$

so the sum is

$$\frac{-\log(1-1/3)}{1/3} = -3\log(2/3).$$

(vi) If

$$f(x) = \sum_{n=0}^{\infty} \frac{(2n+1)x^n}{n!},$$

then

$$f(x^2) = \sum_{n=0}^{\infty} \frac{(2n+1)x^{2n}}{n!}$$

$$= \sum_{n=0}^{\infty} \left(\frac{x^{2n+1}}{n!}\right)'$$

$$= \left(x \sum_{n=0}^{\infty} \frac{x^{2n}}{n!}\right)'$$

$$= \left(x \sum_{n=0}^{\infty} \frac{(x^2)^n}{n!}\right)'$$

$$= \left(xe^{x^2}\right)'$$

$$= 2x^2 e^{x^2} + e^{x^2},$$

so

$$f(x) = 2xe^x + e^x,$$

so the sum is

$$2 \cdot \frac{1}{2} \cdot e^{1/2} + e^{1/2} = 2e^{1/2}.$$

7. (a)

$$(1+x)f'(x) = (1+x)\sum_{n=1}^{\infty} n\binom{\alpha}{n} x^{n-1}$$

$$= \sum_{n=0}^{\infty} \left[n\binom{\alpha}{n} + (n+1)\binom{\alpha}{n+1} \right] x^n$$

$$= \sum_{n=0}^{\infty} \alpha \binom{\alpha}{n} x^n = \alpha f(x).$$

(b)

$$g'(x) = \frac{(1+x)^\alpha f'(x) - f(x)\alpha(1+x)^{\alpha-1}}{(1+x)^{2\alpha}} = 0 \qquad \text{by part (a),}$$

so g is a constant c. Thus $f(x) = c(1+x)^\alpha$. Since $f(0) = 1$, we have $c = 1$.

8. The maximum and minimum of

$$g(x) = \frac{x}{n(1+nx^2)}$$

occur when

$$0 = n(1 + nx^2) - 2n^2x^2$$
$$x = \pm 1/\sqrt{n}$$

so the maximum and minimum have absolute value

$$\frac{1}{2n^{3/2}}.$$

Since

$$\sum_{n=1}^{\infty} \frac{1}{n^{3/2}}$$

converges, by the Integral Test, we can apply the Weierstrass M-test.

9. (a) For $x \geq a$ we have

$$2^n \sin \frac{1}{3^n x} = \frac{1}{x} \left(\frac{2}{3} \right)^n \frac{\sin \left(\dfrac{1}{3^n x} \right)}{\dfrac{1}{3^n x}}$$

$$\leq \frac{1}{a} \left(\frac{2}{3} \right)^n \frac{\sin \left(\dfrac{1}{3^n x} \right)}{\dfrac{1}{3^n x}}.$$

Since $x \geq a$, and since $\lim\limits_{h \to 0} (\sin h)/h = 1$, for sufficiently large n we will have

$$\left| \frac{\sin \left(\dfrac{1}{3^n x} \right)}{\dfrac{1}{3^n x}} \right| \leq 2,$$

hence

$$\left| 2^n \sin \frac{1}{3^n x} \right| \leq \frac{2}{a} \left(\frac{2}{3} \right)^n.$$

Since $\sum\limits_{n=1}^{\infty} \left(\frac{2}{3} \right)^n$ converges, we can apply the Weierstrass M-test.

(b) For $x = 2/(\pi 3^N)$ the terms

$$\sin \frac{1}{3^N x}, \quad \sin \frac{1}{3^{N+1} x}, \quad \sin \frac{1}{3^{N+2} x}, \quad \cdots$$

are

$$\sin \frac{\pi}{2}, \quad \sin \frac{\pi}{6}, \quad \sin \frac{\pi}{18}, \quad \cdots$$

all ≥ 0 with $\sin \pi/2 = 1$. So

$$\sum_{n=N}^{\infty} 2^n \sin \frac{1}{3^n x} \geq 2^N, \qquad \text{when } x = \frac{2}{\pi 3^N}.$$

So $\sum_{n=N}^{\infty}$ cannot be made arbitrarily small on $(0, \infty)$ simply by choosing N sufficiently large, i.e., the sum does not converge uniformly.

10. (a) The maximum of $x/(1 + n^4 x^2)$ on $[0, \infty)$ occurs at $x = 1/n^2$, and the function decreases to 0 after that. So if $a > 0$, then for all but a finite number of n we have

$$\frac{nx}{1 + n^4 x^2} \leq \frac{na}{1 + n^4 a^2} \leq \frac{1}{n^3 a} \qquad \text{on } [a, \infty).$$

Since $\sum 1/n^3$ converges we can apply the Weierstrass M-test.

(b) For $f(x) = \sum_{n=0}^{\infty} nx/(1 + n^4 x^2)$ we have

$$f\left(\frac{1}{N}\right) \geq \sum_{n \geq \sqrt{N}} \frac{n\dfrac{1}{N}}{1 + n^4 \dfrac{1}{N^2}} \qquad \text{since all terms are positive.}$$

Moreover, for $n \geq \sqrt{N}$ we have

$$n^4 \frac{1}{N^2} \geq 1$$

so

$$1 + n^4 \frac{1}{N^2} \leq 2n^4 \frac{1}{N^2},$$

so

$$f\left(\frac{1}{N}\right) \geq \sum_{n \geq \sqrt{N}} \frac{n\dfrac{1}{N}}{2n^4 \dfrac{1}{N^2}} = \frac{N}{2} \sum_{n \geq \sqrt{N}} \frac{1}{n^3}.$$

If M is the smallest integer $\geq \sqrt{N}$, then

$$\sum_{n \geq \sqrt{N}} \frac{1}{n^3} \geq \sum_{n=M}^{\infty} \frac{1}{n^3} \geq \int_M^{\infty} \frac{dx}{x^3} = \frac{1}{2M^2}$$

so

$$\frac{N}{2} \sum_{n \geq \sqrt{N}} \frac{1}{n^3} \geq \frac{N}{2} \cdot \frac{1}{2M^2} = \frac{1}{4} \frac{N}{M^2},$$

rather than $1/4$ as stated. But this is hardly significant: obviously $N/M^2 > 1/2$ for large enough N, so we always have $f(1/N) \geq 1/8$ for large enough N. Consequently, the series cannot converge uniformly to 0 on $[0, \infty)$.

(c) This series is much easier. The maximum and minimum of $nx/(1 + n^5x^2)$ occur at $n = \pm n^{-5/2}$, and the maximum absolute value is $\frac{1}{2}n^{-3/2}$. Since $\sum n^{-3/2}$ converges, the series converges uniformly on all of **R**.

11. (a) Problem 15-33(c) shows that for x in $[\varepsilon, 2\pi - \varepsilon]$ we have

$$|\sin x + \cdots + \sin nx| \leq \frac{1}{\left|\sin \dfrac{x}{2}\right|} \leq \frac{1}{\left|\sin \dfrac{\varepsilon}{2}\right|}.$$

Problem 19-35(b) then shows that

$$\left|\sum_{i=k}^{n} \frac{\sin kx}{k}\right| \leq \frac{1}{k\left|\sin \dfrac{\varepsilon}{2}\right|}$$

for x in $[\varepsilon, 2\pi - \varepsilon]$. So we just need the following result:

Suppose that $\sum\limits_{n=0}^{\infty} f_n(x)$ satisfies a "uniform Cauchy condition" on an interval $[a, b]$, i.e., for every $\varepsilon > 0$ there is some N such that

$$(*) \qquad\qquad |f_{n+1}(x) + \cdots + f_m(x)| < \varepsilon$$

for all $n, m > N$ and all x in $[a, b]$. Then $\sum\limits_{n=0}^{\infty} f_n(x)$ converges uniformly to some f on $[a, b]$

Proof: Certainly $\sum\limits_{n=0}^{\infty} f_n(x)$ converges to some $f(x)$ for each x in $[a, b]$ since $(*)$ shows that for each x the sum $\sum\limits_{n=0}^{\infty} f_n(x)$ satisfies the Cauchy criterion. Now given $\varepsilon > 0$, choose N so that $(*)$ holds for $\varepsilon/2$ and all x in $[a, b]$. Then for $n > N$ we have

$$\left|f(x) - \sum_{k=0}^{n} f_k(x)\right| = \left|\sum_{k=n+1}^{\infty} f_k(x)\right| \leq \varepsilon$$

for all x in $[a, b]$, which shows that $\sum\limits_{k=0}^{\infty} f_k(x)$ converges uniformly to f on $[a, b]$.

(b) The terms

$$\sin\left(k\frac{\pi}{N}\right) \qquad k = N, \ldots, 2N$$

can be written as

$$\sin\left([N + k]\frac{\pi}{N}\right) = \sin\left(\pi + k\frac{\pi}{N}\right) \qquad k = 0, \ldots, N$$

and hence are the negatives of the positive terms

$$\sin k\frac{\pi}{N} \qquad k = 0, \ldots, N.$$

Now for $x = \pi/N$ we have

$$\sum_{k=0}^{N} \sin kx = \frac{\sin\left(\dfrac{N+1}{2} \cdot \dfrac{\pi}{N}\right) \sin\left(\dfrac{N}{2} \cdot \dfrac{\pi}{N}\right)}{\sin \dfrac{\pi}{2N}} \qquad \text{by Problem 15-33.}$$

For N large this is

$$\geq \frac{1}{2} \cdot \frac{1 \cdot 1}{\dfrac{\pi}{2N}} = \frac{N}{\pi}.$$

So for $x = \pi/N$ and N large we have

$$\left| \sum_{k=N}^{2N} \frac{\sin kx}{k} \right| = \sum_{k=N}^{2N} \frac{-\sin kx}{k}$$

$$\geq \frac{1}{2N} \sum_{k=N}^{2N} -\sin kx$$

$$\geq \frac{1}{2N} \sum_{k=0}^{N} \sin kx$$

$$\geq \frac{1}{2N} \cdot \frac{N}{\pi} = \frac{1}{2\pi}.$$

This shows that we *cannot* have a uniform Cauchy condition for the series on $[0, 2\pi]$, so it cannot converge uniformly.

12. (a) $a_n = f^{(n)}(0)/n! = 0$.

(b) $a_0 = f(0) = \lim_{n \to \infty} f(x_n) = 0$, since f is continuous at 0. Thus

$$f(x) = \sum_{n=1}^{\infty} a_n x^n = x\left(\sum_{n=1}^{\infty} a_n x^{n-1}\right) = x\left(\sum_{n=0}^{\infty} a_{n+1} x^n\right) = xg(x).$$

Now $g(x_n) = 0$ (for all $x_n \neq 0$), so by the result just proved, $a_1 = 0$. Thus

$$f(x) = x^2 \sum_{n=0}^{\infty} a_{n+2} x^n,$$

so $a_2 = 0$, etc.

(c) Apply part (b) to $f - g$.

13. If f is even, then $f^{(n)}$ is odd for n odd, so $a_n = f^{(n)}(0)/n! = 0$ for n odd. If f is odd, then $f^{(n)}$ is odd for n even, so $a_n = 0$ for n even.

14. The power series for $f(x) = \log(1-x)$ is $\sum_{n=0}^{\infty}(-1)^n a_n x^n$, where $\sum_{n=0}^{\infty} a_n x^n$ is the power series for $h(x) = \log(1+x)$. Since $\sum_{n=0}^{\infty} a_n x^n$ converges only for

$-1 < x \le 1$, the power series $\sum_{n=0}^{\infty}(-1)^n a_n x^n$ converges only for $-1 \le x < 1$. Since $g(x) = f(x) - h(x)$, its power series converges only for $-1 < x < 1$.

15. (a) Clearly $a_{n-1} \le a_n$. Hence

$$\frac{a_{n+1}}{a_n} \le \frac{2a_n}{a_n} \le 2.$$

(b)

$$\frac{|a_{n+1}x^n|}{|a_n x^{n-1}|} \le 2|x| < 1 \qquad \text{for } |x| < 1/2.$$

(c) We have

$$f(x) = \sum_{n=1}^{\infty} a_n x^{n-1} = 1 + x + 2x^2 + 3x^3 + \cdots,$$

$$xf(x) = \sum_{n=1}^{\infty} a_n x^n = x + x^2 + 2x^3 + \cdots,$$

$$x^2 f(x) = \sum_{n=1}^{\infty} a_n x^{n+1} = x^2 + x^3 + \cdots,$$

so

$$f(x) = 1 + xf(x) = x^2 f(x).$$

(d) Let $\alpha = (-1 - \sqrt{5})/2$ and $\beta = (-1 + \sqrt{5})/2$. Then

$$f(x) = \frac{-1}{x^2 + x - 1} = \frac{1/\sqrt{5}}{x - \left(\dfrac{-1 - \sqrt{5}}{2}\right)} + \frac{-1/\sqrt{5}}{x - \left(\dfrac{-1 + \sqrt{5}}{2}\right)}$$

$$= \frac{1/\sqrt{5}}{x - \alpha} - \frac{1/\sqrt{5}}{x - \beta}$$

$$= \frac{1}{\sqrt{5}}\left(-\frac{1}{\alpha} - \frac{x}{\alpha^2} - \frac{x^2}{\alpha^3} - \cdots\right)$$

$$\quad - \frac{1}{\sqrt{5}}\left(-\frac{1}{\beta} - \frac{x}{\beta^2} - \frac{x^2}{\beta^3} - \cdots\right).$$

(e) Consequently,

$$a_n = \frac{1}{\sqrt{5}} \left(\frac{1}{\beta^n} - \frac{1}{\alpha^n} \right)$$

$$= \frac{1}{\sqrt{5}} \left[\frac{1}{\left(\dfrac{-1 + \sqrt{5}}{2} \right)^n} - \frac{1}{\left(\dfrac{-1 - \sqrt{5}}{2} \right)^n} \right]$$

$$= \frac{1}{\sqrt{5}} \left[\frac{\left(\dfrac{-1 - \sqrt{5}}{2} \right)^n - \left(\dfrac{-1 + \sqrt{5}}{2} \right)^n}{\left(\dfrac{-1 - \sqrt{5}}{2} \right)^n \cdot \left(\dfrac{-1 + \sqrt{5}}{2} \right)^n} \right]$$

$$= \frac{1}{\sqrt{5}} \left[\frac{\left(\dfrac{-1 - \sqrt{5}}{2} \right)^n - \left(\dfrac{-1 + \sqrt{5}}{2} \right)^n}{\left(\dfrac{1 - 5}{4} \right)^n} \right]$$

$$= \frac{1}{\sqrt{5}} \left[\left(\dfrac{-1 + \sqrt{5}}{2} \right)^n - \left(\dfrac{-1 - \sqrt{5}}{2} \right)^n \right].$$

16. If we have

$$f(x)g(x) = \sum_{n=0}^{\infty} c_n x^n,$$

then

$$c_n = \frac{1}{n!} (f \cdot g)^{(n)}(0)$$

$$= \frac{1}{n!} \sum_{k=0}^{n} \binom{n}{k} f^{(k)}(0) \cdot g^{(n-k)}(0)$$

$$= \frac{1}{n!} \sum_{k=0}^{n} \frac{n!}{k!(n-k)!} f^{(k)}(0) \cdot g^{(n-k)}(0)$$

$$= \sum_{k=0}^{n} \frac{f^{(k)}(0)}{k!} \cdot \frac{g^{(n-k)}(0)}{(n-k)!}$$

$$= \sum_{k=0}^{n} a_k b_{n-k}.$$

17. (a) We have

$$|b_n x_0{}^n| = \left| \sum_{k=0}^{n-1} b_k a_{n-k} x_0{}^n \right| \le \sum_{k=0}^{n-1} \left| b_k x_0{}^k a_{n-k} x_0{}^{n-k} \right|$$

$$\le M \sum_{k=0}^{n-1} |b_k x_0{}^k|.$$

(b) By induction on n. It is clear for $n = 0$. Suppose true for all numbers $< n$. Then by part (a)

$$|b_n x_0{}^n| \le M \sum_{k=0}^{n-1} M^{2k}$$

$$= M \sum_{k=0}^{n-1} (M^2)^k$$

$$= M \cdot \frac{M^{2n-2} - 1}{M^2 - 1}.$$

Since $M \ge \sqrt{2}$ we have $M^2 \ge 2$, so $M^2 - 1 \ge 1$, so

$$|b_n x_0{}^n| \le M(M^{2n-2} - 1)$$
$$\le M^{2n-1}$$
$$\le M^{2n}.$$

(c) We have

$$|b_n x^n| = |b_n x_0{}^n| \cdot \frac{|x|^n}{|x_0|^n}$$

$$\le \left(\frac{M^2 |x|}{|x_0|} \right)^n,$$

so if

$$\frac{|x|}{|x_0|} \le \frac{1}{2M^2}$$

we have

$$|b_n x^n| \le \frac{1}{2^n},$$

so $\sum |b_n x^n|$ converges.

18. On $[-a, a]$ the series

$$-\log(1 - x) = x + \frac{x^2}{2} + \frac{x^3}{3} + \cdots$$

converges uniformly and absolutely, so the same is true of the series

$$\sum_{n=0}^{\infty} \frac{x^{2n+1}}{2n+1} \quad \text{and} \quad \sum_{n=0}^{\infty} \frac{x^{2n+2}}{2n+2}$$

consisting of the odd powers, and even powers, respectively. Hence the same is true of

$$\sum_{n=0}^{\infty} \frac{x^{n+1}}{2n+2},$$

which is just the value of the second series at $\sqrt{x}$. It is easy to check that because convergence is uniform and absolute, we can rearrange the series

$$\sum_{n=0}^{\infty} \frac{x^{2n+1}}{2n+1} - \frac{x^{n+1}}{2n+2}.$$

The odd powers can be paired as

$$\frac{x^{2n+1}}{2n+1} - \frac{x^{(2n)+1}}{2(2n)+2} = \frac{x^{2n+1}}{2n+1}\left(1 - \frac{1}{2}\right)$$
$$= \frac{1}{2} \cdot \frac{x^{2n+1}}{2n+1},$$

while the even powers appear once as

$$-\frac{x^{n+1}}{2n+2} = -\frac{1}{2} \cdot \frac{x^{2n+1}}{n+1} \qquad (n \text{ odd}),$$

thus giving altogether $1/2$ the sum of the terms in the series for $\log(1+x)$. But for $x = 1$ we have

$$\sum_{n=0}^{\infty} \frac{1}{2n+1} - \frac{1}{2n+2} = \frac{1}{1} - \frac{1}{2} + \frac{1}{3} - \frac{1}{4} + \frac{1}{5} - \frac{1}{6} \cdots$$
$$= \log 2.$$

19. If $0 \leq x \leq 1$, then $x \geq x^2 \geq x^3 \geq \cdots \geq 0$. Consequently Abel's Lemma shows that $|a_m x^m + \cdots + a_n x^n| < \varepsilon$ if $|a_m + \cdots + a_n| < \varepsilon$. The latter condition is true for sufficiently large m and n. Consequently, $|a_m x^m + a_{m+1} x^{m+1} + \cdots | < \varepsilon$ for sufficiently large m and all x in $[0, 1]$. This means that for all x in $[0, 1]$,

$$\left| \sum_{n=0}^{\infty} a_n x^n - (a_0 + \cdots + a_{m-1} x^{m-1}) \right| < \varepsilon$$

for sufficiently large m. This is precisely the assertion that $\sum_{n=0}^{\infty} a_n x^n$ converges uniformly in $[0, 1]$.

20. Let $a_n = (-1)^n$. For $0 \le x < 1$, we have

$$\sum_{n=0}^{\infty} a_n x^n = 1 - x + x^2 - x^3 + \cdots = \frac{1}{1+x},$$

so

$$\lim_{x \to 1^-} \sum_{n=0}^{\infty} a_n x^n = \lim_{x \to 1^-} \frac{1}{x+1} = \frac{1}{2}.$$

21. (a) (i) By Problem 4(iii) we have

$$(1+x)\log(1+x) - x = \frac{x^2}{2 \cdot 1} - \frac{x^3}{3 \cdot 2} + \frac{x^4}{4 \cdot 3} - \cdots \qquad \text{for } |x| < 1.$$

Now the series on the right does converge for $x = 1$, so by Problem 19,

$$\frac{1}{2 \cdot 1} - \frac{1}{3 \cdot 2} + \frac{1}{4 \cdot 3} - \cdots = 2\log 2 - 2.$$

(ii) If

$$f(x) = x - \frac{x^4}{4} + \frac{x^7}{7} - \frac{x^{10}}{10} + \cdots,$$

then for $|x| < 1$ we have

$$f'(x) = 1 - x^3 + x^6 - x^9 + \cdots = \frac{1}{1+x^3} \qquad \text{by Problem 4(ii)}$$

$$= \frac{1}{(x+1)(x^2 - x + 1)}$$

$$= \frac{1/3}{x+1} + \frac{-x/3 + 2/3}{x^2 - x + 1},$$

$$= \frac{1/3}{x+1} - \frac{1}{6} \cdot \frac{2x - 1}{x^2 - x + 1} + \frac{1}{2} \cdot \frac{1}{x^2 - x + 1},$$

so

$$f(x) = \frac{1}{3} \log(x+1) - \frac{1}{6} \log(x^2 - x + 1)$$

$$+ \frac{3\sqrt{3}}{16} \arctan \frac{x - 1/2}{\sqrt{3/4}} + \frac{3\sqrt{3}}{16} \arctan \frac{1/3}{\sqrt{3/4}}$$

$$= \frac{1}{3} \log(x+1) - \frac{1}{6} \log(x^2 - x + 1)$$

$$+ \frac{3\sqrt{3}}{16} \arctan \frac{x - 1/2}{\sqrt{3/4}} + \frac{3\sqrt{3}}{16} \frac{\pi}{6}.$$

Consequently,

$$1 - \frac{1}{4} + \frac{1}{7} - \frac{1}{10} + \cdots = \lim_{x \to 1} f(x) = \frac{\log 2}{3} + \frac{\sqrt{3}}{16} \pi.$$

(b) Let

$$f(x) = \sum_{n=0}^{\infty} a_n x^n \qquad |x| < 1$$

$$g(x) = \sum_{n=0}^{\infty} b_n x^n \qquad |x| < 1.$$

Then, as on page 505 of the text,

$$\sum_{n=0}^{\infty} c_n x^n = \left(\sum_{n=0}^{\infty} a_n x^n \right) \cdot \left(\sum_{n=0}^{\infty} b_n x^n \right) \qquad |x| < 1.$$

It follows from Problem 19 that

$$\sum_{n=0}^{\infty} c_n = \lim_{x \to 1^-} \left(\sum_{n=0}^{\infty} a_n x^n \right) \cdot \left(\sum_{n=0}^{\infty} b_n x^n \right)$$

$$= \sum_{n=0}^{\infty} a_n \cdot \sum_{n=0}^{\infty} b_n.$$

22. (a) Choose N so that if $n \geq N$, then $|f(x) - f_n(x)| \leq 1$ for all x in $[a, b]$. Since f_N is bounded, there is some M such that $|f_N(x)| \leq M$ for all x in $[a, b]$. Then $|f(x)| \leq |f_N(x)| + 1 \leq M + 1$.

(b) Let $f_n(x) = nx$ for $0 \leq x \leq 1/\sqrt{n}$, and $f(x) = 1/x$ for $1/\sqrt{n} \leq x \leq 1$. Then $\lim_{n \to \infty} f_n(0) = 0$ and $\lim_{n \to \infty} f_n(x) = 1/x$ for $0 < x \leq 1$.

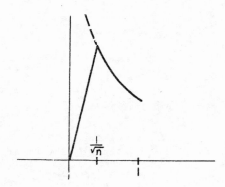

23. Let $f_n(x) = [f(x + 1/n) - f(x)]/(1/n)$.

24. Let $\{a_n\}$ be the sequence

$$0, \ 1, \ \tfrac{1}{2}, \ \tfrac{1}{3}, \ \tfrac{2}{3}, \ \tfrac{1}{4}, \ \tfrac{3}{4}, \ \tfrac{1}{5}, \ \tfrac{2}{5}, \ \tfrac{3}{5}, \ \tfrac{4}{5}, \ \tfrac{1}{6}, \ \tfrac{5}{6}, \dots.$$

Let $f_n(x) = 0$ if $x \neq a_1, \ldots, a_n$, and let $f_n(a_j) = 1$ for $1 \leq j \leq n$.

25. (a) Given $\varepsilon > 0$ choose f_n so that

$$(*) \qquad |f_n(x) - f(x)| < \frac{\varepsilon}{3(b-a)} \qquad \text{for } x \text{ in } [a, b].$$

Also choose a partition P of $[a, b]$ such that

$$(1) \qquad U(f_n, P) - L(f_n, P) < \frac{\varepsilon}{3}.$$

It follows from $(*)$ that

$$(2) \qquad |U(f_n, P) - U(f, P)| \leq \frac{\varepsilon}{3},$$

$$(3) \qquad |L(f_n, P) - L(f, P)| \leq \frac{\varepsilon}{3}.$$

From (1), (2) and (3) we obtain

$$|U(f, P) - L(f, P)| < \varepsilon.$$

(b) The hypotheses of Theorem 3 say that $\{f_n'\}$ converges uniformly to g, and the proof shows that $g = f'$. Thus, $\{f_n'\}$ converges uniformly to f'. Now since

$$f(x) = f(a) + \int_a^x f'(x)\, dx$$

$$f_n(x) = f_n(a) + \int_a^x f_n'(x)\, dx$$

we have

$$|f(x) - f_n(x)| \leq |f_n(a) - f(a)| + \int_a^x |f_n'(x) - f'(x)|\, dx$$

$$\leq \frac{\varepsilon}{2} + \int_a^x \frac{\varepsilon}{2(b-a)}\, dx \qquad \text{for large enough } n$$

$$< \varepsilon.$$

(c) Since we are still assuming that f_n' converges uniformly to g we have for any x in $[a, b]$

$$\int_{x_0}^x g = \lim_{n \to \infty} \int_{x_0}^x f_n'$$

$$= \lim_{n \to \infty} [f_n(x) - f_n(x_0)].$$

Since $l = \lim_{n \to \infty} f_n(x_0)$ exists, it follows that

$$f(x) = \lim_{n \to \infty} f_n(x) \text{ exists}$$

and

$$\int_{x_0}^x g = f(x) - l,$$

so $f' = g$. Notice that this proof works even if $[a, b]$ is replaced by an infinite interval.

(d) If

$$s_n(x) = \sum_{j=1}^n \frac{(-1)^j}{x + j},$$

then $\{s_n(0)\}$ converges to $\sum (-1)^j/j$. Moreover,

$$s_n'(x) = \sum_{j=1}^n \frac{(-1)^{j+1}}{(x + j)^2}$$

and $\{s_n'\}$, i.e., the series

$$\sum_{j=1}^\infty \frac{(-1)^{j+1}}{(x + j)^2},$$

converges uniformly on $[0, \infty)$ by the Weierstrass M-test. So by part (c), $\{s_n\}$ converges uniformly on $[0, \infty)$, i.e., the series

$$\sum_{j=1}^\infty \frac{(-1)^j}{x + j}$$

converges uniformly on $[0, \infty)$.

26. Since we have

$$\lim_{n \to \infty} \int_0^1 f_n = \int_0^1 f,$$

we just have to show that

$$\lim_{n \to \infty} \int_{1-1/n}^1 f_n = 0.$$

First of all, choose $\delta > 0$ so that $|f(x) - f(1)| < 1$ on $[1-\delta, 1]$. Then for sufficiently large n we also have $|f_n(x) - f(1)| < 2$ on $[1 - \delta, 1]$. Then $|f_n(x)| < f(1) + 2$ on $[1 - \delta, 1]$, so for $1 - \delta \le x_0 < 1$ we have

$$\left| \int_{x_0}^1 f_n(x)\, dx \right| \le \int_{x_0}^1 |f_n(x)|\, dx$$

$$\le \int_{1-\delta}^1 |f_n(x)|\, dx$$

$$\le \delta \cdot (f(1) + 2).$$

So we just have to choose δ so that this product is $< \varepsilon$, and then n so large that $1 - \delta < 1 - 1/n$.

This is not true if convergence is not uniform. If $\{f_n\}$ is the sequence of functions in Figure 4 on page 492 of the text, then

$$g_n(x) = f_n(1 - x)$$

give a counterexample.

27. (a) If convergence is not uniform, then for some $\varepsilon > 0$ there are arbitrarily large n with $f_n(x) \geq \varepsilon$ for some x on $[a, b]$. So we can choose distinct $x_1, x_2, x_3, \ldots$ and $n_1 < n_2 < n_3 < \cdots$ with

$$f_{n_k}(x_k) \geq \varepsilon.$$

Some subsequence of the $\{x_k\}$ converges to a point x in $[a, b]$; simply by throwing away terms from the original sequence and renumbering we can assume that the original sequence $x_k \to x$. Now $f_n(x) \to f(x) = 0$ so there is some n such that $f_n(x) < \varepsilon$. Since f_n is continuous we have $f_n(y) < \varepsilon$ for all y close enough to x. Hence, in particular,

$$f_n(x_k) < \varepsilon$$

for large enough k. But if k is also large enough so that $n_k > n$ then

$$f_{n_k}(x_k) \leq f_n(x_k) < \varepsilon,$$

a contradiction.

(b) Apply part (a) to the functions $\{f_n - f\}$.

(c) The functions in Figure 1 on page 491 of the text give a counterexample on $[0, 1]$ when f isn't continuous. They also give a counterexample on the open interval $(0, 1)$, with $f = 0$.

28. (a) Since $x_n \to x$ and f is continuous, for any $\varepsilon > 0$ for large enough n we have

(1) $$|f(x) - f_n(x)| < \varepsilon/2.$$

Moreover, for large enough n we have

$$|f(y) - f_n(y)| < \varepsilon/2$$

for all y on $[a, b]$, and in particular

(2) $$|f(x_n) - f_n(x_n)| < \varepsilon/2.$$

Adding (1) and (2) we obtain

$$|f(x) - f_n(x_n)| < \varepsilon.$$

(b) No, in fact, just choose all f_n to be some function f which is not continuous at x, and choose $x_n \to x$ such that $f(x_n) \to f(x)$ is false.

(c) Choose x_n as in the hint, so that

(*) $|f_n(x_n) - f(x_n)| > \varepsilon.$

By the Bolzano-Weierstrass theorem some subsequence of the $\{x_n\}$ converges. By throwing away terms of the sequence and renumbering, assume that $x_n \to x$. Then by assumption $f_n(x_n) \to f(x)$, so (*) gives

$$\varepsilon \le \lim_{n\to\infty} |f_n(x_n) - f(x_n)| = |f(x) - f(x)| = 0.$$

29. (a) Suppose $\{u_0, \ldots, u_m\}$ contains $\{t_0, \ldots, t_n\}$. For each i we have

$$t_i = u_\alpha < u_{\alpha+1} < \cdots < u_\beta = t_{i+1}$$

for some $u_\alpha, \ldots, u_\beta$. Then f has the constant value s_i on each $(u_{\alpha+j-1}, u_{\alpha+j})$. Thus the sum $\sum_{i=1}^{n} s_i (t_i - t_{i-1})$ is the sum $\sum_{j=1}^{m} s_j'(u_j - u_{j-1})$ where s_j' is the constant value of f on (u_{j-1}, u_j). To deal with the general case, consider a partition containing both $\{u_0, \ldots, u_m\}$ and $\{t_0, \ldots, t_n\}$.

(b) Choose N so that for $n > N$ we have $|f(x) - s_n(x)| < \varepsilon/2$ for all x in $[a, b]$.

(c) From $|s_n(x) - s_m(x)| < \varepsilon$ it follows easily that

$$\left| \int_a^b s_n - \int_a^b s_m \right| < \varepsilon(b - a).$$

(d) Choose N so that for $n > N$ we have both $|f(x) - s_n(x)| < \varepsilon/2$ and $|f(x) - s_m(x)| < \varepsilon/2$ for all x in $[a, b]$.

(e) For any $\varepsilon > 0$, choose N so that if $n > N$, then

$$\left| \lim_{n\to\infty} \int_a^b s_n - \int_a^b s_n \right| < \frac{\varepsilon}{3},$$

$$\left| \lim_{n\to\infty} \int_a^b t_n - \int_a^b t_n \right| < \frac{\varepsilon}{3},$$

$$|s_n(x) - t_n(x)| < \frac{\varepsilon}{3(b - a)} \qquad \text{for all } x \text{ in } [a, b].$$

The last equation implies that

$$\left| \int_a^b s_n - \int_a^b t_n \right| < \frac{\varepsilon}{3}.$$

It follows that $\left| \lim_{n\to\infty} \int_a^b s_n - \lim_{n\to\infty} \int_a^b t_n \right| < \varepsilon.$

(f) Let

$A = \{\, y : a \le y \le b$ and there is a step function s

on $[a, y]$ such that $|f(x) - s(x)| < \varepsilon$ for all x in $[a, y]\,\}$.

Let $\alpha = \sup A$. Since f is continuous at α, there is a $\delta > 0$ such that $|f(x) - f(\alpha)| < \varepsilon$ for $|x - a| < \delta$. There is some y in A with $\alpha - \delta < y < \alpha$. Thus there is a step function s defined on $[a, y]$ with $|f(x) - s(x)| < \varepsilon$ for all x in $[a, y]$. Define $s_1(x) = s(x)$ for x in $[a, y]$ and $s_1(x) = f(\alpha)$ for $y < x \le \alpha$. Then s_1 is a step function defined on $[a, \alpha]$ with $|f(x) - s_1(x)| < \varepsilon$ for all x in $[a, \alpha]$. This shows that α is in A. Similarly, if $\alpha < b$, then pick δ as before, and let s be a step function defined on $[a, \alpha]$ with $|f(x) - s(x)| < \varepsilon$ for x in $[a, \alpha]$. If $s_1(x)$ is defined as $s(x)$ for x in $[a, \alpha]$ and as $f(\alpha)$ for $\alpha < x \le \alpha + \delta/2$, then $|f(x) - s_1(x)| < \varepsilon$ for x in $[a, \alpha + \delta/2]$. So $\alpha + \delta/2$ is in A, contradicting the definition of α. So $\alpha = b$, which completes the proof.

[The class of regulated functions can be determined more explicitly, as follows. A step function S has the property that $\lim\limits_{x \to a^+} s(x)$ and $\lim\limits_{x \to a^-} s(x)$ exist for all a. It is not hard to show that a uniform limit of step functions must have the same property (the proof is a simple modification of the proof of Theorem 2). The converse is also true—if f has right and left hand limits at every point, then f is regulated. Notice that the class of regulated functions is smaller than the class of integrable functions. For example, if $f(0) = 0$ and $f(x) = \sin 1/x$ for $0 < x \le 1$, then f is integrable on $[0, 1]$ (by Problem 13-19, for example), but is not regulated.]

30. The function f_n is shown below. The length of each f_n is 2, since two sides of an equilateral triangle have a total length of twice the other side.

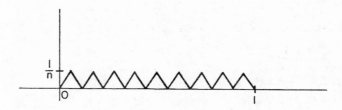

CHAPTER 25

1. (ii) $|(3+4i)^{-1}| = 1/|3+4i| = 1/5;\ \theta = -\text{argument of } 3+4i = -\arctan 4/3.$

(iv) $|\sqrt[7]{3+4i}| = \sqrt[7]{|3+4i|} = \sqrt[7]{5};\ \theta = (\arctan 4/3)/7.$

2. (ii) $(x^2)^2 + x^2 + 1 = 0$, so

$$x^2 = \frac{-1 \pm \sqrt{1-4}}{2}$$

$$= \frac{-1+\sqrt{3}i}{2} \quad \text{or} \quad \frac{-1-\sqrt{3}i}{2}$$

$$= \cos\frac{2\pi}{3} + i\sin\frac{2\pi}{3} \quad \text{or} \quad \cos\frac{4\pi}{3} + i\sin\frac{4\pi}{3}.$$

So x is one of the square roots of these numbers, so x is one of

$$\cos\frac{\pi}{3} + i\sin\frac{\pi}{3} = \frac{1}{2} + \frac{\sqrt{3}}{2}i,$$

$$\cos\frac{4\pi}{3} + i\sin\frac{4\pi}{3} = -\frac{1}{2} - \frac{\sqrt{3}}{2}i,$$

$$\cos\frac{2\pi}{3} + i\sin\frac{2\pi}{3} = -\frac{1}{2} + \frac{\sqrt{3}}{2}i,$$

$$\cos\frac{5\pi}{3} + i\sin\frac{5\pi}{3} = \frac{1}{2} - \frac{\sqrt{3}}{2}i.$$

(iv)

$$x = \frac{7}{3} - \frac{4i}{3},$$

$$y = \frac{1}{3} + 2i.$$

3. (ii) All z with $|z| = 1$.

(iv) The ellipse consisting of all points the sum of whose distances from a and b is c, if $c > |a-b|$; the line segment between a and b if $|a-b| = c$; $\emptyset$ if $|a-b| > c$.

6. For one value of $\sqrt{-i}$ the point $z\sqrt{-i}$ is obtained by rotating z by an angle of $-\pi/2$, so the diagonal goes into the real axis under multiplication by $\sqrt{-i}$. Similarly, for one value of $\sqrt{i}$, multiplication by $\sqrt{i}$ is rotation by $\pi/2$. So $\sqrt{i} \cdot z\sqrt{-i}$ is obtained by rotating the plane until the diagonal lies along the real axis, then reflecting through the real axis, and then rotating back by the same amount. Hence

z and $\sqrt{i} \cdot z\sqrt{-i}$ are reflections of each other through the diagonal; we obtain the negative of this answer for the other choices of $\sqrt{-i}$ and $\sqrt{i}$.

7. (a) Since $a_0, \ldots, a_{n-1}$ are real, we have

$$0 = \overline{(a + bi)^n + a_{n-1}(a + bi)^{n-1} + \cdots + a_0}$$
$$= \left(\overline{a + bi}\right)^n + a_{n-1}\left(\overline{a + bi}\right)^{n-1} + \cdots + a_0$$
$$= (a - bi)^n + a_{n-1}(a - bi)^{n-1} + \cdots + a_0.$$

(b) Since $a + bi$ and $a - bi$ are roots, $z^n + a_{n-1}z^{n-1} + \cdots + a_0$ is divisible by $z - (a + bi)$ and $z - (a - bi)$, and by their product

$$[z - (a + bi)] \cdot [z - (a - bi)] = z^2 - 2az + (a^2 + b^2).$$

8. (a) Suppose that $a + b\sqrt{c} = a' + b'\sqrt{c}$. If $b = b'$, then also $a = a'$. If $b \neq b'$, then we would have $\sqrt{c} = (a - a')/(b - b')$, contradicting the fact that $\sqrt{c}$ is irrational (Problem 2-17).

(b) The proofs are almost exactly the same as parts (1)–(6) of Theorem 1.

(c) Since $a_0, \ldots, a_{n-1}$ are integers, we have

$$0 = \overline{\left(a + b\sqrt{c}\right)^n + a_{n-1}\left(a + b\sqrt{c}\right)^{n-1} + \cdots + a_0}$$
$$= \left(\overline{a + b\sqrt{c}}\right)^n + a_{n-1}\left(\overline{a + b\sqrt{c}}\right)^{n-1} + \cdots + a_0.$$

9. The 4th roots of i are

$$\cos\theta + i\sin\theta$$

for

$$\theta = \frac{\pi}{8}, \quad \frac{\pi}{8} + \frac{\pi}{2}, \quad \frac{\pi}{8} + \pi, \quad \frac{\pi}{8} + \frac{3\pi}{2}.$$

We have

$$\cos\frac{\pi}{4} = \sin\frac{\pi}{4} = \frac{\sqrt{2}}{2};$$

using Problem 15-15(b) we then have

$$\cos\frac{\pi}{8} = \sqrt{\frac{1 + \sqrt{2}/2}{2}} = \frac{\sqrt{2 + \sqrt{2}}}{2}$$

$$\sin\frac{\pi}{8} = \sqrt{\frac{1 - \sqrt{2}/2}{2}} = \frac{\sqrt{2 - \sqrt{2}}}{2},$$

so

$$\sqrt[4]{i} = \frac{\sqrt{2 + \sqrt{2}}}{2} + \frac{i\sqrt{2 - \sqrt{2}}}{2}.$$

10. (a) If $\omega^n = 1$, then $(\omega^k)^n = \omega^{nk} = (\omega^n)^k = 1$.

(b) There are two primitive 3rd roots and 4 primitive 5th roots (in each case, all roots except 1); there are two primitive 4th roots (i and $-i$) and six primitive 9th roots (if ω is the root with smallest argument, then 1, ω^3, and ω^6 are not primitive). [In general, the number of primitive n^{th} roots is the number of numbers from 1 to $n - 1$ that have no factor in common with n.]

(c) By Problem 2-5,

$$1 + \omega + \cdots + \omega^{n-1} = \frac{1 - \omega^n}{1 - \omega} = 0.$$

11. (a) The assertion is clear if the line is the real axis, because in that case the imaginary parts of $z_1, \ldots, z_k$ are either all positive or all negative, so the same is true for the sum. In general, let θ be the angle between the line and the real axis, and let $w = \cos\theta + i\sin\theta$. Then $z_1 w^{-1}, \ldots, z_k w^{-1}$ all lie on one side of the real axis, so the same is true of $z_1 w^{-1} + \cdots + z_k w^{-1} = (z_1 + \cdots + z_k)w^{-1}$, which shows that $z_1 + \cdots + z_k$ lies on the corresponding side of the original line.

(b) z^{-1} is above the real axis if and only if z is below the real axis, and conversely. This proves the assertion when the line is the real axis. The general case then follows as in part (a).

12. The hypotheses remain true when each z_j is multiplied by the same w. So we can assume that z_1 is real, in fact, that $z_1 = 1$. It follows that $z_2 + z_3$ is real, so $z_2 = a + bi$, $z_3 = a - bi$. Moreover, $2a + 1 = 0$, so $a = -1/2$; since $a^2 + b^2 = 1$, we have $b = \sqrt{3}/2$. The points 1, $-1/2 + i\sqrt{3}/2$, and $-1/2 - i\sqrt{3}/2$ do lie on the vertices of an equilateral triangle.

CHAPTER 26

1. (a) If $|x - x_0| < \delta$, then $|\alpha(x) - \alpha(x_0)| < \delta$. Similarly, if $|y - y_0| < \delta$, then $|\beta(y) - \beta(y_0)| < \delta$.

(b) $g = f \circ \alpha$; $h = f \circ \beta$.

2. (a) g is a continuous real-valued function on $[0, 1]$ with $g(0) = f(z)$ and $g(1) = f(w)$. So g takes on all values between $f(z)$ and $f(w)$ on $[0, 1]$.

(b) Let $f(x + iy) = x + i(y + x^2)$, on $[0, 1] \times [-1, 0]$.

3. (a) There is, by the Fundamental Theorem of Algebra, some number z_1 such that $z_1{}^n + a_{n-1}z_1{}^{n-1} + \cdots + a_0 = 0$. Then

$$z^n + a_{n-1}z^{n-1} + \cdots + a_0 = (z - z_1)(z^{n-1} + b_{n-2}z^{n-2} + \cdots + b_0)$$

for some numbers $b_0, \ldots, b_{n-2}$ (as in Problem 3-7). Using an inductive argument, we can assume that

$$z^{n-1} + b_{n-2}z^{n-2} + \cdots + b_0 = \prod_{i=2}^{n}(z - z_i)$$

for some numbers $z_2, \ldots, z_n$.

(b) According to Problem 25-7, the non-real numbers $z_1, \ldots, z_n$ from part (a) occur in paris which are conjugates of each other, and $(z - z_i)(z - \bar{z}_i)$ has real coefficients.

4. (a) is obvious.

(b) If $f = \sum_{i=1}^{n} h_i{}^2$ and $g = \sum_{j=1}^{m} k_j{}^2$, then

$$fg = \sum_{i=1}^{m}\sum_{j=1}^{m}(h_i k_j)^2.$$

(c) If $f(a) = 0$, then $f(x) = (x - a)f_1(x)$ for some polynomial function f_1. Then $f_1(x) \geq 0$ for $x \geq a$, and $f_1(x) \leq 0$ for $x \leq a$. So $f_1(a) = 0$. Thus every root of

393

f is a double root, so $f(x) = \displaystyle\prod_{i=1}^{k} (x - a_i)^2 g(x)$ where $g(x) > 0$ for all x. Since g has no roots, Problem 3 shows that g is a product of quadratic factors $x^2 + ax + b$ without roots. Thus $a^2 - 4b < 0$, so we can write

$$x^2 + ax + b = \left(x + \frac{a}{2}\right)^2 + \left(\sqrt{b - a^2/4}\right)^2,$$

which is a sum of squares. So f is a product of sums of squares, so f is a sum of squares.

5. (a) Follow the procedure given in the hint, to obtain a decreasing sequence of rectangles $[a_i, b_i] \times [c_i, d_i]$, each containing infinitely many points of A. By the Nested Intervals Theorem (Problem 8-14), there is a point x in all $[a_i, b_i]$ and a point y in all $[c_i, d_i]$. Then $z = (x, y) = x + iy$ is in all $[a_i, b_i] \times [c_i, d_i]$. If $\varepsilon > 0$, then for some i the set $[a_i, b_i] \times [c_i, d_i]$ is contained in $\{a : |z - a| < \varepsilon\}$, so there are infinitely many points of A in $\{a : |z - a| < \varepsilon\}$.

(b) If f were not bounded on $[a, b] \times [c, d]$, then there would be points a_n in $[a, b] \times [c, d]$ with $|f(a_n)| \geq N$. If z is a limit point of $\{a_n : n$ in $\mathbf{N}\}$, then for every $\varepsilon > 0$ there are points a_n with $|a_n - z| < \varepsilon$, so $f(a_n) \geq N$. This contradicts the fact that f is continuous at z.

(c) Let $\alpha = \sup\{f(z) : z$ in $[a, b] \times [c, d]\}$; this exists by part (b). If $\alpha \neq f(z)$ for all z in $[a, b] \times [c, d]$, then $g(z) = 1/(f(z) - \alpha)$ would be a continuous unbounded function on $[a, b] \times [c, d]$.

6. (a) If $c = \alpha + \beta i$, then $z = a + bi$ satisfies $z^2 = c$ if and only if

$$a^2 - b^2 = \alpha,$$
$$2ab = \beta,$$

which can be solved to give

$$a = \sqrt{2\alpha + 2\sqrt{\alpha^2 + \beta^2}} \\ b = \dfrac{\beta}{2\sqrt{2\alpha + 2\sqrt{\alpha^2 + \beta^2}}} \quad \text{or} \quad a = -\sqrt{2\alpha + 2\sqrt{\alpha^2 + \beta^2}} \\ b = \dfrac{-\beta}{2\sqrt{2\alpha + 2\sqrt{\alpha^2 + \beta^2}}}$$

(b) If $n = 2k$, then a solution of $z^k - \sqrt{c} = 0$ will be a solution of $z^{2n} - c = 0$. (If k is even, we can continue until we reach an odd number.)

(c) For this f we have

$$g(z) = f(z + z_0) = (z + z_0)^n - c = (z_0{}^n - x) + (nz_0)z + \cdots.$$

(d) Suppose, for example, that $-c = \alpha + \beta i$ with $\alpha, \beta > 0$. If $\delta^n < \alpha$, then $|-c - \delta^n| < |-c|$. The same argument works for all other cases.

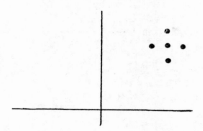

7. (a)

$$f'(x) = \sum_{\alpha=1}^{k} m_\alpha (z - z_1)^{m_1} \cdots (z - z_\alpha)^{m_\alpha - 1} \cdots (z - z_k)^{m_k}$$

$$= \sum_{\alpha=1}^{k} m_\alpha (z - z_1)^{m_1} \cdots (z - z_\alpha)^{m_\alpha} \cdots (z - z_k)^{m_k} (z - z_\alpha)^{-1}$$

$$= (z - z_1)^{m_1} \cdots (z - z_k)^{m+k} \cdot \sum_{\alpha=1}^{k} m_\alpha (z - z_\alpha)^{-1}.$$

(b) If $z_1, \ldots, z_k$ did all lie on the same side of a straight line through z, then $z - z_1, \ldots, z - z_k$ would all lie on the same side of a straight line through 0. The same would then be true of $m_1^{-1}(z - z_1), \ldots, m_k^{-1}(z - z_k)$, since $m_1, \ldots m_k > 0$. By Problem 25-11, this would imply that $g(z) \neq 0$, a contradiction.

(c) If z satisfied $f'(z) = 0$ but z were not in the convex hull of the set $\{z_1, \ldots, z_k\}$, then there would be a straight line through z containing the points $z_1, \ldots, z_k$. This contradicts part (b).

8. The proof is exactly the same as for real-valued functions defined on **R**.

9. (a) Let $z_0 = x_0 + iy_0$. Since

$$\alpha + i\beta = f'(z_0) = \lim_{z \to 0} \frac{f(z_0 + z) - f(z_0)}{z},$$

it must be true, in particular, that for real δ we have

$$\alpha + i\beta = \lim_{\delta \to 0} \frac{f(z_0 + \delta) - f(z_0)}{\delta}$$

$$= \lim_{\delta \to 0} \left[\frac{g(x_0 + \delta) - g(x_0)}{\delta} + i\frac{h(x_0 + \delta) - h(x_0)}{\delta} \right]$$

$$= g'(x_0) + ih'(x_0),$$

so $\alpha = g'(x_0)$ and $\beta = h'(x_0)$.

(b) We also have

$$
\begin{aligned}
\alpha + i\beta &= \lim_{\delta \to 0} \frac{f(x_0 + \delta i) - f(z_0)}{\delta i} \\
&= \lim_{\delta \to 0} \left[\frac{k(y_0 + \delta) - k(y_0)}{\delta i} + i\frac{l(y_0 + \delta) - l(y_0)}{\delta i} \right] \\
&= \frac{k'(y_0)}{i} + l'(y_0),
\end{aligned}
$$

so $l'(y_0) = \alpha$ and $k'(y_0) = -\beta$.

(c) Part (b) shows that u and v are constant along horizontal and vertical lines.

10. (a)

$$
f^{(k)}(x) = \frac{1}{2i} \left(\frac{(-1)^k k!}{(x - i)^{k+1}} - \frac{(-1)^k k!}{(x + i)^{k+1}} \right).
$$

(b)

$$
\begin{aligned}
\arctan^{(k)}(0) &= f^{(k-1)}(0) \\
&= \frac{(-1)^{k-1}(k - 1)!}{2i} \left(\frac{1}{(-1)^k} - \frac{1}{i^k} \right) \\
&= \frac{(-1)^{k-1}(k - 1)!}{2i} i^k [1 + (-1)^{k-1}].
\end{aligned}
$$

[If k is even, then $\arctan^{(k)} = 0$. If $k = 2l + 1$, then $\arctan^{(2l+1)}(0) = (2l)!(-1)^l$.]

CHAPTER 27

1. (ii) Absolutely convergent.

(iv) Absolutely convergent, since $|1/2 + i/2| = \sqrt{2}/2 < 1$.

2. (ii) The limit

$$\lim_{n\to\infty} \frac{|z|^{n+1}/(n+1)}{|z|^n/n} = |z| \lim_{n\to\infty} \frac{n}{n+1} = |z|$$

is < 1 for $|z| < 1$ and > 1 for $|z| > 1$.

(iv) The limit

$$\lim_{n\to\infty} \frac{|z|^{n+1}(n+1+2^{-n-1})}{|z|^n(n+2^{-n})} = |z| \lim_{n\to\infty} \frac{n+1+2^{-n-1}}{n+2^{-n}} = |z|$$

is < 1 for $|z| < 1$ and > 1 for $|z| > 1$.

3. (ii) Since

$$\lim_{n\to\infty} \frac{\sqrt[n]{n!}\,|z|}{\sqrt[n]{n^n}} = |z| \lim_{n\to\infty} \frac{\sqrt[n]{n!}}{n} = \frac{|z|}{e} \qquad \text{(by Problem 22-13)},$$

the radius of convergence is e.

(iv) Since

$$\lim_{n\to\infty} \frac{\sqrt[n]{n^2}\,|z|}{\sqrt[n]{2^n}} = |z| \lim_{n\to\infty} \frac{\left(\sqrt[n]{n}\right)^2}{2} = \frac{|z|}{2} \qquad \text{(by Problem 22-1(vi))},$$

the radius of convergence is 2.

4. (a) Since $\overline{\lim}_{n\to\infty} \sqrt[n]{|a^n|}\,|z| = \overline{\lim}_{n\to\infty} \sqrt[n]{|a_n z^n|}$, Problem 23-7 shows that the series $\sum_{n=0}^{\infty} a_n z^n$ converges (absolutely).

(b) If $\overline{\lim}_{n\to\infty} \sqrt[n]{|a_n z^n|} = 1 + \varepsilon$ for $\varepsilon > 0$, then there are infinitely many n with $\sqrt[n]{|a_n z^n|} \geq 1 + \varepsilon/2$, so $|a_n z^n| \geq (1 + \varepsilon/2)^n$ for infinitely many n, so the terms $a_n z^n$ are unbounded.

(c) Since the terms $\sqrt[n]{|a_n|}$ are unbounded, the same is true for $\sqrt[n]{|a_n z^n|}$ for $z \neq 0$. This is all the more true for $|a_n z^n|$, so $\sum_{n=0}^{\infty} a_n z^n$ diverges.

5. If z is on the unit circle, then $|z^n/n^2| \leq 1/n^2$, so $\sum_{n=1}^{\infty} |z_n|/n^2$ converges by the comparison test.

The series $\sum_{n=1}^{\infty} z^n$ certainly diverges for $z = 1$. If $z \neq 1$, then by Problem 2-5

$$\sum_{n=1}^{N} z^n = z \cdot \frac{1-z^N}{1-z}.$$

If $\lim\limits_{N \to \infty} \sum_{n=1}^{N} z^n$ existed, then $\lim\limits_{N \to \infty} z^N$ would exist, which is impossible, since $l = \lim\limits_{N \to \infty} z^N$ would imply that $zl = \lim\limits_{N \to \infty} z^{N+1} = l$, or $z = 1$.

The series $\sum_{n=1}^{\infty} z^n/n$ diverges for $z = 1$ and converges for $z = -1$.

6. (a) We have the absolutely convergent series

$$e^z = \sum_{k=0}^{\infty} \frac{z^k}{k!}, \qquad e^w = \sum_{k=0}^{\infty} \frac{w^k}{k!}.$$

Theorem 23-9 holds just as well for complex series, so $e^z \cdot e^w$ is given by any sum containing all pairs of products. In particular, we can choose the Cauchy product $\sum_{n=0}^{\infty} c_n$, where

$$c_n = \sum_{k=0}^{n} \frac{z^k}{k!} \frac{w^{n-k}}{(n-k)!}.$$

But this is exactly the power series for

$$e^{z+w} = \sum_{n=0}^{\infty} \frac{(z+w)^n}{n!},$$

since

$$\frac{(z+w)^n}{n!} = \sum_{k=0}^{n} \frac{\binom{n}{k} z^k w^{n-k}}{n!}$$

$$= \sum_{k=0}^{\infty} \frac{n!}{k!(n-k)!} \frac{z^k w^{n-k}}{n!}$$

$$= \sum_{k=0}^{\infty} \frac{z^k}{k!} \frac{w^{n-k}}{(n-k)!}.$$

(b)

$\sin z \cos w + \cos z \sin w$

$$= \left(\frac{e^{iz} - e^{-iz}}{2i} \right) \left(\frac{e^{iw} + e^{-iw}}{2} \right) + \left(\frac{e^{iz} + e^{-iz}}{2} \right) \left(\frac{e^{iw} - e^{-iw}}{2i} \right)$$

$$= \frac{e^{i(z+w)} - e^{-i(z+w)}}{2i} = \sin(z + w)$$

$$\cos z \cos w - \sin z \sin w$$

$$= \left(\frac{e^{iz} - e^{-iz}}{2}\right)\left(\frac{e^{iw} + e^{-iw}}{2}\right) - \left(\frac{e^{iz} + e^{-iz}}{2i}\right)\left(\frac{e^{iw} - e^{-iw}}{2i}\right)$$

$$= \frac{e^{i(z+w)} + e^{-i(z+w)}}{2i} = \cos(z + w)$$

7. (a) Since $e^{iy} = \cos y + i \sin y$, the problem is just a restatement of Problem 15-22.

(b)

$$|e^{x+iy}| = |e^x \cdot e^{iy}| = |e^x| \cdot |\cos y + i \sin y| = |e^x|.$$

8. (a) If $z \neq 0$, then $z = r(\cos \theta + i \sin \theta)$ for some $r > 0$. Then $\exp(\log r + i\theta) = z$.

(b) We have $\sin z = w$ when

$$\frac{e^{iz} - e^{-iz}}{2i} = w,$$

$$(e^{iz})^2 - 2iwe^{iz} - 1 = 0,$$

$$e^{iz} = \frac{2iw \pm \sqrt{-4w + 4}}{2}$$

$$= iw \pm \sqrt{1 - w}.$$

This equation can always be solved for z, by part (a), since $iw \pm \sqrt{1 - w} \neq 0$.

9. (i) We have

$$\cos z = 1 - \frac{z^2}{2!} + \frac{z^4}{4!} - \cdots$$

and if we write

$$\frac{1}{\cos z} = 1 + a_2 z^2 + a_4 z^4 + \cdots$$

then we find

$$a_2 - \frac{1}{2!} = 0$$

$$a_4 - \frac{a_2}{2!} + \frac{1}{4!} = 0$$

leading to

$$\frac{1}{\cos z} = 1 + \frac{z^2}{2} + \frac{5}{24}z^4 + \cdots .$$

Hence

$$\tan z = \left(1 + \frac{z^2}{2} + \frac{5}{24}z^4 + \cdots\right)\left(z - \frac{z^3}{6} + \frac{z^5}{120} + \cdots\right)$$

$$= z + \frac{1}{3}z^3 + \frac{2}{15}z^5 + \cdots.$$

(ii) If we write

$$\sqrt{1-z} = 1 + a_1 z + a_2 z^2 + \cdots$$

then

$$1 - z = 1 + 2a_1 z + (2a_2 + a_1{}^2)z^2 + \cdots$$

so $2a_1 = -1$ and $2a_2 + a_1{}^2 = 0$, hence

$$\sqrt{1-z} = 1 - \frac{1}{2}z - \frac{1}{8}z^2 + \cdots.$$

Then if we write

$$\frac{1}{\sqrt{1-z}} = 1 + b_1 z + b_2 z^2 + \cdots$$

we have

$$b_1 - \frac{1}{2} = 0$$

$$b_2 - \frac{b_1}{2} - \frac{1}{8} = 0,$$

so

$$\frac{1}{\sqrt{1-z}} = 1 + \frac{1}{2}z + \frac{3}{8}z^2 + \cdots.$$

We could also get this from the binomial series (Problem 24-7), which holds for $|z| < 1$:

$$(1-z)^{-1/2} = 1 + \binom{-1/2}{1}(-z) + \binom{-1/2}{2}z^2 + \cdots$$

$$= 1 + \frac{1}{2}z + \frac{(-1/2)(-1/2 - 1)}{2}z^2 + \cdots.$$

Finally, we have

$$z(1-z)^{-1/2} = z + \frac{1}{2}z^2 + \frac{3}{8}z^3 + \cdots.$$

(iii)

$$e^{\sin z} - 1 = \left(z - \frac{z^3}{6} + \cdots\right) + \frac{\left(z - \frac{z^3}{6} + \cdots\right)^2}{2} + \frac{(z - \cdots)^3}{6} + \cdots$$

$$= z - \frac{z^3}{6} + \frac{z^2}{2} - \frac{z^4}{6} + \frac{z^3}{6} + \cdots$$

$$= z + \frac{z^2}{2} - \frac{z^4}{6} + \cdots$$

so

$$\frac{e^{\sin z} - 1}{z} = 1 + \frac{z}{2} - \frac{z^3}{6} + \cdots.$$

(iv) We have

$$\log(1 + z) = z - \frac{z^2}{2} + \frac{z^3}{3} - \cdots$$

(we know there will be some power series for $\log(1 + z)$, so it must be this one, since this works for z real). Hence

$$\log(1 - z) = -z - \frac{z^2}{2} - \frac{z^3}{3} - \cdots$$

so

$$\log(1 - z^2) = -z^2 - \frac{z^4}{2} - \frac{z^6}{3} - \cdots.$$

(v)

$$\sin^2 z = \left(z - \frac{z^3}{6} + \frac{z^5}{120} - \cdots\right)\left(z - \frac{z^3}{6} + \frac{z^5}{120} + \cdots\right)$$

$$= z^2 - \frac{z^4}{3} + \frac{2z^6}{45} + \cdots$$

so

$$\frac{\sin^2 z}{z^2} = 1 - \frac{z^2}{3} + \frac{2z^4}{45} + \cdots.$$

(vi) From (i) we have

$$\frac{1}{\cos z} = 1 + \frac{z^2}{2} + \frac{5}{24}z^4 + \cdots$$

so

$$\frac{1}{\cos^2 z} = 1 + z^2 + \frac{2}{3}z^4 + \cdots$$

so

$$\frac{\sin(z^2)}{z\cos^2 z} = \left(z - \frac{z^5}{6} + \frac{z^9}{120} + \cdots\right)\left(1 + z^2 + \frac{2}{3}z^4 + \cdots\right)$$

$$= z + z^3 + \frac{1}{2}z^5 + \cdots.$$

(vii) We have

$$\frac{1}{z^4 - 2z^2 + 3} = \frac{1}{3} \cdot \frac{1}{1 - \left(\dfrac{2z^2}{3} - \dfrac{z^4}{3}\right)}$$

$$= \frac{1}{3}\left[1 + \left(\frac{2z^2}{3} - \frac{z^4}{3}\right) + \left(\frac{2z^2}{3} - \frac{z^4}{3}\right)^2 + \cdots\right]$$

$$= \frac{1}{3} + \frac{2z^2}{9} + \frac{z^4}{27} + \cdots.$$

(viii) We have

$$\sqrt{1+z} = (1+z)^{1/2} = 1 + \binom{1/2}{1}z + \binom{1/2}{2}z^2 + \binom{1/2}{3}z^3 + \cdots$$

$$= 1 + \frac{1}{2}z - \frac{1}{8}z^2 + \frac{1}{16}z^3 - \frac{5}{128}z^4 + \cdots$$

so

$$e^{(\sqrt{1+z}-1)} = \left(\frac{1}{2}z - \frac{1}{8}z^2 + \frac{1}{16}z^3 - \frac{5}{128}z^4 + \cdots\right)$$

$$+ \frac{\left(\dfrac{1}{2}z - \dfrac{1}{8}z^2 + \cdots\right)^2}{2} + \frac{\left(\dfrac{1}{2}z - \dfrac{1}{8}z^2 + \cdots\right)^3}{6} + \cdots$$

$$= \frac{1}{2}z + \left(-\frac{1}{8} + \frac{1}{8}\right)z^2 + \left(\frac{1}{16} - \frac{1}{16} + \frac{1}{48}\right)z^3$$

$$+ \left(-\frac{5}{128} + \frac{1}{128} - \frac{1}{64}\right)z^4 + \cdots$$

$$= \frac{1}{2}z + \frac{1}{48}z^3 - \frac{3}{64}z^4 + \cdots$$

so

$$\frac{1}{z}\left[e^{(\sqrt{1+z}-1)}\right] = \frac{1}{2} + \frac{1}{48}z^2 - \frac{3}{64}z^3 + \cdots.$$

10. (a) This follows from Problem 26-9 with $y_0 = 0$.

(b) This follows from (a) by induction.

(c) Applying (b) we have, with $a_n = 1$, for real x,

$$\sum_{k=0}^{n} a_k \left[\bar{u}^{(k)}(x) + i\bar{v}^{(k)}(x) \right] = 0.$$

Since $\bar{u}^{(k)}$ and $\bar{v}^{(k)}$ are real, the real and imaginary parts of the left side are

$$\sum_{k=0}^{n} a_k \bar{u}^{(k)}(x) \qquad \text{and} \qquad \sum_{k=0}^{n} a_k \bar{v}^{(k)}(x),$$

so these must both be 0.

(d) If $a = b + ci$ is a complex root of $z^n + a_{n-1}z^{n-1} + \cdots + a_0 = 0$, then as in Problem 18-42, the function $f(z) = e^{az} = e^{bz} \cos cz + ie^{bz} \sin cz$ satisfies (∗). So (assuming the a_i are real) it follows form (c) that $g(x) = e^{bx} \cos cx$ and $h(x) = e^{bk} \sin cx$ also satisfy (∗).

11. (a), (b) $e^z = e^{x+iy} = e^x(\cos y + i \sin y) = w$ means that $e^x = |w|$, so $x = \log |w|$, and y is an argument of w. In particular $e^{x+iy_0} = e^{x+iy_1}$ if y_0 and y_1 are arguments for w (for example, $e^0 = e^{2\pi i}$), so exp is not one-one.

(c) Suppose that there were a continuous function log defined for $|z| = 1$ such that $\exp(\log(z)) = z$ for all z with $|z| = 1$. Then we could write $\log(z) = \alpha(z) + i\beta(z)$ for continuous real-valued functions α and β. We must have $\alpha(z) = 0$ for $|z| = 1$, and $\beta(z)$ an argument of z for $|z| = 1$. This contradicts the fact that there is no continuous argument function.

(d) If θ is an argument for a, then one logarithm of a is

$$\log |a| + i\theta.$$

It is easy to see that

$$e^{m[\log |a| + i\theta]}$$

is indeed the product of $e^{\log |a| + i\theta}$ with itself m times, i.e., a^m. Moreover, any other logarithm of a is

$$\log |a| + i\theta + i \cdot 2k\pi$$

for some k, and

$$e^{m[\log |a| + i\theta + i \cdot 2k\pi]} = a^m \cdot e^{2km\pi i}$$

$$= a^m, \qquad \text{since } m \text{ is an integer and } e^{2\pi i} = 1.$$

(e) As we know from the proof of Theorem 25-2, the nth roots of a have absolute values $\sqrt[n]{|a|}$ and arguments

$$\frac{1}{n}(\theta + 2k\pi)i, \qquad k = 0, \ldots, n - 1.$$

So they are

$$e^{\log \sqrt[n]{|a|} + \frac{1}{n}(\theta + 2k\pi)i} = e^{\frac{1}{n}[\log |a| + (\theta + 2k\pi)i]}, \qquad k = 0, \ldots, n - 1.$$

The logarithms of a consist of

$$\log |a| + (\theta + 2k\pi)i.$$

So $a^{m/n}$ consists of all values

$$e^{\frac{m}{n}[\log |a| + (\theta + 2k\pi)i]},$$

where we can clearly consider only $k = 0, \ldots, n - 1$. Then these numbers are b^m for

$$b = e^{\frac{1}{n}[\log |a| + (\theta + 2k\pi)i]},$$

i.e., for b an nth root of a.

(f) The logarithms of a are

$$\log |a| + \begin{cases} 2k\pi i, & a > 0 \\ (2k + 1)\pi i, & a < 0. \end{cases}$$

So the values of a^b are $|a|^b$ times the numbers

$$e^{2kb\pi i} \qquad \left(\text{or } e^{(2k+1)b\pi i} \right).$$

Since b is irrational no two exponents differ by an integral multiple of $2\pi i$, so all these numbers are distinct.

(g) The logarithms of i are numbers of the form $i(2k\pi + \pi/2)$, and the values of i^i are the real numbers

$$e^{-(2k\pi + \pi/2)}.$$

(h) 1^i has the values $e^{i(2k\pi i)} = e^{-2k\pi}$. The logarithms of these real numbers have the values $-2k\pi + 2l\pi i$. So $(1^i)^i$ has the values

$$e^{i(-2k\pi + 2l\pi i)} = e^{-2l\pi}.$$

But 1^{-1} has only the values $e^{-(2k\pi i)} = 1$.

(i) The elements of $a^{b \cdot c}$ are e^{bcz}, where z is a logarithm of a, so that

$$e^z = a.$$

But then

$$e^{bz} = (e^z)^b = a^b,$$

so bz is a logarithm of a^b, so $e^{bcz} = e^{bz \cdot c}$ is an element of $(a^b)^c$. It is not generally true that $a^{b \cdot c} = \left(a^b \right)^c \cap (a^c)^b$, attractive as that hypothesis might seem. It fact, part (h) shows this is false when $a = 1$, $b = c = i$.

12. **(a)** $|x + i| = 1 + x^2$, and an argument for $x + i$ is $\arctan 1/x = \pi/2 - \arctan x$, while an argument for $x - i$ is $\arctan(-1/x) = -\arctan 1/x = -(\pi/2 - \arctan x)$.

(b) From part (a) we obtain

$$\frac{1}{2i}[\log(x-i)-\log(x+i)] = \frac{1}{2i}\cdot -2i\left(\frac{\pi}{2}-\arctan x\right)$$

$$= \arctan x - \pi/2,$$

which differs by a constant from the usual answer, $\arctan x$.

13. (a) Since $a_n - a_m = (b_n - b_m) + i(c_n - c_m)$, we have $|b_n - b_m| \le |a_n - a_m|$ and $|c_n - c_m| \le |a_n - a_m|$, so $\{b_n\}$ and $\{c_n\}$ are Cauchy if $\{a_n\}$ is. Since we also have $|a_n - a_m| \le |b_n - b_m| + |c_n - c_m|$, it follows that $\{a_n\}$ is Cauchy if $\{b_n\}$ and $\{c_n\}$ are.

(b) If $\{a_n\}$ is Cauchy, then $\{b_n\}$ and $\{c_n\}$ are, so $\{b_n\}$ and $\{c_n\}$ converge to α and β, respectively. Thus $\{a_n\}$ converges to $\alpha + i\beta$, by Theorem 1.

(c) The hint is the answer. Since Cauchy sequences of complex numbers are the same as convergent sequences of complex numbers, there is a Cauchy criterion for convergence of complex series: $\sum_{n=1}^{\infty} a_n$ converges if and only if $\lim_{m,n\to\infty} |a_{n+1} + \cdots + a_m| = 0$. Now write down the proofs for the first halves of both Theorems 23-5 and 23-8, interpreting all numbers as complex numbers.

14. (a) We have

$$\sum_{k=1}^{n} e^{ikx} = e^{ix} \cdot \sum_{k=0}^{n-1} (e^{ix})^k$$

$$= e^{ix} \cdot \frac{1 - e^{inx}}{1 - e^{ix}}$$

$$= e^{ix} \cdot \frac{e^{inx/2} \cdot e^{-inx/2}(1 - e^{inx})}{e^{ix/2} \cdot e^{-ix/2}(1 - e^{ix})}$$

$$= e^{ix/2} e^{inx/2} \cdot \frac{e^{-inx/2} - e^{inx/2}}{e^{-ix/2} - e^{ix/2}}$$

$$= e^{i(n+1)x/2} \cdot \frac{\sin\left(\frac{n}{2}x\right)}{\sin\frac{x}{2}},$$

by the formulas on page 555 of the text.

(b) The real and imaginary parts of this equation give

$$\cos x + \cdots + \cos nx = \frac{\sin\left(\frac{n}{2}x\right)}{\sin\frac{x}{2}} \cdot \cos\left(\frac{n+1}{2}x\right)$$

$$\sin x + \cdots + \sin nx = \frac{\sin\left(\frac{n}{2}x\right)}{\sin\frac{x}{2}} \cdot \sin\left(\frac{n+1}{2}x\right).$$

To transform the first, we note that by Problem 15-14 we have

$$2\sin\left(\frac{n}{2}x\right)\cos\left(\frac{n+1}{2}x\right) = \sin\left(\frac{[n+\frac{1}{2}]x + [-\frac{x}{2}]}{2}\right)\cos\left(\frac{[n+\frac{1}{2}]x - [-\frac{x}{2}]}{2}\right)$$

$$= \sin(x + \tfrac{1}{2})x - \sin\frac{x}{2}$$

so

$$\frac{\sin\left(\frac{n}{2}x\right)}{\sin\frac{x}{2}}\cos\left(\frac{n+1}{2}x\right) = \frac{\sin(n+\frac{1}{2})x - \sin\frac{x}{2}}{2\sin\frac{x}{2}}$$

$$= \frac{\sin(n+\frac{1}{2})x}{2\sin\frac{x}{2}} - \frac{1}{2}.$$

15. (a)

$$r_{n+1} = \frac{a_{n+2}}{a_{n+1}} = \frac{a_{n+1} + a_n}{a_{n+1}} = 1 + \frac{a_n}{a_{n+1}} = 1 + \frac{1}{r_n}.$$

(b) If $r = \lim_{n\to\infty} r_n$ exists, then

$$r = \lim_{n\to\infty} r_n = \lim_{n\to\infty} 1 + \frac{1}{r_n} = 1 + \frac{1}{\lim_{n\to\infty} r_n} = 1 + \frac{1}{r},$$

so $r - (1+\sqrt5)/2$ (clearly $r > 0$). To prove that the limit actually exists, note that if $r_n < (1+\sqrt5)/2$, then $r_n{}^2 - r_n - 1 < 0$, so

$$r_n < \frac{2r_n + 1}{r_n + 1} = r_{n+2}.$$

Thus $r_1 < r_3 < r_5 < \cdots < 2$, so $\lim_{n\to\infty} r_{2n+1}$ exists. Similarly, $\lim_{n\to\infty} r_{2n}$ exists. Moreover, the equation $r_{n+2} = (2r_n + 1)/(r_n + 1)$ leads, as before, to the fact that both limits are $(1+\sqrt5)/2$.

(c) The limit

$$\lim_{n\to\infty}\frac{|a_{n+1}z^{n+1}|}{|a_n z^n|}=\lim_{n\to\infty}\frac{a_{n+1}}{a_n}|z|=\frac{1+\sqrt5}{2}|z|$$

is <1 for $|z|<2/(1+\sqrt5)$ and >1 for $|z|>2/(1+\sqrt5)$.

16. (a)

$$-\frac{z}{2}+\frac{z}{2}\cdot\frac{e^z+1}{e^z-1}=\frac{z}{2}\left(-1+\frac{e^z+1}{e^z-1}\right)$$
$$=\frac{z}{e^z-1};$$
$$\frac{e^{-z}+1}{e^{-z}-1}=\frac{(e^{-z}+1)e^z}{(e^{-z}-1)e^z}=\frac{1+e^z}{1-e^z}=\frac{e^z+1}{e^z-1}.$$

These formulas show that $z/(e^z-1)=-z/2+h(z)$ where h is even. Consequently, the power series for h contains only even powers of z. Thus $-1/2=b_1$ is the coefficient of z in the power series for $z/(e^z-1)$, and $b_n=0$ for odd $n>1$.

(b) If $n>1$, then the coefficient of z^n must be 0. But this coefficient is $\sum_{i=0}^{n-1}\binom{n}{i}b_1$.

(c)

$$z\cot z=z\frac{\cos z}{\sin z}=z\cdot\frac{(e^{iz}+e^{-iz})/2}{(e^{iz}-e^{-iz})/2i}$$
$$=\frac{-2iz}{2}\cdot\frac{e^{2iz/2}+e^{-2iz/2}}{e^{2iz/2}-e^{-2iz/2}}$$
$$=\sum_{n=1}^{\infty}\frac{b_{2n}}{(2n)!}(2iz)^{2n}$$
$$=\sum_{n=1}^{\infty}\frac{b_{2n}}{(2n)!}(-1)^n 2^{2n}z^{2n}.$$

(d) From the formula $\tan 2z=2\tan z(1-\tan^2 x)$ (Problem 15-9) we have

$$\cot z-2\cot 2z=\frac{1}{\tan z}-\frac{1-\tan^2 z}{\tan z}=\frac{\tan^2 z}{\tan z}=\tan z.$$

(e)

$$\tan z=\cot z-2\cot 2z$$
$$=\sum_{n=1}^{\infty}\frac{b_{2n}}{(2n)!}(-1)^n 2^{2n}z^{2n-1}-\sum_{n=1}^{\infty}\frac{b_{2n}}{(2n)!}(-1)^n 2^{2n}z^{2n-1}$$
$$=\sum_{n=1}^{\infty}\frac{b_{2n}}{(2n)!}(-1)^{n-1}2^{2n}(2^{2n}-1)z^{2n-1}.$$

17. (a) Applying $(*)$ to $f^{(k)}$ we obtain

$$f^{(k)}(x+1) - f^{(k)}(x) = \sum_{n=1}^{\infty} \frac{f^{(k+n)}(x)}{n!}.$$

Thus

$$\sum_{k=0}^{\infty} \frac{b_k}{k!}[f^{(k)}(x+1) - f^{(k)}(x)] = \sum_{k=0}^{\infty} \sum_{n=1}^{\infty} \frac{b_k}{k!} \frac{f^{(k+n)}(x)}{n!}.$$

The coefficient of $f^{(1)}(x)$ is $b_0/0!1! = 1$. The coefficient of $f^{(j)}(x)$ for $j > 1$ is

$$\sum_{k=0}^{j-1} \frac{b_k}{k!(j-k)!} = \frac{1}{j!} \sum_{k=0}^{j-1} \binom{j}{k} b_k = 0, \qquad \text{by Problem 16(b).}$$

(b)

$$f'(0) + \cdots + f'(n) = \sum_{x=0}^{n} \left(\sum_{k=0}^{\infty} \frac{b_k}{k!}[f^{(k)}(x+1) - f^{(k)}(x)] \right)$$

$$= \sum_{k=0}^{\infty} \frac{b_k}{k!} \sum_{x=0}^{n}[f^{(k)}(x+1) - f^{(k)}(x)]$$

$$= \sum_{k=0}^{\infty} \frac{b_k}{k!}[f^{(k)}(n+1) - f^{(k)}(0)].$$

(c) Let f be a polynomial function with $f' = g$. Then $f(n+1) - f(0) = \int_0^{n+1} g$. Since $b_0 = 1$, part (b) becomes

$$g(0) + \cdots + g(n) = \int_0^{n+1} g(t)\, dt + \sum_{k=1}^{\infty} \frac{b_k}{k!}[g^{(k-1)}(n+1) - g^{(k-1)}(0)].$$

(d) $g^{(k)}(x) = p! x^{p-k}/(p-k)!$ for $k \le p$, so part (c) applied with $n-1$ instead of n gives

$$\sum_{k=1}^{n-1} k^p = \int_0^{n+1} x^p\, dx + \sum_{k=1}^{p+1} \frac{b_k}{k!} \frac{p!}{(p-k+1)!} n^{p-k+1}$$

$$= \frac{n^{p+1}}{p+1} + \sum_{k=1}^{p+1} \frac{b_k}{k} \binom{p}{k-1} n^{p-k+1}.$$

Thus

$$\sum_{k=1}^{n} k^p = \frac{n^{p+1}}{p+1} + n^p + b_1 n^p + \sum_{k=1}^{p+1} \frac{b_k}{k} \binom{p}{k-1} n^{p-k+1}$$

$$= \frac{n^{p+1}}{p+1} + \frac{n^p}{2} + \sum_{k=1}^{p+1} \frac{b_k}{k} \binom{p}{k-1} n^{p-k+1}.$$

18. (a) Clearly $\phi_n(0) = b_n$. If $n > 1$, then

$$\phi_n(1) = \sum_{k=0}^{n} \binom{n}{k} b_{n-k} = \sum_{k=0}^{n} \binom{n}{k} b_k, \qquad \text{since } \binom{n}{k} = \binom{n}{n-k}$$

$$= \sum_{k=0}^{n-1} \binom{n}{k} b_k + b_n = b_n, \qquad \text{by Problem 16(b).}$$

$$\phi_n' = \sum_{k=1}^{n} k \binom{n}{k} b_{n-k} x^{k-1}$$

$$= \sum_{k=0}^{n-1} (k+1) \binom{n}{k+1} b_{n-1+k} x^k$$

$$= \sum_{k=0}^{n-1} n \binom{n-1}{k} b_{n-1+k} x^k$$

$$= n\phi_{n-1}(x).$$

To prove the last equation, note first that

$$\phi_2(1-x) = (1-x)^2 - (1-x) + \frac{1}{6} = x^2 - 2x + 1 - 1 + x + \frac{1}{6}$$

$$= x^2 - x + \frac{1}{6} = \phi_2(x).$$

Now suppose that $\phi_n(x) = (-1)^n \phi_n(1-x)$ for some $n > 1$. Then the function $g(x) = \phi_{n+1}(1-x)$ satisfies

$$g'(x) = -\phi_{n+1}'(1-x) = -(n+1)\phi_n(1-x)$$

$$= (-1)^{n+1}(n+1)\phi_n(x) = (-1)^{n+1}\phi_n'(x).$$

Moreover, $g(0) = \phi_{n+1}(1) = b_{n+1} = \phi_{n+1}(0)$, so $g(x) = (-1)^{n+1}\phi_{n+1}(x)$ for all x.

(b) Substituting from $(*)$ we have

$$\sum_{k=0}^{N} \frac{b_k}{k!} [f^{(k)}(x+1) - f^{(k)}(x)] = \sum_{k=0}^{N} \frac{b_k}{k!} \sum_{n=1}^{N-k} \frac{f^{(k+n)}(x)}{n!} + \sum_{k=0}^{N} \frac{b_k}{k!} R_{N-k}{}^k(x).$$

The coefficient of $f'(x)$ in the double sum is $b_0/0!1! = 1$. For $1 < j \le N$, the coefficient of $f^{(j)}(x)$ is

$$\sum_{k=0}^{j-1} \frac{b_k}{k!(j-k)!} = 0 \qquad \text{by Problem 16(b).}$$

(c) The term $R_{N-k}{}^k(x)$ is the remainder $R_{N-k,x}(x+1)$ for the function $f^{(k)}$. Thus

$$R_{N-k}{}^k(x) = \int_{x}^{x+1} \frac{f^{(k+N-k+1)}(t)}{(N-k)!} (x+1-t)^n \, dt.$$

So

$$\sum_{k=0}^{N} \frac{b_k}{k!} R_{N-k}{}^k(x) = \int_x^{x+1} \sum_{k=0}^{N} \frac{b_k}{k!(N-k)!} (x+1-t)^n f^{(N+1)}(t)\, dt$$

$$= \int_x^{x+1} \frac{\phi_N(x+1-t)}{N!} f^{(N+1)}(t)\, dt.$$

(d) From parts (b) and (c) we obtain

$$f'(x) + \cdots + f'(x+n) = \sum_{k=0}^{N} \frac{b_k}{k!} [f^{(k)}(x+n+1) - f^{(k)}(x)]$$

$$- \sum_{j=0}^{n} \int_{x+j}^{x+j+1} \frac{\phi_N(x+j+1-t)}{N!} f^{(N+1)}(t)\, dt.$$

Applying this to $g = f'$ we obtain the desired formula.

(e) If t is in $[x+j, x+j+1]$, then $x-t$ is in $[-j-1, -j]$. Therefore, by definition, $\psi_N(x-t) = \phi_N(x+j+1-t)$.

19. (a) Apply Problem 18(d) with $g = \log$, $x = 1$, $n-2$ for n and $N = 2$. We obtain

$$\log(n-1)! = \log 1 + \cdots + \log(n-1)$$

$$= \int_1^n \log t\, dt + \frac{b_1}{1!}(\log n - \log 1) + \frac{b_2}{2!}\left(\frac{1}{n} - 1\right)$$

$$+ (-1)^3 \int_1^n \frac{\psi_2(t)}{2!} \cdot -\frac{1}{t^2}\, dt$$

$$= \int_1^n \log t\, dt - \frac{1}{2}\log n + \frac{1}{12}\left(\frac{1}{n} - 1\right) + \int_1^n \frac{\psi_2(t)}{2t^2}\, dt$$

$$= n\log n - n + 1 - \frac{1}{2}\log n + \frac{1}{12}\left(\frac{1}{n} - 1\right) + \int_1^n \frac{\psi_2(t)}{2t^2}\, dt.$$

(b) Consequently,

$$\log n! = \log(n-1)! + \log n = \left(n + \frac{1}{2}\right)\log n - n + \frac{1}{12n} + \frac{11}{12} + \int_1^n \frac{\psi_2(t)}{2t^2}\, dt.$$

So

$$\log\left(\frac{n!}{n^{n+1/2}e^{-n+1/12n}}\right) = \frac{11}{12} + \int_1^n \frac{\psi_2(t)}{2t^2}\, dt.$$

(c) Since ψ_2 is periodic, it is bounded. Thus $\int_1^\infty \psi_2(t)/2t^2\,dt$ exists, since $\int_1^\infty 1/t^2\,dt$ exists. So we have

$$\log\left(\frac{n!}{n^{n+1/2}e^{-n+1/12n}}\right) = \frac{11}{12} + \int_1^\infty \frac{\psi_2(t)}{2t^2}\,dt - \int_n^\infty \frac{\psi_2(t)}{2t^2}\,dt$$

$$= \beta + \frac{11}{12} - \int_n^\infty \frac{\psi_2(t)}{2t^2}\,dt$$

$$= \log\alpha - \int_n^\infty \frac{\psi_2(t)}{2t^2}\,dt,$$

or

$$\log\left(\frac{n!}{\alpha n^{n+1/2}e^{-n+1/12n}}\right) = -\int_n^\infty \frac{\psi_2(t)}{2t^2}\,dt.$$

(d) Part (c) implies that

$$0 = \lim_{n\to\infty} -\int_n^\infty \frac{\psi_2(t)}{2t^2}\,dt = \lim_{n\to\infty} \log\left(\frac{n!}{\alpha n^{n+1/2}e^{-n+1/12n}}\right)$$

so

$$1 = \lim_{n\to\infty} \frac{n!}{\alpha n^{n+1/2}e^{-n+1/12n}} = \lim_{n\to\infty} \frac{n!}{\alpha n^{n+1/2}e^{-n}}.$$

Thus

$$\lim_{n\to\infty} \frac{(n!)^2}{\alpha^2 n^{2n+1}e^{-2n}} = 1.$$

Replacing n by $2n$ and taking square roots we have

$$\lim_{n\to\infty} \frac{(2n)!}{\alpha(2n)^{2n+1/2}e^{-2n}} = 1,$$

so

$$\sqrt{\pi} = \lim_{n\to\infty} \frac{(n!)^2}{(2n)!} \cdot \frac{2^{2n}}{\sqrt{n}}$$

$$= \lim_{n\to\infty} \frac{\alpha^2 n^{2n+1}e^{-2n}2^{2n}}{\alpha(2n)^{2n+1/2}e^{-2n}\sqrt{n}}$$

$$= \alpha \lim_{n\to\infty} \frac{n^{2n+1/2}\sqrt{n}\,2^{2n}}{2^{2n}\sqrt{2}\,n^{2n+1/2}\sqrt{n}}$$

$$= \frac{\alpha}{\sqrt{2}}.$$

(e) By Problem 18(a) we have $\phi_3' = 2\phi_2$. Also $\phi_3(0) = b_3 = 0$, and

$$\phi_3(x) = -\phi_3(1-x),$$

from which it follows that

$$\phi_3(1) = 0$$
$$\phi_3(\tfrac{1}{2}) = 0.$$

It follows immediately that

$$\int_0^{1/2} \phi_2(t)\,dt = \int_0^1 \phi_2(t)\,dt = 0.$$

Clearly

$$\bar{\psi}(x) = \int_0^x \psi_2(t)\,dt \quad \begin{cases} \geq 0 & \text{for } 0 \leq x \leq 1/2 \\ \leq 0 & \text{for } 1/2 \leq x \leq 1, \end{cases}$$

with $\bar{\psi}(n) = 0$ for all n.

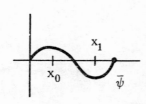
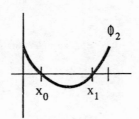

Moreover, $\bar{\psi}(x) = -\bar{\psi}(1 - x)$ on $[0, 1]$, since $\bar{\psi} = \phi_3/3$, so

$$\bar{\bar{\psi}}(x) = \int_0^x \bar{\psi}(t)\,dt \geq 0 \qquad \text{on } [0, 1], \text{ and hence everywhere,}$$

and $\bar{\bar{\psi}}(n) = 0$ for all n. Now we have

$$\int_n^\infty \psi_2(t) \cdot \frac{1}{2t^2}\,dt = \bar{\psi}(t) \cdot \frac{1}{2t^2}\Big|_n^\infty + \int_n^\infty \bar{\psi}(t) \cdot \frac{1}{t^3}\,dt$$

$$= \int_n^\infty \bar{\psi}(t) \cdot \frac{1}{t^3}\,dt$$

$$= \bar{\bar{\psi}}(t) \cdot \frac{1}{t^3}\Big|_n^\infty + 3\int_n^\infty \bar{\bar{\psi}}(t) \cdot \frac{1}{t^4}\,dt$$

$$= 3\int_n^\infty \bar{\bar{\psi}} \cdot \frac{1}{t^4}\,dt$$

$$> 0.$$

(f) The minimum of $\phi_2(x)$ for x in $[0, 1]$ occurs at $x = 1/2$, where $\phi_2(x) = -1/12$, and the maximum occurs at $x = 0$ and $x = 1$, where $\phi_2(x) = 1/6$. Clearly

$$\left| \int_n^\infty \frac{\psi_2(t)}{2t^2}\,dt \right| < \int_n^\infty \frac{|\psi_2(t)|}{2t^2}\,dt \leq \int_n^\infty \frac{1}{12t^2}\,dt = \frac{1}{12n}.$$

(g) From parts (c) and (d) we have

$$-\frac{1}{12n} < \log\left(\frac{n!}{\sqrt{2\pi}\, n^{n+1/2} e^{-n+1/12n}}\right) < 0$$

so

$$e^{-1/12n} < \frac{n!}{\sqrt{2\pi}\, n^{n+1/2} e^{-n+1/12n}} < 1$$

or

$$\sqrt{2\pi}\, n^{n+1/2} e^{-n} < n! < \sqrt{2\pi}\, n^{n+1/2} e^{-n+1/12n}.$$

1. It is clear that $a + b = b + a$, since the table for $+$ is symmetric. Clearly $a + 0 = a$ for all a, and condition 2(ii) is satisfied because each row in the table contains 0. To check that $(a + b) + c = a + (b + c)$ it suffices to consider only cases where $a, b, c \neq 0$. Because $x + y = y + x$, this equation clearly holds when $a = c$. This leaves the cases

$$(1 + b) + 2 = 1 + (b + 2),$$
$$(2 + b) + 1 = 2 + (b + 1),$$

which are equivalent to each other, either of which can be checked by letting $b = 1$ and 2. Conditions (4)–(6) are checked similarly. Finally, (7) is clear if $a = 0$ or 1. For $a = 2$ we can assume $b, c \neq 0$ and the condition is clear if $b = c = 1$. This leaves only the cases $a = 2$, $b = 2$, $c = 2$ and $a = 2$, $b = 1$, $c = 2$ and $a = 2$, $b = 2$, $c = 1$, the last two being equivalent.

F cannot be made into an ordered field because $1 = 1^2$ would have to be positive, but $1 + 1 + 1 = 0$.

2. F will not be a field because we will have $2 \cdot 2 = 0$.

3. As in Problem 1, conditions (2), (3), (5), and (6) are clear. Condition (1) can be checked case-by-case. To check (5) we can assume that $a, b, c \neq 0, 1$. This leaves only the cases $(\alpha \cdot \beta) \cdot \alpha = (\alpha \cdot \beta) \cdot \alpha$ and $(\alpha \cdot \beta) \cdot \beta = \alpha \cdot (\beta \cdot \beta)$.

4. (a) $a + a = a \cdot (1 + 1) = a \cdot 0 = 0$.

(b) $0 = a^{-1} \cdot 0 = a^{-1}(a + a) = 1 + 1$.

5. (a) The assertion is obvious when $n = 1$. Suppose it is true for n. Then

$$\underbrace{(1 + \cdots + 1)}_{m \text{ times}} \cdot \underbrace{(1 + \cdots + 1)}_{n+1 \text{ times}} = \underbrace{(1 + \cdots + 1)}_{m \text{ times}} \cdot (\underbrace{[1 + \cdots + 1]}_{n \text{ times}} + 1)$$

$$= \underbrace{(1 + \cdots + 1)}_{m \text{ times}} \cdot \underbrace{(1 + \cdots + 1)}_{n \text{ times}} + \underbrace{(1 + \cdots + 1)}_{m \text{ times}}$$

$$= \underbrace{(1 + \cdots + 1)}_{mn \text{ times}} + \underbrace{(1 + \cdots + 1)}_{m \text{ times}} = \underbrace{1 + \cdots + 1}_{\substack{mn+m=m(n+1) \\ \text{times}}} \ .$$

(b) If m were not prime, so that $m = kl$ for some $k, l < m$, then

$$0 = \underbrace{(1 + \cdots + 1)}_{m=kl \text{ times}} = \underbrace{(1 + \cdots + 1)}_{k \text{ times}} \cdot \underbrace{(1 + \cdots + 1)}_{l \text{ times}} = a \cdot b.$$

Therefore either $a = 0$ or $b = 0$, contradicting the assumption about m.

6. (a) If this were not true, then P would have infinitely many distinct elements, namely, those of the form

$$\underbrace{1 + \cdots + 1}_{n \text{ times}}$$

for all n.

(b) Suppose $m > n$. Then

$$\underbrace{1 + \cdots + 1}_{m-n \text{ times}} = \left(\underbrace{1 + \cdots + 1}_{m \text{ times}}\right) - \left(\underbrace{1 + \cdots + 1}_{n \text{ times}}\right) = 0.$$

7. The solutions are

$$x + (\alpha d - \beta b) \cdot (ad - bc)^{-1},$$
$$y = (\alpha c - \beta a) \cdot (bc - ad)^{-1}.$$

8. (a) One, namely 0.

(b) If a has one square root b, then it also has the square root $-b$. Moreover, if $c^2 - a = b^2$, then $(c - b) \cdot (c + b) = 0$, so $c = b$ or $c = -b$. Consequently, b and $-b$ are the only square roots; these are distinct precisely when $1 + 1 \neq 0$.

9. (a) is a straightforward check

(b) In part (a), the symbol 2 means $1 + 1$, which is 0 in F_2; the solution in part (a) is correct only if $1 + 1 \neq 0$.

10. (a) Most conditions require only a straightforward check. The element $(0, 0)$ will play the role of 0 and $(1, 0)$ will play the role of 1. To verify 5(ii), note that if $(x, y) \neq (0, 0)$, so that $x \neq 0$ and $y \neq 0$, then $x^2 - ay^2 \neq 0$, since a does not have a square root. Then

$$(x, y) \odot \left(\frac{x}{x^2 - ay^2}, \frac{-y}{x^2 - ay^2}\right) = (1, 0).$$

(b) is a straightforward check.

(c) $(0, 1)$ is a square root of a.

11. (a) The inverse of (a_1, a_2, a_3, a_4) is

$$\left(\frac{a_1}{\gamma}, -\frac{a_2}{\gamma}, -\frac{a_3}{\gamma}, -\frac{a_4}{\gamma}\right),$$

where $\gamma = a_1^2 + a_2^2 + a_3^2 + a_4^2$.

(b) Each entry in the following table is the product $a \cdot b$ where a is on the left and b is above.

$$
\begin{array}{c|ccc}
 & i & j & k \\
\hline
i & -1 & k & -j \\
j & -k & -1 & i \\
k & j & -i & -1
\end{array}
$$

[If we denote 1, i, j and k by v_1, v_2, v_3, and v_4, then the definition of multiplication can be written

$$\left(\sum_{i=1}^{4} a_i v_i\right) \cdot \left(\sum_{j=1}^{4} b_j v_j\right) = \sum_{i,j=1}^{4} a_i b_j (v_i v_j).$$

This allows a simpler proof that multiplication is associative, by first checking that it is associative for ± 1, $\pm i$, $\pm j$, $\pm k$.]

CHAPTER 29

Since the detailed examination of other constructions of the real numbers was recommended only for masochists, detailed answers to the two problems in this chapter will not be given.

CHAPTER 30

1. (a) $f(0) = f(0+0) = f(0) + f(0)$, so $f(0) = 0$. Since f is an isomorphism and $0 \neq 1$, it follows that $f(1) \neq 0$. Consequently, the equation $f(1) = f(1 \cdot 1) = f(1) \cdot f(1)$ implies that $f(1) = 1$.

(b) $0 = f(0) = f(a + -a) = f(a) + f(-a)$, so $f(-a) = -f(a)$. Similarly, $1 = f(1) = f(a \cdot a^{-1}) = f(a) \cdot f(a^{-1})$, so $f(a^{-1}) = f(a)^{-1}$.

2. As an example, a proof for (a) will be given. If $a^2 + 1 = 0$ for some α in F_1, then by Problem 1, $0 = f(0) = f(\alpha^2 + 1) = f(\alpha \cdot \alpha) + f(1) = f(\alpha)^2 + 1$, so $f(\alpha)$ is a solution of the equation $x^2 + 1 = 0$ in F_2.

3. (1) If $x \neq y$, then $f(x) \neq f(y)$, so $g(f(x)) \neq g(f(y))$, so $(g \circ f)(x) \neq (g \circ f)(y)$.
(2) If z is in F_3, then $z = g(y)$ for some y in F_2, and $y = f(x)$ for some x in F_1. Then $z = (g \circ f)(x)$.
(3)

$$(g \circ f)(x + y) = g(f(x + y)) = g(f(x) + f(y)) = g(f(x)) + g(f(y))$$
$$= (g \circ f)(x) + (g \circ f)(y),$$
$$(g \circ f)(x \cdot y) = g(f(x \cdot y)) = g(f(x) \cdot f(y)) = g(f(x)) \cdot g(f(y))$$
$$= (g \circ f)(x) \cdot (g \circ f)(y).$$

(4) If $x < y$, then $f(x) < f(y)$, so $g(f(x)) < g(f(y))$, i.e., $(g \circ f)(x) < (g \circ f)(y)$.

4. $g^{-1} \circ f$ is an isomorphism form **R** to **R**, so $g^{-1} \circ f = I$, so $g = f$.

5. Let $f(x + iy) = x - iy$. [Since $i^2 = -1$ we must have $f(i)^2 = f(-1) = -1$, so $f(i) = i$ or $-i$, which suggests the answer. This particular isomorphism is the only one, aside from the identity, which any one can write down, but there are actually infinitely many others. This is one of those facts which requires, aside from a knowledge of algebra, some of the sophisticated theorems from set theory which will be found in references [8] and [9] of the Suggested Reading.]